Basic System Analysis

Basic System Analysis

Imran Ullah Khan
Assistant Professor
Department of Electronics and Communication Engineering
Kanpur Institute of Technology
Kanpur, U.P. (India)

Jaishri Panjwani
Senior Lecturer
Department of Electrical & Electronics Engineering
Kanpur Institute of Technology
Kanpur, U.P. (India)

I.K. International Publishing House Pvt. Ltd.

NEW DELHI • BANGALORE

Published by

I.K. International Publishing House Pvt. Ltd.
S-25, Green Park Extension
Uphaar Cinema Market
New Delhi–110 016 (India)
E-mail: info@ikinternational.com
Website: www.ikbooks.com

ISBN: 978-93-81141-16-8

10 9 8 7 6 5 4 3 2 1

Published by Krishan Makhijani for I.K. International Publishing House Pvt. Ltd., S-25, Green Park Extension, Uphaar Cinema Market, New Delhi–110 016 and printed by Rekha Printers Pvt. Ltd., Okhla Industrial Area, Phase II, New Delhi–110 020.

Dedicated to our Parents

Mr. S.U. Khan, Mrs. Shafiya Khan

&

Mr. K.C. Panjwani, Mrs. Kavita Panjwani

Preface

This book deals with the properties and classification of Basic System Analysis. The several chapters introduce the important principles and properties of Fourier transform (Continuous time), Laplace transform, Z-transform, Fourier series, Linear time invariant (LTI) systems, Convolution, etc. This subject covers basic concepts required for Digital signal processing, Communication systems, Networks and similar subjects.

Difficulties faced by an average student in learning and understanding this subject has motivated us to write this book. This book is being written, keeping average and below average students in mind. Difficult topics have been simplified using an easy to understand language.

To get an edge in competitive market, Objective type questions with answers, Question bank with answers (Question bank includes important questions) have been included. This book consists of five chapters in all and covers the complete syllabii of Basic System Analysis of U.P. Technical University and various other universities.

To make it different and better from few available books, we have used solved examples, solved questions from previous year Question papers and Model Question papers at the end of book. The book has been made adequately self-sufficient for students to avoid the need for another support book.

Chapter 1 deals with commonly used Continuous-time signals, Types of signals and their representation, Basic system properties and their detailed analysis. Different types of analogous systems are discussed in detail.

Chapter 2 deals with Fourier series representation of continuous-time and discrete-time periodic signals and their properties. It also includes detailed discussion about Continuous-time Fourier transform (CTFT) and its properties. This chapter also presents Application of Fourier Transform to Network Analysis.

Chapter 3 deals with Laplace transform and its use in signals and systems analysis. Applications of Laplace transform to the solution of differential equations with initial conditions are discussed. Applications of Laplace Transform to Network Analysis and Laplace Transform of complex waveforms is discussed in detail.

Chapter 4 is devoted to detailed analysis of State space analysis of linear system, State Transition Matrix, Solution of state equations for homogeneous and not homogeneous systems and application of State variable technique is discussed in detail.

Chapter 5 is devoted to Z-transform and its use in signals and systems analysis. The properties of Z-transform are presented in detail. Application of Z-transform for solving the differential equations are presented.

Authors

Acknowledgements

It is our great pleasure to thank Mr. Rakesh Kumar Pandey (HOD-EN, KIT), Mr. K.K. Mishra (TPO) who encouraged us to write this book.

We are deeply indebted to Mr. Somendra Shukla (HOD, KIT) and Brajesh Varshney (Dean), Vaibhav Purwar (COE) and Mr. D.S. Kushwaha (HO, PSIT), Ms. Akanksha Gupta, Mrs. Neha Sharma, Mr. Ashish Agarwal and Mr. Shashikant Singh for valuable comments and suggestions.

Special thanks are due to Mr. Sujeet Singh, Mr. Mohd. Rashid, Mr. Harvir Singh (HOD, NCET), Mr. Himanshu Vijay and Mr. Tanveer Qamar.

Finally, we pay sincere gratitude to Injila, Amil and Amreen, Rehan and Gopal Panjwani, Shyam Panjwani for their patience and constant motivation during the preparation of this book.

Authors

Contents

1

Introduction to Continuous-Time Signals and Systems

1.1 SIGNALS

Definition. A signal is a function of one or more independent variables that contain information about the nature and behaviour of that phenomenon.

In a RC circuit signal may be represented as voltage across the capacitor or current flowing through the resistor. A signal is represented as a function of independent variable t, thus a signal is denoted by $x(t)$. One of the definitions of signal in the Merrian-Webster dictionary is:

"A detectable physical quantity or impulse (as a voltage, current, or magnetic field strength) by which messages or information can be transmitted."

One-dimensional signals: Variation of room temperature, ac power signal, electrocardiogram, speech signal, etc.

Two-dimensional signals: X-rays, sonograms, pictures, etc.

1.2 TYPES OF SIGNALS AND THEIR REPRESENTATION

The following are the types of signals which will be of interest in this book. We will focus on these broad classes of signals.

- Continuous-time signals and discrete-time signals
- Periodic and non-periodic signals
- Even and odd signals
- Energy and power signals
- Deterministic and random signals
- One-dimensional signals and two-dimensional signals

1.2.1 Continuous-time signals

Continuous-time signals are defined for every value (instant) of time t and are represented by $x(t)$. A continuous-time signal is also called analog signal. Most of the signals encountered in practice are continuous-time signals.

Example 1: The temperature recorded over an interval of time as shown in Fig. 1.1. It is defined at every time instant.

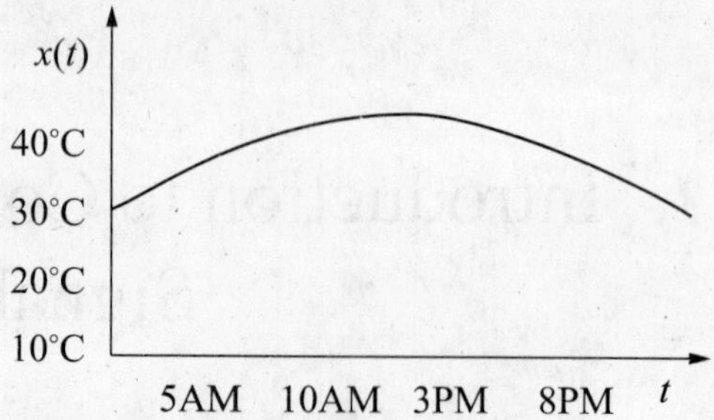

Fig. 1.1.

- Waveform of ac power supply.

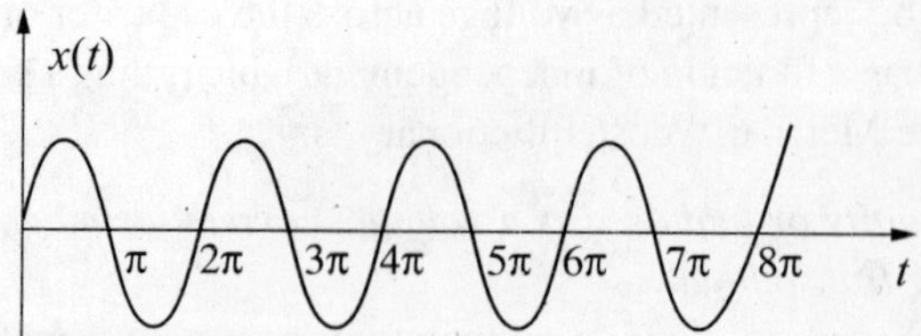

Fig. 1.2.

Discrete-time signal

The discrete-time signals are defined at discrete-instant of time and are represented by $x(n)$, where n is index. Some signals are discrete in nature and some signals may be discrete representation of continuous-time signals.

Example 2: The amount deposited every month in a savings account is discrete.

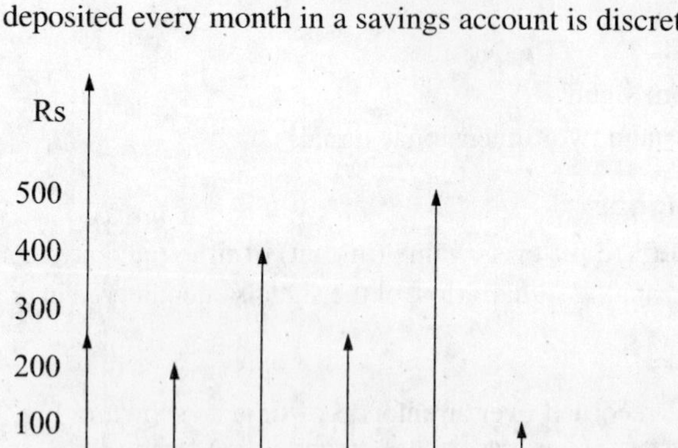

Fig. 1.3.

Population survey of India by government

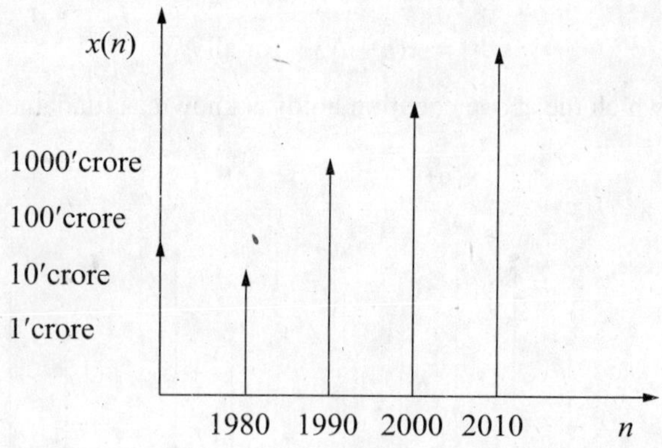

Fig. 1.4.

The other type of discrete-time signal may be obtained by sampling a continuous-time signal at regular intervals (variations of room temperature).

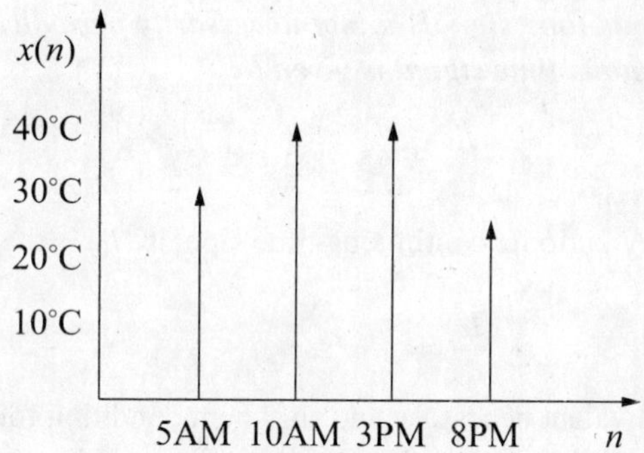

Fig. 1.5.

1.2.2 Periodic and non-periodic signal

A continuous-time signal $x(t)$ is said to be periodic if it satisfies the condition

$$x(t) = x(t+T) \quad \text{for all } t$$

The smallest value of T that satisfies the above condition is known as fundamental period.

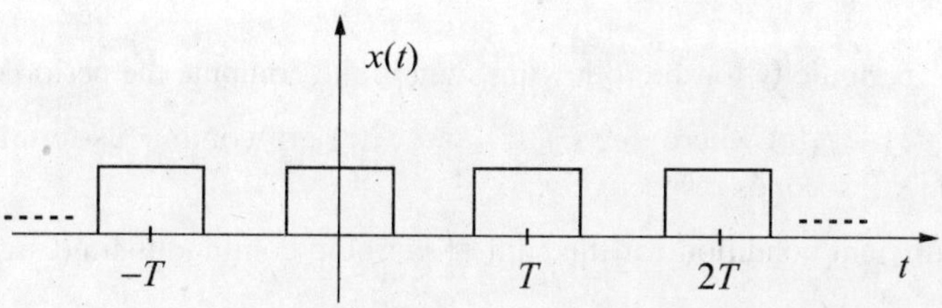

Fig. 1.6. A periodic signal with period T.

A signal is non-periodic if the condition is not satisfied for at least one value of t.

A discrete-time signal $x(n)$ is said to be periodic if it satisfies the condition

$$x(n) = x(n+N) \quad \text{for all } N$$

The smallest value of N for which the above equation holds is known as fundamental period.

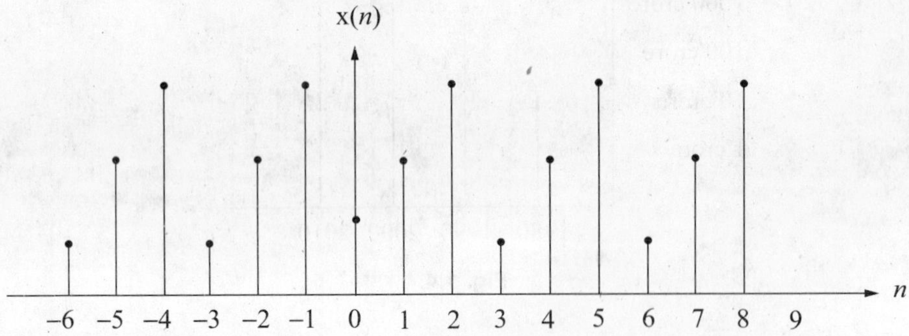

Fig. 1.7. A discrete-time periodic signal with period $n = 3$.

The sum of N periodic continuous-time signals is not necessarily periodic. The condition of periodicity for the sum of N periodic continuous-time signal is given by

$$\frac{T_0}{T_{0i}} = n_i \quad 1 \le i \le N$$

where T_0 is the period of sum of N periodic continuous-time signals, T_{0i} is the period of ith signal in the sum and n_i is an integer.

$$\frac{T_0}{T_{0i}} = \frac{n_i}{n_1} \quad 2 \le i \le N$$

The above equation is an equivalent necessary and sufficient condition for a sum of periodic continuous-time signals to be periodic. This condition states that the sum of N periodic continuous-time signal is periodic if all the ratios of the period of the first signal in the sum to the period of another signal in the sum be rational. It means it is the ratio of integers.

For determining the period of a sum signal, if it is periodic, first convert each ratio from the above equation to a ratio of integers. If this ratio is not rational then the sum signal is not periodic. If it is rational, then find the factor that is the highest common divisor from the numerator and denominator of each individual ratio. Now find the least common multiple of denominators of all the factors. Assume that as n_1, so the period of sum signal is $T_0 = T_{01} \cdot n_1$

The above equations are also applicable for discrete-time signals.

Example 3: Determine periodicity for the following signals and compute the period of sum signals.

- $F_1(t) = f_1(t) + f_2(t) + f_3(t)$ where $f_1(t), f_2(t)$ and $f_3(t)$ are continuous-time periodic signals with period 2/3, 1.5 and $\sqrt{2}$ seconds respectively.

A necessary and sufficient condition for the sum of periodic continuous-time signals to be periodic is given by

$$\frac{T_{01}}{T_{0i}} = \frac{n_i}{n_1} \quad 2 \le i \le N$$

Given $T_{01} = 2/3$ seconds, $T_{02} = 1.5 = 3/2$ seconds, $T_{03} = \sqrt{2}$

$$\text{for } i = 2 \qquad \frac{T_{01}}{T_{02}} = \frac{n_2}{n_1} = \frac{2/3}{3/2} = \frac{4}{9} \quad \text{Rational}$$

$$\text{for } i = 3 \qquad \frac{T_{01}}{T_{03}} = \frac{n_3}{n_1} = \frac{2/3}{\sqrt{2}} = \frac{2}{3\sqrt{2}} \quad \text{Not rational}$$

Signal $F_1(t)$ is not periodic since T_{01}/T_{03} is not rational

- $F_2(t) = f_1(t) + f_2(t) + f_3(t)$ where $f_1(t), f_2(t)$ and $f_3(t)$ are continuous-time periodic signals with period 2/3, 1.5 and 3/4 seconds respectively.

A necessary and sufficient condition for the sum of periodic continuous-time signals to be periodic is given by

$$\frac{T_{01}}{T_{0i}} = \frac{n_i}{n_1} \quad 2 \leq i \leq N$$

Given $T_{01} = 2/3$ seconds, $T_{02} = 1.5 = 3/2$ seconds, $T_{03} = 3/4$

$$\text{for } i = 2 \qquad \frac{T_{01}}{T_{02}} = \frac{n_2}{n_1} = \frac{2/3}{3/2} = \frac{4}{9} \quad \text{Rational}$$

$$\text{for } i = 3 \qquad \frac{T_{01}}{T_{03}} = \frac{n_3}{n_1} = \frac{2/3}{3/4} = \frac{8}{9} \quad \text{Rational}$$

The least common multiple of denominators of all the factors is 9 so the period is $(2/3) * 9 = 6$.

Example 4: Find if the following sequences are periodic or not, if yes, find their fundamental time period.

(1) $x(t) = e^{j(\pi/4)t}$

(2) $x_e(t) = \cos(50\pi t) + \cos(100\pi t)$

(3) $x(n) = 3\sin\left(\frac{1}{8}n\right)$

(4) $x(t) = 10\sin 12\pi t + 4\sin 18\pi t$

(5) $x(t) = 3\tan 4t$

(6) $x(t) = 3 + t^2$

(7) $x(t) = je^{j5t}$

(8) $x(t) = 3\cos 4t + 2\sin 2\pi t$

(9) $x(t) = 2\cos(10t + 1) - \sin(4t - 1)$

(10) $x(t) = 2u(t) + 2\sin 2t$

(11) $x(n) = \cos(2\pi n)$

(12) $x(n) = \sin\left(\frac{6\pi}{7}n + 1\right)$

(13) $x(n) = e^{j\frac{2\pi}{3}n} + e^{j\frac{3\pi}{4}n}$

(14) $x(n) = 12\cos(20n)$

(15) $x(t) = e^{-|t|}$

SOLUTION:

(1)
$$x(t) = e^{j\frac{\pi}{4}t} = \cos\left(\frac{\pi}{4}\right)t + j\sin\left(\frac{\pi}{4}\right)t$$
$$w_{01} = \frac{\pi}{4} \quad w_{02} = \frac{\pi}{4}$$
$$T_1 = \frac{2\pi}{w_{01}} = 8 \quad T_2 = \frac{2\pi}{w_{02}} = 8$$
$$\frac{T_1}{T_2} = 1, \text{rational number, signal is periodic.}$$

Fundamental time period $T = T_1 = T_2 = 8$

(2)
$$x(t) = \cos(50\pi t) + \cos(100\pi t)$$
$$\text{Here } w_{01} = 50\pi \quad \text{and } w_{02} = 100\pi$$
$$T_1 = \frac{2\pi}{w_{01}} = \frac{1}{25} \quad \text{and } T_2 = \frac{2\pi}{100\pi} = \frac{1}{50}$$
$$\frac{T_1}{T_2} = 2 \text{ rational number, hence periodic.}$$

F.T.P. $T = T_1 = 2T_2 = \dfrac{1}{25}$

(3)
$$x(n) = 3\sin\left(\frac{1}{8}n\right)$$
$$w_0 = \frac{1}{8}$$
$$N = \frac{2\pi m}{w_0} = \frac{2\pi}{1/8}m = 16\pi m$$

For any value of m, N is not a integer, so $x(n)$ is non-periodic.

(4)
$$x(t) = 10\sin(12\pi t) + 4\sin(18\pi t)$$
$$\text{Here } w_{01} = 12\pi \quad \text{and } w_{02} = 18\pi$$
$$T_1 = \frac{2\pi}{w_{01}} = \frac{1}{6} \quad \text{and } T_2 = \frac{2\pi}{w_{02}} = \frac{2\pi}{18\pi} = \frac{1}{9}$$
$$\frac{T_1}{T_2} = \frac{3}{2} \text{ rational, signal is periodic.}$$

Fundamental period $T = 2T_1 = 3T_2 = \frac{1}{3}$

(5)
$$x(t) = 3\tan 4t$$

The tangent function is periodic, it has a period of π. The above function can be expressed as,

$$x(t) = 3\tan(4t \pm n\pi) = 3\tan\left\{4\left(t \pm \frac{n\pi}{4}\right)\right\}$$

Thus, the period is $T = \pi/4$

(6)
$$x(t) = 3 + t^2$$

This is not a periodic function since its value continuously increases with time.

(7)
$$x(t) = je^{j5t}$$
$$= j(\cos 5t + j\sin 5t) = j\cos 5t - \sin 5t$$
$$w_{01} = 5 \quad \text{and} \quad w_{02} = 5$$
$$T_1 = \frac{2\pi}{5} \quad T_2 = \frac{2\pi}{w_{02}} = \frac{2\pi}{5}$$
$$\frac{T_1}{T_2} = 1, \text{ rational number, hence periodic.}$$

F.T.P. $T = T_1 = T_2 = \frac{2\pi}{5} = 0.4\pi$ sec

(8)
$$x(t) = 3\cos 4t + 2\sin 2\pi t$$

The period of the signal $3\cos 4t$ is $T_1 = \frac{2\pi}{w_{01}} = \frac{2\pi}{4} = \frac{\pi}{2}$ sec and the period of the signal $2\sin 2\pi t$ is $T_2 = \frac{2\pi}{w_{02}} = \frac{2\pi}{2\pi} = 1$ sec. The ratio of time period $\frac{T_1}{T_2} = \frac{\pi/2}{1} = \pi/2$ is not a rational number. Therefore, the signal is not periodic.

(9)
$$x(t) = 2\cos(10t + 1) - \sin(4t - 1)$$

Time period of the signal $2\cos(10t + 1)$ is $T_1 = \frac{2\pi}{w_{01}} = \frac{2\pi}{10} = \frac{\pi}{5}$ sec

Time period of the signal $\sin(4t - 1)$ is $T_2 = \frac{2\pi}{w_{02}} = \frac{2\pi}{4} = \frac{\pi}{2}$ sec

The ratio of two periods is $\frac{T_1}{T_2} = \frac{\pi/5}{\pi/2} = \frac{2}{5}$

$\frac{2}{5}$ is a rational number. Therefore, the sum of two signals is periodic and the period is given by $T = 5T_1 = 2T_2 = \pi$ sec.

(10)
$$2u(t) + 2\sin 2t$$

Period of the signal $2\sin 2t$ is $T_1 = \frac{2\pi}{w_{01}} = \frac{2\pi}{2} = \pi$ sec.

The signal $2u(t)$ is non-periodic as shown in figure.

Therefore, the sum of two signals $2u(t) + 2\sin 2t$ is non-periodic

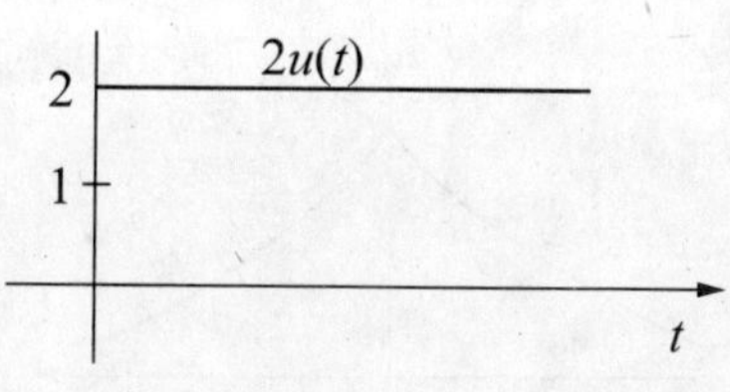

Fig. 1.8. Graphical representation of $2u(t)$

(11) $$x(n) = \cos(2\pi n)$$

Period of the signal $x(n) N = \dfrac{2\pi m}{w_0}$

$$N = \frac{2\pi}{2\pi} m = m$$

The smallest possible value of m for which N is an integer is 1. Therefore, $N = 1$ and the sequence is periodic.

(12) $$x(n) = \sin\left(\frac{6\pi}{7} n + 1\right)$$

$$w_0 = 6\pi/7$$

Period of the signal is $N = \dfrac{2\pi m}{w_0} = \dfrac{2\pi m}{6\pi/7} = \left(\dfrac{7}{3}\right) m$

The smallest possible value of m for which m is an integer is 3.

$$N = \left(\frac{7}{3}\right) 3 = 7$$

(13) $$x(n) = e^{j\left(\frac{2\pi}{3}\right)n} + e^{j\left(\frac{3\pi}{4}\right)n}$$

Period of the signal $e^{j\left(\frac{2\pi}{3}\right)n}$ is $N_1 = \dfrac{2\pi m}{w_{01}} = \dfrac{2\pi m}{2\pi/3} = 3m$ for $m = 1$; $N_1 = 3$

Period of the signal $e^{j\left(\frac{3\pi}{4}\right)n}$ is $N_2 = \dfrac{2\pi m}{w_{02}} = \dfrac{2\pi m}{3\pi/4} = \dfrac{8}{3} m$ for $m = 3$; $N_2 = 8$

$$\frac{N_1}{N_2} = \frac{3}{8}$$

$$N = 8N_1 = 3N_2 = 24$$

The given signal is periodic with period 24.

(14) $$x(n) = 12\cos(20n)$$

Period of the signal is $N = \dfrac{2\pi m}{w_0} = \dfrac{2\pi m}{20} = \dfrac{\pi m}{10}$

For all values of m, N is not an integer. Therefore, given sequence is non-periodic.

(15) $$x(t) = e^{-|t|}$$

The signal $e^{-|t|}$ is non-periodic as shown in figure.

$$e^{-|t|} = \begin{cases} e^{-t} & \text{for } t > 0 \\ e^{t} & \text{for } t < 0 \end{cases}$$

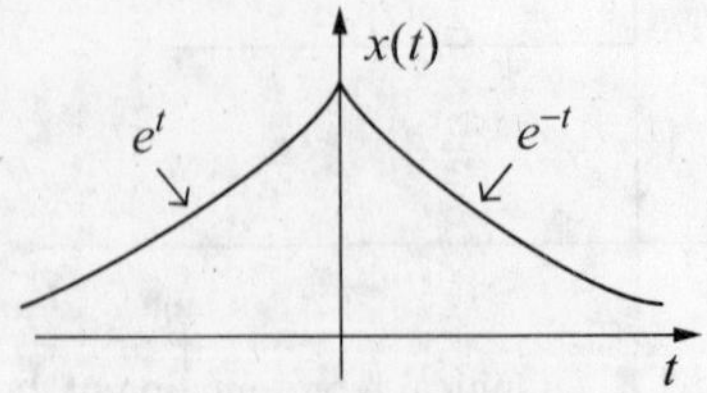

Fig. 1.9. Graphical representation of $e^{-|t|}$

1.2.3 Even (symmetric) and odd (anti-symmetric) signals

Even signal: A CTS $x(t)$, is said to be even if it satisfies the condition

$$x(t) = x(-t) \quad \text{for all } t.$$

e.g.: $\cos t,\ t\sin t,\ t^2$

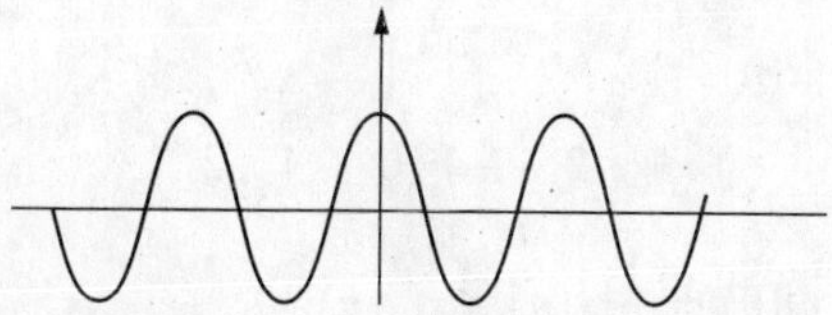

Fig. 1.10. Cosinusoidal signal

Odd signal: A CTS $x(t)$ is said to be odd if

$$x(t) = -x(-t) \quad \text{for all t}$$

e.g.: $\sin t$, $t \cos t$, t^3

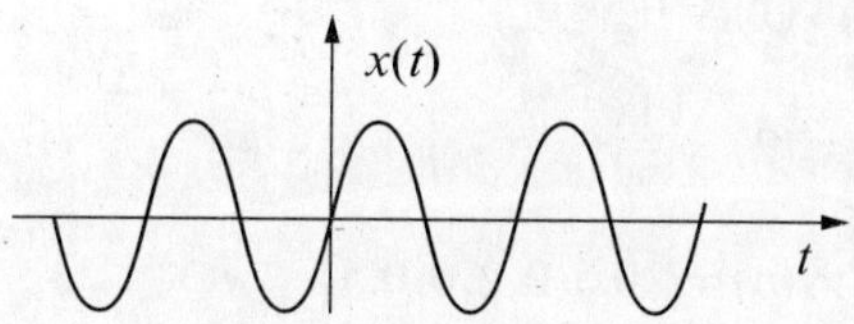

Fig. 1.11. Sinusoidal signal

Any signal $x(t)$ can be expressed as sum of even and odd components

$$x(t) = x_e(t) + x_0(t) \tag{1}$$

Replacing t by $-t$

$$x(-t) = x_e(-t) - x_0(-t)$$

$$\text{or} \quad x(-t) = x_e(t) - x_0(t) \tag{2}$$

$$(1)+(2) \quad \boxed{x_e(t) = \frac{1}{2}[x(t)+x(-t)]}$$

$$(1)-(2) \quad \boxed{x_0(t) = \frac{1}{2}(x(t)-x(-t))}$$

Example 5: Find the even and odd components of the following signals

(i) $x(t) = \cos t + \sin t + \cos t \sin t$

(ii) $x(n) = \underset{\uparrow}{\{-2, 1, 2, -1, 3\}}$

SOLUTION:

(i)
$$x(t) = \cos t + \sin t + \cos t \sin t \tag{1}$$

$$x(-t) = \cos t - \sin t - \cos t \sin t \tag{2}$$

$$x_e(t) = \frac{1}{2}[x(t) + x(-t)] = \cos t$$

$$x_0(t) = \frac{1}{2}[x(t) - x(-t)] = \sin t + \cos t \sin t$$

(ii)

$$x(n) = \begin{matrix} \{-2 & 1 & 2 & -1 & 3\} \\ & & \uparrow & & \\ -2 & -1 & 0 & 1 & 2 \end{matrix}$$

$$x_e(n) = \frac{1}{2}[x(n) + x(-n)]$$

$$x_e(-2) = \frac{1}{2}(-2+3) = \frac{1}{2}$$

$$x_e(-1) = 0$$

$$x_e(0) = 2$$

$$x_e(1) = 0$$

$$x_e(2) = \frac{1}{2}$$

$$x_e(n) = \{0.5, 0, 2, 0, 0.5\}$$

$$x_0(n) = \frac{1}{2}[x(n) - x(-n)]$$

$$x_0(-2) = \frac{1}{2}[x(-2) - x(2)] = \frac{1}{2}(-2-3) = -2.5$$

$$x_0(-1) = 1$$

$$x_0(0) = 0$$

$$x_0(1) = -1$$

$$x_0(2) = 2.5$$

$$x_0(n) = \{-2.5, 1, 0, -1, 2.5\}$$

Example 6: Sketch and label the even and odd components of the signals shown in figure.

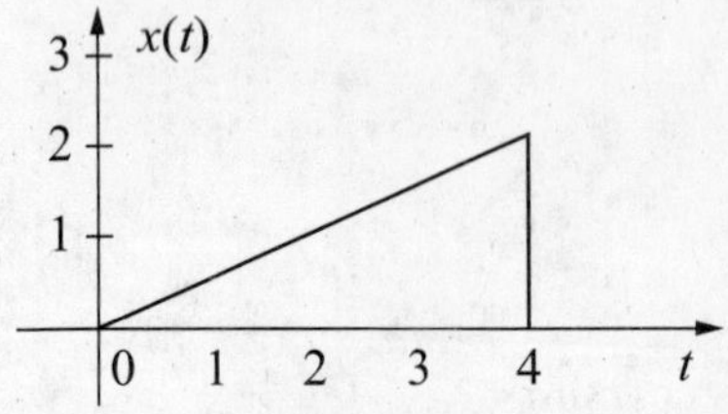

Fig. 1.12.

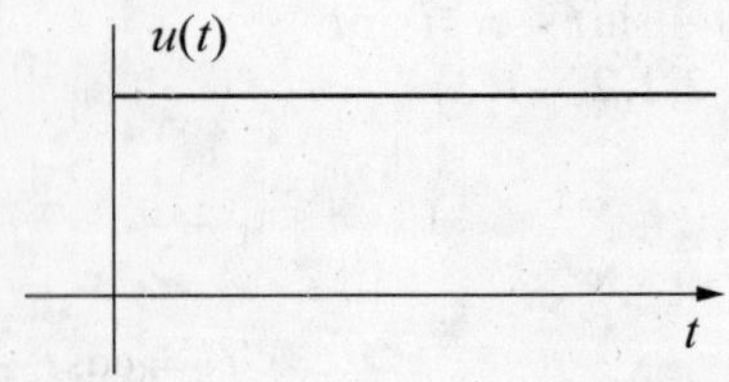

Fig. 1.13.

SOLUTION:

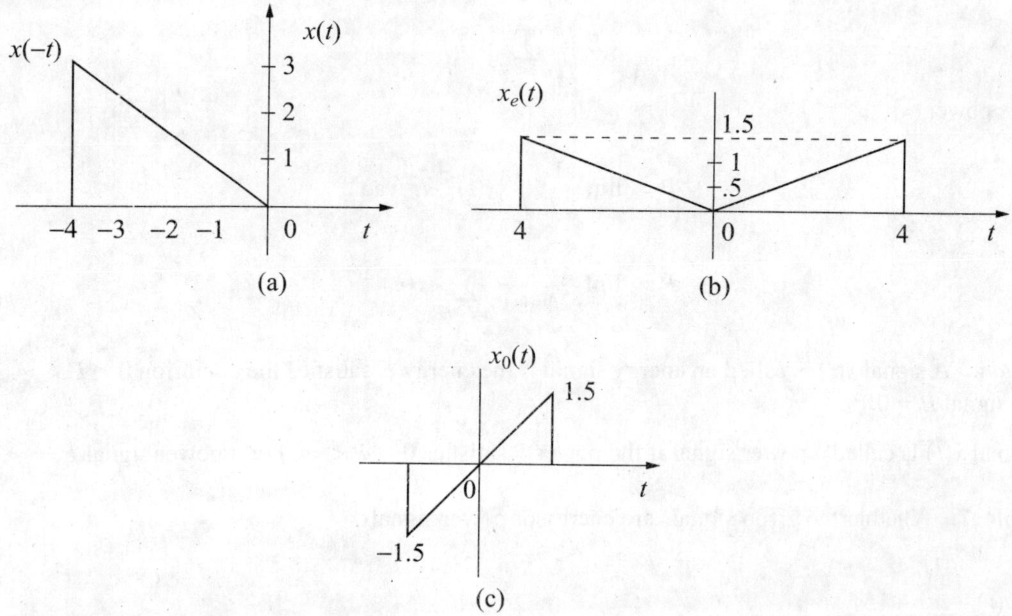

Fig. 1.14. (a) Signal $x(-t)$ (b) Even signal $x_e(t)$ (c) Odd signal $x_0(t)$ of the signal of Fig. 1.12

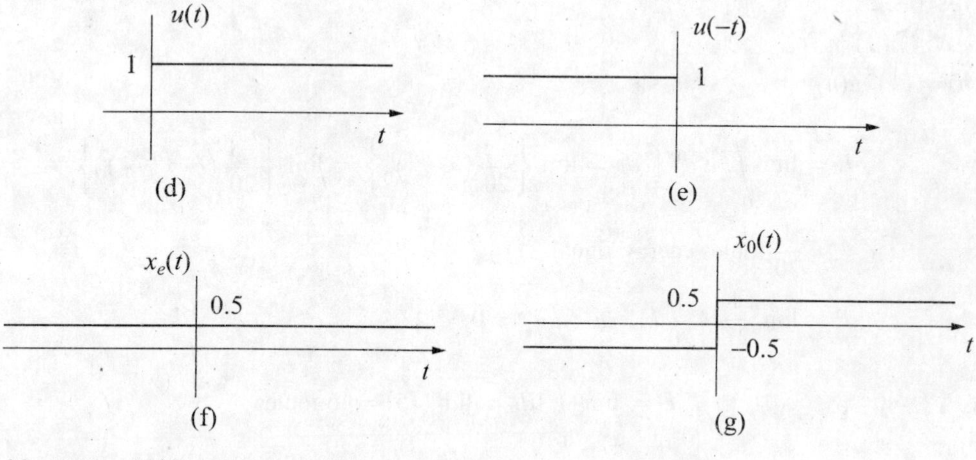

Fig. 1.14. (d) Signal $u(t)$ (e) Signal $u(-t)$ (f) Even signal $x_e(t)$ (g) Odd signal $x_0(t)$ of the signal of Fig. 1.13.

1.2.4 Energy and power signals

For a signal $x(t)$, total energy is defined as

$$E = \lim_{T\to\infty} \int_{-T}^{T} |x(t)|^2 \, dt \text{ joules}$$

For a discrete-time signal $x(n)$, the energy E is defined as

$$E = \sum_{n=-\infty}^{\infty} |x(n)|^2$$

Average power

$$P = \lim_{T\to\infty} \frac{1}{2T} \int_{-T}^{T} |x(t)|^2 \, dt \text{ watts}$$

$$P = \lim_{N\to\infty} \frac{1}{2N+1} \sum_{n=-N}^{N} |x(n)|^2$$

Definition. A signal $x(t)$ is called an energy signal if the energy E satisfies the condition $0 < E < \infty$. For an energy signal $P = 0$.

Signal $x(t)$ is called a power signal if the power P satisfies $0 < P < \infty$. For a power signal $E = \infty$.

Example 7: Whether the given signals are energy or power signals.

(i) $e^{-10t}u(t)$
(ii) $u(t) - u(t-15)$
(iii) $\cos(10\pi t)u(t)u(2-t)$
(iv) $u(t) - u(t-1)$
(v) $r(t) - r(t-2)$
(vi) $(1 + e^{-5t})\,u(t)$
(vii) $x_1(n) = \left(\frac{1}{3}\right)^n u(n)$

(i)
$$E = \lim_{T\to\infty} \int_0^T |e^{-10t}|^2 \, dt = \lim_{T\to\infty} \left\{\frac{-1}{20}\left(e^{-20t}\right)_0^T\right\} = \lim_{T\to\infty} \left\{\frac{-1}{20}\left(e^{-20T} - 1\right)\right\}$$

$$= \frac{1}{20} \text{ joules energy signal}$$

$$P = \lim_{T\to\infty} \frac{1}{2T} \left\{\frac{1}{20}\left(1 - e^{-20T}\right)\right\} = 0$$

(ii)
$$E = \lim_{T\to\infty} \int_0^{15} 1 dt = \lim_{T\to\infty} (15) = 15 \text{ joules}$$

$$P = \lim_{T\to\infty} \frac{1}{2T}[15] = 0$$

(iii)

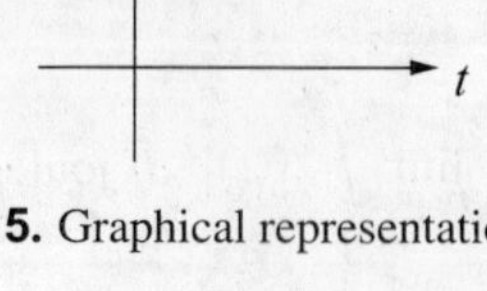

Fig. 1.15. Graphical representation of $u(t)$

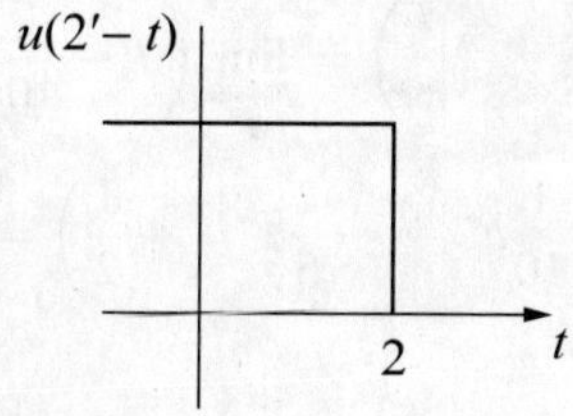

Fig. 1.16. Graphical representation of $u(2-t)$

$$u(t) = \begin{cases} 1 & \text{for} \quad t \geq 0 \\ 0 & \text{Otherwise} \end{cases} \qquad u(2-t) = \begin{cases} 1 & \text{for} \quad t \leq 2 \\ 2 & \text{Otherwise} \end{cases}$$

$$\text{So} \quad u(t)u(2-t) = \begin{cases} 1 & \text{for} \quad 0 \leq t \leq 2 \\ 0 & \text{Otherwise} \end{cases}$$

$$E = \lim_{T\to\infty} \left[\int_0^2 \cos^2 2\pi t\, dt \right] = \lim_{T\to\infty} \left[\frac{1}{2} \int_0^2 (1+\cos 4\pi t)\, dt \right]$$

$$= \lim_{T\to\infty} \frac{1}{2} \left[t + \left(\frac{\sin 4\pi t}{4\pi} \right)_0^2 \right] = \lim_{T\to\infty} \frac{1}{2}[2] = 1 \text{ joule}$$

$$P = \lim_{T\to\infty} \frac{1}{2T} \cdot 1 = 0$$

(iv)

$$E = \lim_{T\to\infty} \int_0^1 dt = 1 \text{ joule}$$

$$P = \lim_{T\to\infty} \frac{1}{2T} \cdot 1 = 0$$

(v)

$$x(t) = r(t) - r(t-2) = tu(t) - (t-2)u(t-2)$$

$$x(t) = t[u(t) - u(t-2)] + 2u(t-2)$$

$$E = \lim_{T\to\infty} \left[\int_0^2 t^2 dt + \int_2^T 4dt \right] = \left[\frac{8}{3} + 4T - 8 \right] = \infty$$

$$P = \lim_{T\to\infty} \frac{1}{2T} \left[\frac{8}{3} + 2T - 4 \right] = 2 \text{ watt}$$

(vi)

$$x(t) = (1+e^{-5t})u(t)$$

$$E = \lim_{T\to\infty} \int_0^T (1+e^{-5t})^2 dt = \lim_{T\to\infty} \int_0^T (1+e^{-10t} + 2e^{-5t})\, dt$$

$$= \lim_{T\to\infty}\left(t - \frac{1}{10}e^{-10t} - \frac{2}{5}e^{-5t}\Big|_0^T\right) = \lim_{T\to\infty}\left(T - \frac{1}{10}e^{-10T} - \frac{2}{5}e^{-5T} + \frac{1}{10} + \frac{2}{5}\right) = \infty$$

$$P = \lim_{T\to\infty}\frac{1}{2T}\left(T - \frac{1}{10}e^{-10T} - \frac{2}{5}e^{-5T} + \frac{1}{2}\right) = \lim_{T\to\infty}\left(\frac{1}{2}+0\right) = \frac{1}{2}\text{ watt}$$

(vii)

$$E = \sum_{n=-\infty}^{\infty}\left(\frac{1}{3}\right)^{2n}u(n) = \sum_{n=0}^{\infty}\left(\frac{1}{9}\right)^{n} = 1 + \frac{1}{9} + \frac{1}{9^2} + \frac{1}{9^3} + \cdots = \frac{1}{1-1/9} = \frac{9}{8}$$

$$P = \lim_{N\to\infty}\frac{1}{2N+1}\sum_{n=-N}^{N}\left(\frac{1}{9}\right)^{n}u(n)$$

$$= \lim_{N\to\infty}\frac{1}{2N+1}\sum_{n=0}^{N}\left(\frac{1}{9}\right)^{n} = \lim_{N\to\infty}\frac{1}{2N+1}\left[\frac{\left(1-\left(\frac{1}{9}\right)^{N+1}\right)}{1-\frac{1}{9}}\right] = 0$$

Example 8: Find if the following signals are energy or power signals; if yes, determine their energy/power.

(i) $x(t) = \begin{cases} A & -T \le t \le T \\ 0 & \text{elsewhere} \end{cases}$

(ii) $x(n) = n$

SOLUTION:

(i) The signal $x(t)$ is time limited, hence this is an energy signal, its energy is given as

$$E = \int_{-\infty}^{\infty}|x(t)|^2\,dt = \int_{-T}^{T}A^2\,dt = 2A^2T$$

Thus, energy is finite. Hence, it is an energy signal.

(ii) The signal $x(n) = n$ grows as value of n increases hence energy of this signal will be infinite. Since it is a non-periodic signal, its power can be calculated as

$$P = \lim_{N\to\infty}\frac{1}{2N+1}\sum_{n=-N}^{N}|x(n)|^2 = \lim_{N\to\infty}\frac{1}{2N+1}\sum_{n=-N}^{N}n^2$$

$$= \lim_{N\to\infty}\frac{1}{2N+1}2\sum_{n=0}^{N}n^2 = \lim_{N\to\infty}\frac{1}{2N+1}2\sum_{n=1}^{\infty}n^2 + 0$$

Here use the formula, $\sum_{n=1}^{N} n^2 = \frac{N(N+1)(2N+1)}{6}$

$$P = \lim_{N\to\infty}\frac{1}{2N+1}2\times\frac{N(N+1)(2N+1)}{6} = \infty$$

Since energy and power both are infinite, the given signal is neither an energy nor a power signal.

Comparison of energy and power signals

S. No.	Energy signal	Power signal
1.	Total normalized energy is finite and non-zero. $0 < E < \infty$	Total normalized average power is finite and non-zero. $0 < P < \infty$
2.	Total normalized energy is given as $E = \lim_{T\to\infty} \int_{-T}^{T} \lvert x(t)\rvert^2 \, dt$	Total power is given as $P = \lim_{T\to\infty} \frac{1}{2T} \int_{-T}^{T} \lvert x(t)\rvert^2 \, dt$
3.	Non-periodic signals are energy signals	Practical periodic signals are power signals
4.	Energy signals are time limited	Power signals can exist over infinite time
5.	Power of the energy signal is zero over infinite time, $P = 0$	Energy of power signals is infinite over infinite time, $E = \infty$
6.	Example: Rectangular pulse 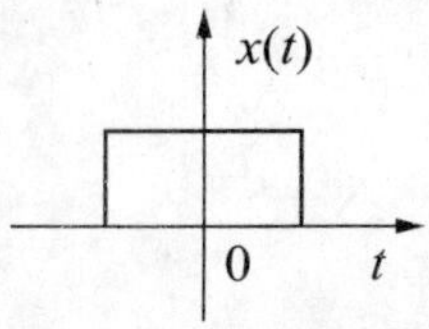	Example: Periodic pulse train 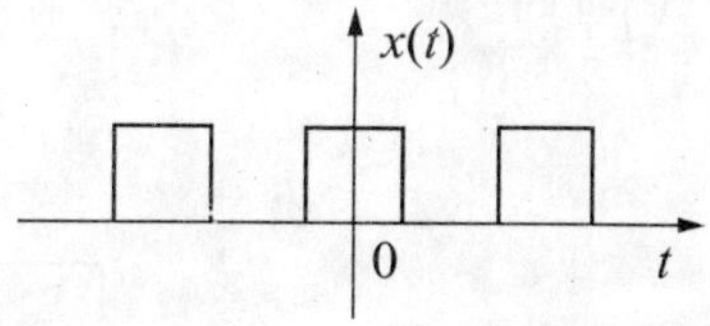

Example 9: Determine whether the given signals are energy signals or power signals, and evaluate their normalized energy or power.

(i) $x(t) = \text{rect}\,(t/T)$

(ii) $x(t) = \text{rect}\,(t/T)\cos\omega_0 t$

(iii) $x(t) = \cos^2\omega_0 t$

SOLUTION:

(i)

$$x(t) = \text{rect}\,(t/T)$$

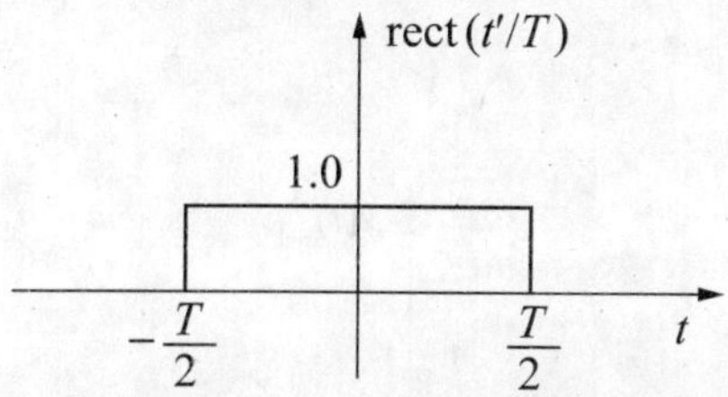

Fig. 1.17. Rectangular pulse

$$\text{rect}\left(\frac{t}{T}\right) = \begin{cases} 1 & \frac{-T}{2} < t < \frac{T}{2} \\ 0 & \text{elsewhere} \end{cases}$$

Total normalized energy $E = \lim_{T\to\infty} \int_{-T}^{T} |x(t)|^2 \, dt$

$$E = \int_{-\infty}^{\infty} |x(t)|^2 \, dt = \int_{-\infty}^{\frac{-T}{2}} |x(t)|^2 \, dt + \int_{\frac{-T}{2}}^{\frac{T}{2}} |x(t)|^2 \, dt + \int_{\frac{T}{2}}^{\infty} |x(t)|^2 \, dt$$

(1) and (3) are zero

$$E = \int_{\frac{-T}{2}}^{\frac{T}{2}} dt = (t)_{\frac{-T}{2}}^{\frac{T}{2}} = T, \text{ which is finite and non-zero.}$$

Hence, this is an energy signal.

(ii) $x(t) = \text{rect}\left(\frac{t}{T}\right) \cos \omega_0 t$

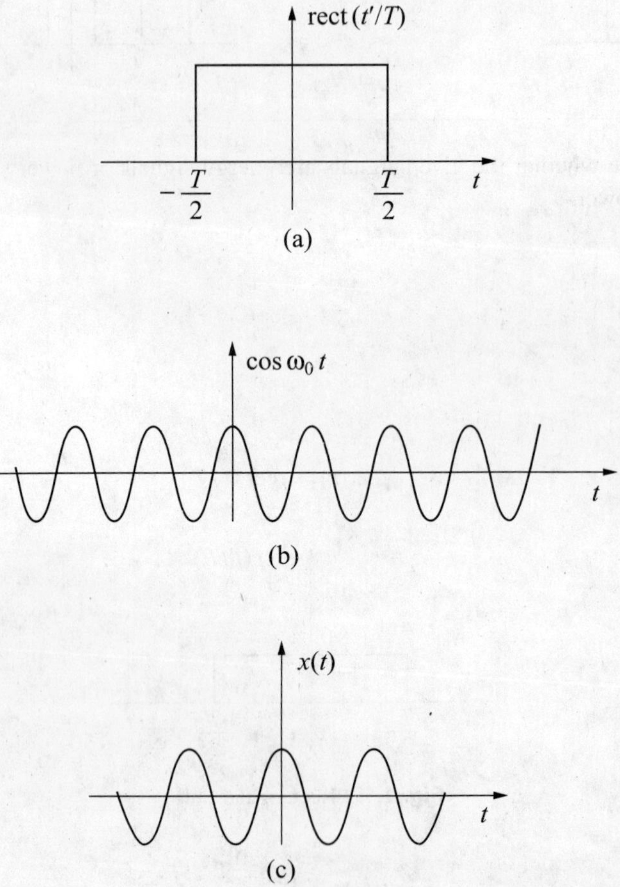

Fig. 1.18. (a) Rectangular pulse (b) Cosinusoidal wave (c) Graphical representation of $\text{rect}\left(\frac{t}{T}\right) \cos \omega_0 t$

$$E = \int_{-\infty}^{\infty} |x(t)|^2 dt = \int_{\frac{-T}{2}}^{\frac{T}{2}} \cos 2\omega_0 t dt = \frac{1}{2}\int_{\frac{-T}{2}}^{\frac{T}{2}} (1+\cos 2\omega_0 t)\, dt$$

$$= \frac{1}{2}\left[\int_{\frac{-T}{2}}^{\frac{T}{2}} dt + \int_{\frac{-T}{2}}^{\frac{T}{2}} \cos 2\,\omega_0 t dt\right]$$

$$= \frac{1}{2}[T] \sin\theta \int_{\frac{-T}{2}}^{\frac{T}{2}} \cos 2\omega_0 t dt$$

Since $\int_{-T/2}^{T/2} \cos 2\omega_0 t dt$ is the integration of "full cycle" of cosine wave which is zero.

$$E = \frac{T}{2}$$

or

$$\int_{\frac{-T}{2}}^{\frac{T}{2}} \cos 2\omega_0\, t dt = \left[\frac{\sin 2\,\omega_0\, t}{2\omega_0}\right]_{\frac{-T}{2}}^{\frac{T}{2}}$$

$$= \frac{1}{2\omega_0}\left[\sin\left(2\cdot\frac{2\pi}{T}\cdot\frac{T}{2}\right) + \sin\left(2\cdot\frac{2\pi}{T}\cdot\frac{T}{2}\right)\right] = \frac{1}{2\omega_0}(2\sin 2\pi) = 0$$

(iii) $x(t) = \cos 2\omega_0 t$.

The given function is a periodic cosine wave. We can say that this is a power signal for when we calculate P it is finite and $E = \infty$

$$P = \lim_{T\to\infty} \frac{1}{2T}\int_{-T}^{T} |x(t)|^2 dt = \lim_{T\to\infty} \frac{1}{2T}\int_{-T}^{T} \cos 4\omega_0 t dt$$

$$= \lim_{T\to\infty} \frac{1}{2T}\int_{-T}^{T} \frac{1}{8}[3+4\cos 2\omega_0 t + \cos 4\omega_0 t] dt$$

$$= \underset{(1)}{\lim_{T\to\infty} \frac{1}{2T}\int_{-T}^{T} \frac{3}{8} dt} + \underset{(2)}{\lim_{T\to\infty} \frac{1}{2T}\int_{-T}^{T} \frac{1}{2}\cos 2\,\omega_0 t dt} + \underset{(3)}{\lim_{T\to\infty} \frac{1}{2T}\int_{-T}^{T} \frac{1}{8}\cos 4\,\omega_0 t dt}$$

$$= \lim_{T\to\infty} \frac{1}{2T}\frac{3}{8}\cdot 2T = \frac{3}{8}.$$

This proves that the power of the given signal is finite and non-zero, hence it is a power signal. Solve (2) and (3). Integrated as stated in the previous example or we can say that terms (2) and (3) are zero since it is integration of cosine wave over "full cycles".

1.2.5 Deterministic and random signals

Deterministic signal: A deterministic signal is a signal exhibiting no uncertainty of its magnitude and phase at any given instant of time, its instantaneous value can be accurately determined by a mathematical equation.

For example, sine wave, cosine wave, triangular wave, exponential pulse, etc. are deterministic signals, since they have unique mathematical equations, i.e.

$$x(t) = 10\sin(50\pi t)$$
$$x(n) = e^{-n}$$

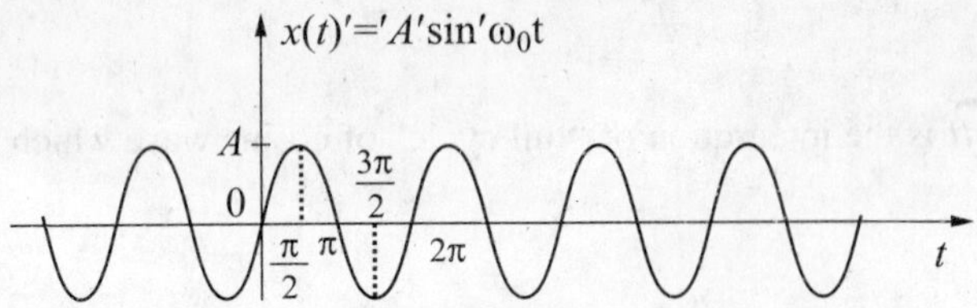

Fig. 1.19. Sine wave represented by $x(t) = A\sin\omega_0 t$ is a deterministic continuous-time signal

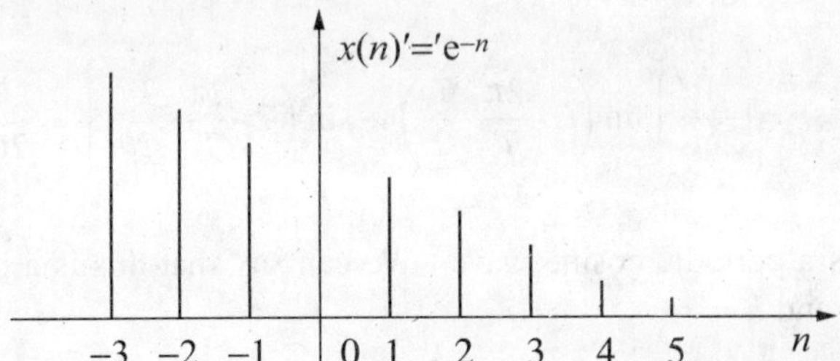

Fig. 1.20. Exponential signal $x(n) = e^{-n}$ is a deterministic discrete time signal

Random signal: A random signal is a signal characterized by uncertainity about its actual occurrence. That means a random signal cannot be defined by some mathematical expression. Mathematical equation cannot be written for these signals to predict their value in future.

Causal and non-causal signals: A continuous-time signal $x(t)$ is said to be causal if $x(t) = 0$ for $t < 0$, otherwise the signal is non-causal for a non-causal signal $x(t) = 0$ for $t > 0$.

Similarly, a discrete-time signal $x(n)$ is said to be causal if $x(n) = 0$ for $n < 0$; otherwise the signal is non-causal.

One-dimensional signals: A signal is said to be a one-dimensional signal if it depends on only one variable. For example, variation of room temperature, ac power signal, electrocardiogram, speech signal, etc.

Two-dimensional signals: A signal is said to be multidimensional signal if it depends on more than one variable. For example, x rays, sonograms, pictures, etc.

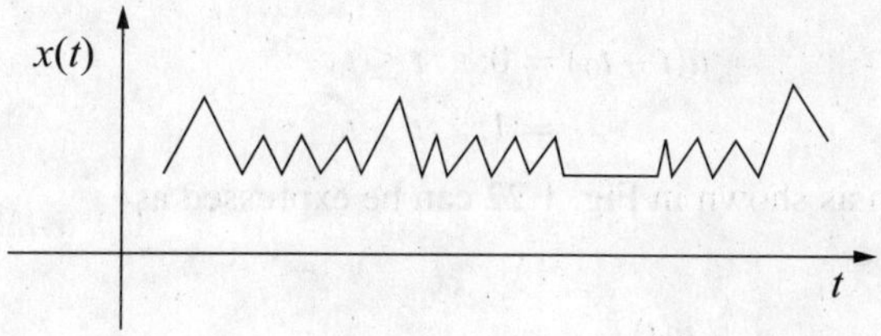

(a)

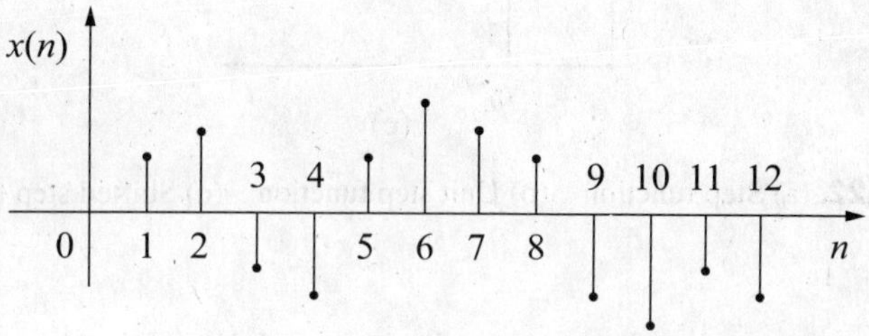

(b)

Fig. 1.21. (a) & (b) Random signals

1.3 COMMONLY USED SIGNALS

1.3.1 Unit step function

A step function is shown in figure.

(a)

The mathematical representation of step function is

$$V(t) = 0; \quad t \leq 0$$
$$= k; \quad t > 0$$

An unit step function, denoted by $u(t)$, is defined as

$$u(t) = 0; \quad t \leq 0$$
$$= 1; \quad t > 0$$

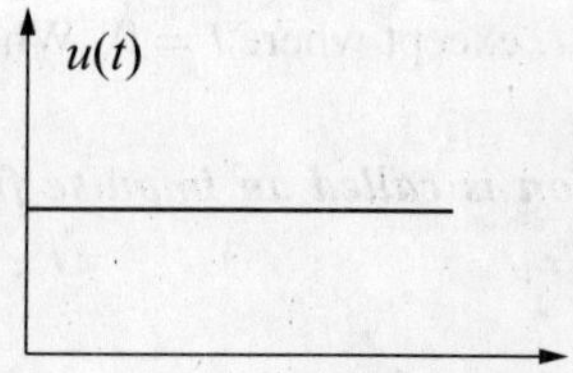

(b)

$$u(t-t_0) = 0; \quad t \le t_0$$
$$= 1; \quad t > t_0$$

A shifted or delayed step function as shown in Fig. 1.22 can be expressed as

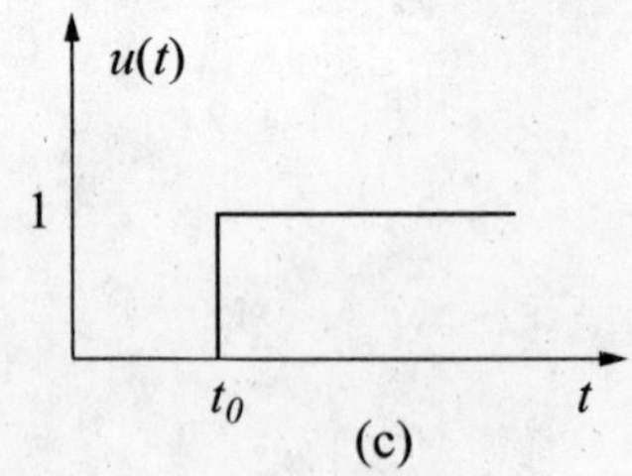

Fig. 1.22. (a) Step function (b) Unit step function (c) Shifted step function

Time inversion

If we want to reverse the unit step function, we can flip it around the y axis as such: $u(-t)$. With a little bit of manipulation, we can come to an important result:

$$u(-t) = 1 - u(t)$$

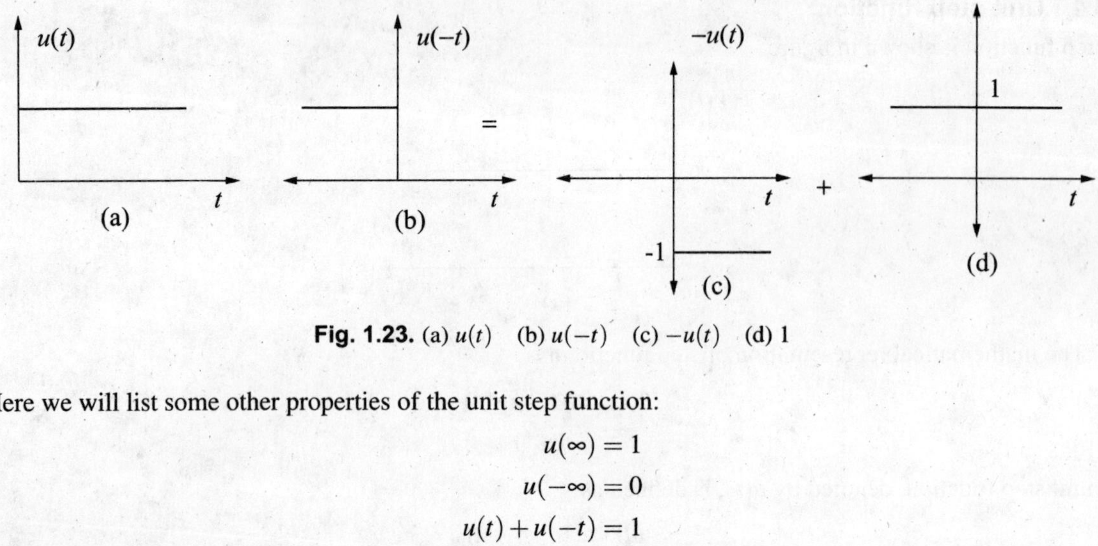

Fig. 1.23. (a) $u(t)$ (b) $u(-t)$ (c) $-u(t)$ (d) 1

Here we will list some other properties of the unit step function:

$$u(\infty) = 1$$
$$u(-\infty) = 0$$
$$u(t) + u(-t) = 1$$

Derivative

The unit step function is level in all places except for a discontinuity at $t = 0$. For this reason, the derivative of the unit step function is 0 at all points t, except where $t = 0$. Where $t = 0$, the derivative of the unit step function is infinite.

The derivative of a unit step function is called an impulse function. The impulse function will be described in more detail next.

Gate Function:

A **gate function** $G(t)$ can be written as the sum of two step functions.

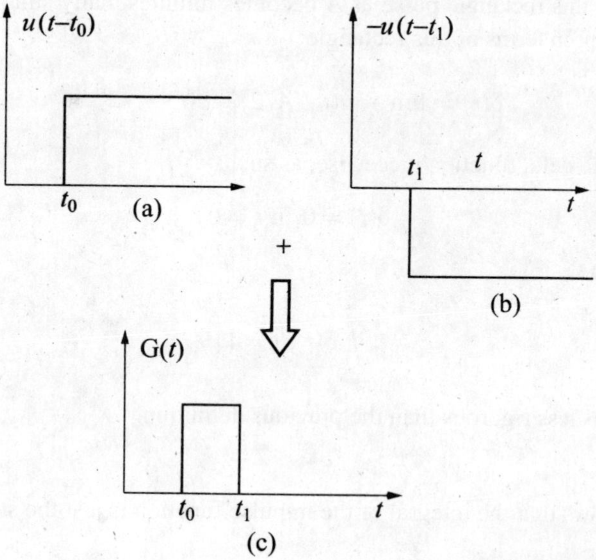

Fig. 1.24. (a) $u(t-t_0)$ (b) $u(t-t_1)$ (c) $u(t-t_0)-u(t-t_1)$

$$G(t) = u(t-t_0) + (-u(t-t_1)) = u(t-t_0) - u(t-t_1)$$

1.3.2 Impulse function

An impulse function is a special function that is often used by engineers to model certain events. An impulse function is not realizable, in that by definition the output of an impulse function is infinity at certain values. An impulse function is also known as a "delta function". Specifically, the unit-impulse function is known as the Dirac delta function. The term "impulse function" is unambiguous, because there is only one definition of the term "impulse".

Let's start by drawing out a rectangle pulse function, $\delta(t)$, as such:

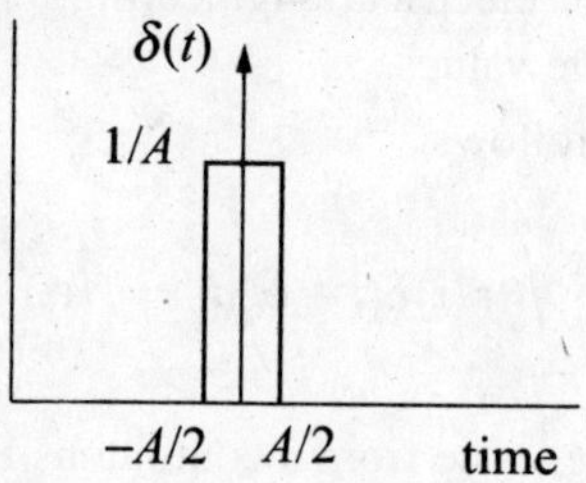

Fig. 1.25. Rectangular pulse as an approximation to $\delta(t)$

We can define this rectangle pulse in terms of the unit step function:

$$D(t) = \frac{1}{A}[u(t+A/2) - u(t-A/2)]$$

Important points related to rectangle pulse that make it impulse

- The width of the pulse is zero.
- The height of the pulse goes to infinity.
- The area under the pulse-curve is always unity.

Now, we want to analyze this rectangle pulse as A becomes infinitesimally small. We can define this new function, the delta function, in terms of this rectangle:

$$\delta(t) = \lim_{A \to 0} \frac{1}{A}[u(t + A/2) - u(t - A/2)]$$

We can similarly define the delta function piecewise, as such:

$$\delta(t) = 0, \text{ if } t \neq 0.$$

$\delta(t)$ is undefined for $t = 0$.

$$\int_{-\infty}^{\infty} \delta(t)dt = 1.$$

Although, this definition is less rigorous than the previous definition.

Integration
From its definition it follows that the integral of the impulse function is just the step function:

$$\int \delta(t)dt = u(t)$$

Thus, defining the derivative of the unit step function as the impulse function is justified.

Sapling Theorem (1)
Furthermore, for an integrable function f:

$$\int_{-\infty}^{\infty} \delta(t - A)f(t)dt = f(A)$$

This is known as the sampling theorem because, it effectively samples the value of the function f, at location A.

The delta function has many uses in electrical engineering, and one of the most important uses is to sample a continuous function into discrete values.

The sampling theorem is written as follows:

$$\int_{-\infty}^{\infty} f(t)\delta(t - \alpha)dt = f(\alpha)$$

Using this theory, we can extract a single value from a continuous function by multiplying with an impulse, and then integrating.

1.3.3 Ramp function

A Ramp function, as shown in figure (a), is described as

$$\begin{aligned} r(t) &= 0; \quad t \leq 0 \\ &= \tan\theta; \quad t > 0 \end{aligned}$$

A shifted or delayed ramp function is shown in figure (b)

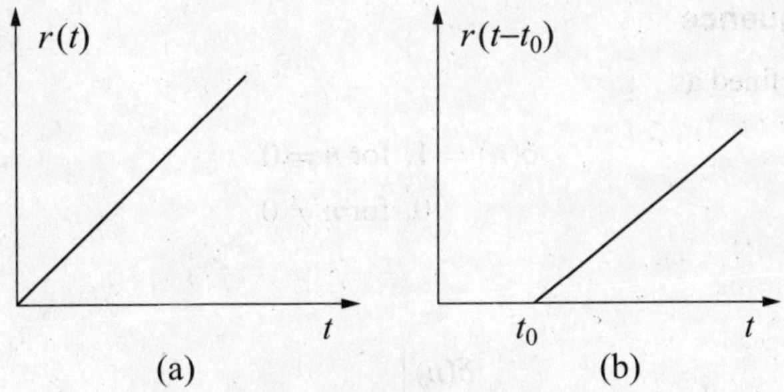

Fig. 1.26. (a) Ramp function (b) Shifted ramp function

$$r(t) = 0; \quad t \leq t_0$$
$$= \tan\theta; \quad t > t_0$$

Discrete-time signals

A discrete-time signal is defined only at discrete instants of time, thus in this case the independent variable takes on only discrete values, which are usually uniformly spaces.

1.3.4 Unit step sequence

Unit step sequence is defined as

$$u(n) = 1 \text{ for } n \geq 0$$
$$0 \text{ for } n < 0$$

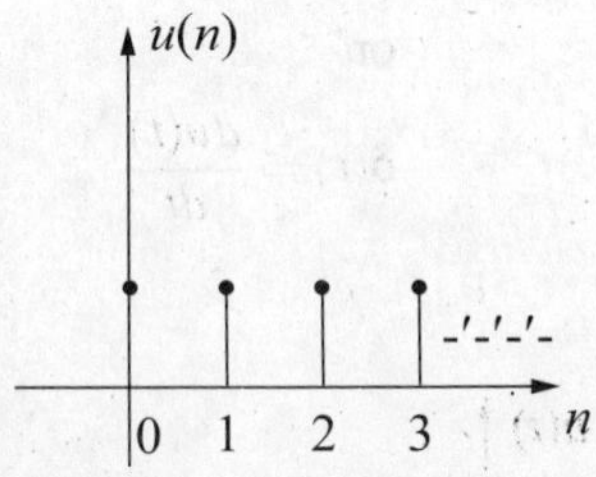

Fig. 1.27. Unit step sequence

1.3.5 Unit ramp sequence

Unit ramp function is defined as

$$r(n) = n \text{ for } n \geq 0$$
$$0 \text{ for } n < 0$$

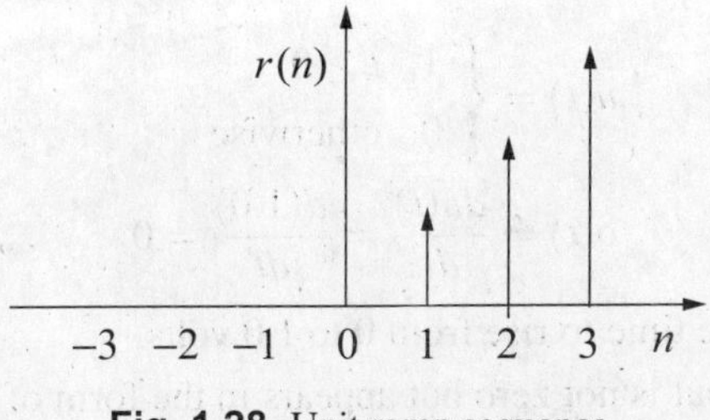

Fig. 1.28. Unit ramp sequence

1.3.6 Unit sample sequence

Unit sample sequence is defined as

$$\delta(n) = 1, \text{ for } n = 0$$
$$\quad 0, \text{ for } n \neq 0$$

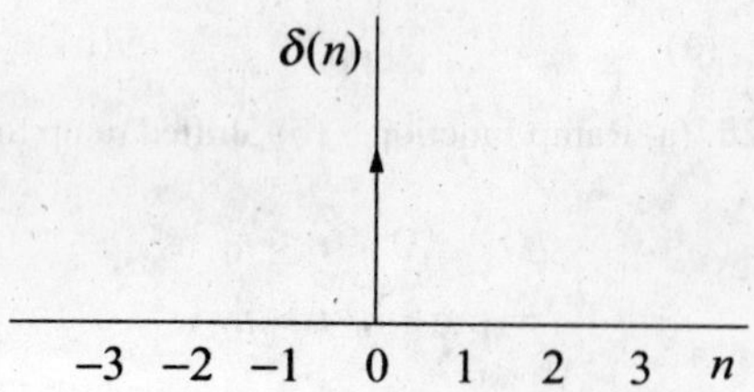

Fig. 1.29. Unit sample sequence

1.3.7 Interrelationship between unit step, unit impulse and unit ramp

From its definition, it follows that the integral of the impulse function is just the step function.

$$\int \delta(t)dt = u(t)$$

or

$$\delta(t) = \frac{du(t)}{dt}$$

Explanation:

Fig. 1.30. Unit step function

$$u(t) = \begin{cases} 1, & t \geq 0 \\ 0, & \text{otherwise} \end{cases}$$

$$\delta(t) = \frac{du(t)}{dt} = \frac{d(1.0)}{dt} = 0$$

Actually the step input takes a finite time to rise from 0 to 1.0 volts.

Due to this finite time, the output is not zero but appears in the form of a spike at $t = 0$.

The ramp function can be obtained by applying unit step function to an integrator

$$r(t) = \int u(t)dt$$

or

$$u(t) = \frac{dr(t)}{dt}$$

$$\boxed{\delta(t) = \frac{du(t)}{dt} = \frac{d^2r(t)}{dt^2}}$$

Example 10: Prove the following relationships.

(a) $u(n) = \sum_{k=-\infty}^{n} \delta(k)$

(b) $u(n) = \sum_{k=0}^{\infty} \delta(n-k)$

(c) $\delta(n) = u(n) - u(n-1)$

SOLUTION:

(a)

$$u(n) = \sum_{k=-\infty}^{n} \delta(k)$$

$$\text{R.H.S. } \sum_{k=-\infty}^{n} \delta(k)$$

Let n be $= -1$, so $\sum_{k=-\infty}^{-1} \delta(k) = \ldots\delta(-2) + \delta(-1)$

$$= 0 \qquad\qquad \text{Since } \delta(k) = \begin{cases} 1 & k = 0 \\ 0 & k \neq 0 \end{cases}$$

Now $n = 0$, so $\sum_{k=-\infty}^{0} \delta(k) = \ldots\delta(-2) + \delta(-1) + \delta(0)$

$$= 1$$

$n = 1$, so $\sum_{k=-\infty}^{1} \delta(k) = \ldots\delta(-2) + \delta(-1) + \delta(0) + \delta(1)$

$$= \ldots 0 + 0 + 1 + 0$$

$$= 1$$

$n = 2$, so $\sum_{k=-\infty}^{1} \delta(k) = \ldots\delta(-2) + \delta(-1) + \delta(0) + \delta(1) + \delta(2)$

$$= \ldots 0 + 0 + 1 + 0 + 0$$

$$= 1$$

$$\vdots$$

so and so forth

Therefore, $\sum_{k=-\infty}^{n} \delta(k) = \begin{cases} 0 & \text{for} \quad k < 0 \\ 1 & \text{for} \quad k \geq 0 \end{cases} = u(n)$

(b)

$$u(n) = \sum_{k=0}^{\infty} \delta(n-k)$$

R.H.S. $\quad \sum_{k=0}^{\infty} \delta(n-k) = \delta(n) + \delta(n-1) + \delta(n-2) + \delta(n-3) + \ldots$

Graphical representation

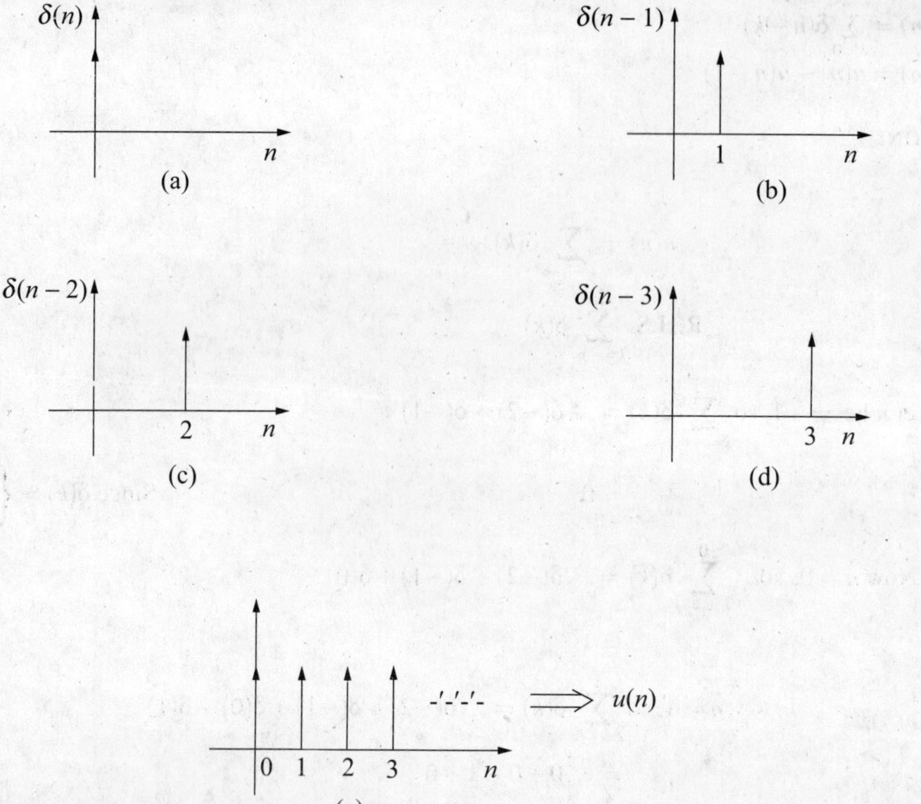

Fig. 1.31. (a) $\delta(n)$ (b) $\delta(n-1)$ (c) $\delta(n-2)$ (d) $\delta(n-3)$ (e) Summation of (a), (b), (c), (d) ...

On adding all these waveforms, we get

$$u(n) = \begin{cases} 1 & n \geq 0 \\ 0 & n < 0 \end{cases}$$

(c) $\delta(n) = u(n) - u(n-1)$

Graphical representation

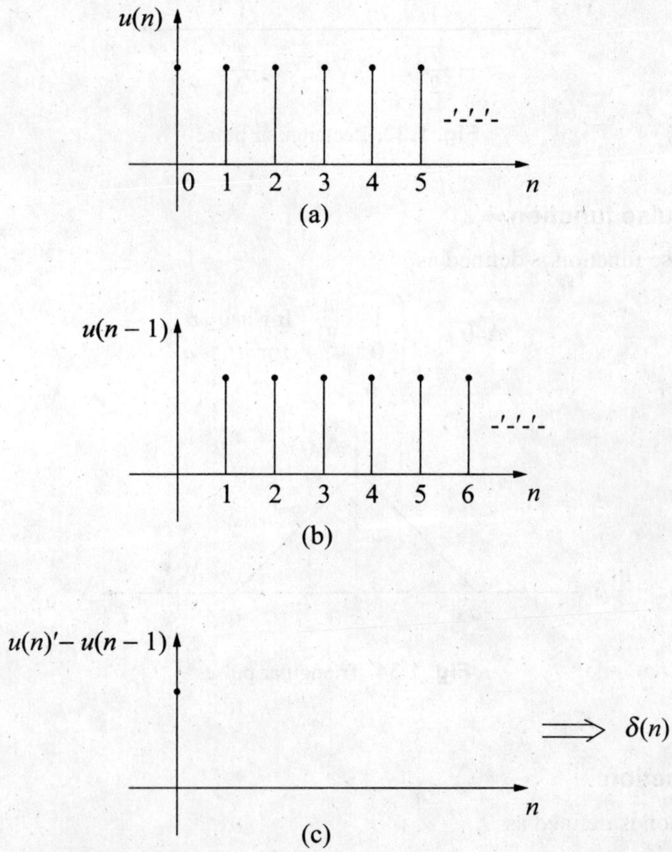

Fig. 1.32. (a) $u(n)$ (b) $u(n-1)$ (c) $u(n) - u(n-1)$

1.3.8 Rectangular function

Also known as rectangle function, rect function, unit pulse, or the normalized boxcar function.

Consider a rectangular pulse of duration T and amplitude A, as shown in figure to define this pulse mathematically in convenient form, we use the following notation

$$\prod(t) = \text{rect}\left(\frac{t}{T}\right) = \left\{ \begin{array}{ll} 1 & \text{for } -\frac{T}{2} \le t < \frac{T}{2} \\ 0 & \text{elsewhere} \end{array} \right\}$$

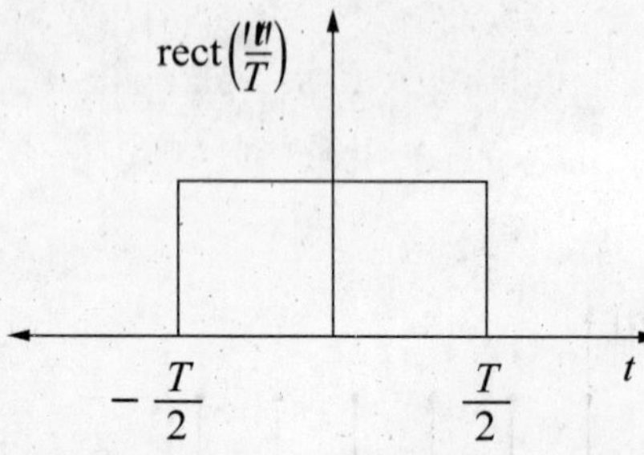

Fig. 1.33. Rectangular pulse

1.3.9 Triangular pulse function

The unit triangular pulse function is defined as

$$\Delta_a(t) = \begin{cases} 1 - \frac{|t|}{a} & \text{for } |t| \le a \\ 0 & \text{for } |t| > a \end{cases}$$

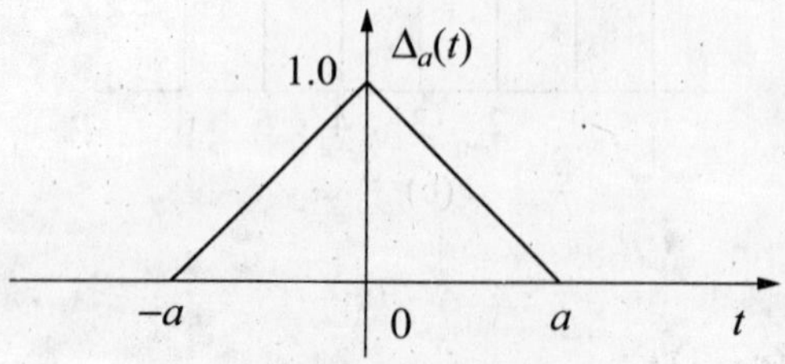

Fig. 1.34. Triangular pulse

1.3.10 Signum function

The unit signum function is defined as

$$\operatorname{sgn}(t) = \begin{cases} 1 & t > 0 \\ 0 & t = 0 \\ -1 & t < 0 \end{cases}$$

Any real number can be expressed as the product of its absolute value and its sgn function:

$$x = (\operatorname{sgn} x)|x|$$

sgn function is related to unit step function

$$\operatorname{sgn} t = 2u(t) - 1$$

Explanation:

$$\operatorname{sgn}(t) = 2u(t) - 1$$

$$x_1(t) \quad x_2(t)$$

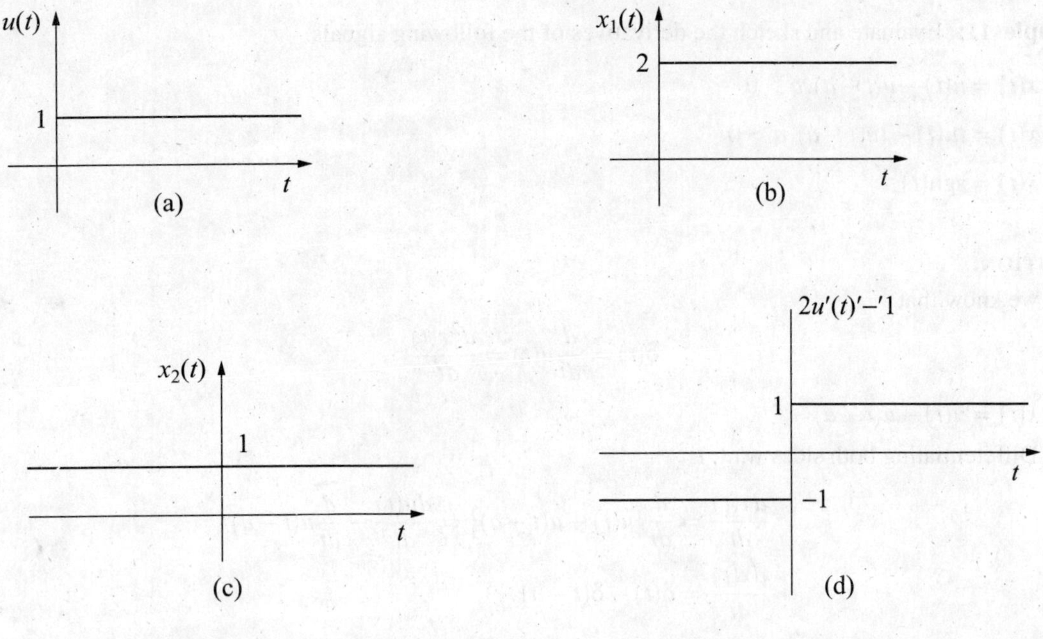

Fig. 1.35. Graphical representation of (a) $u(t)$ (b) $2u(t)$ (c) 1 (d) $2u(t) - 1$

1.3.11 Sinc function

The sinc function, also called the "sampling function", is a function that arises frequently in signal processing and the theory of Fourier transforms. The full name of the function is "sine cardinal," but it is commonly referred to by its abbreviation, "sinc". The "sinc" function is defined by

$$\text{sinc}(x) = \frac{\sin(\pi x)}{\pi x} \quad \text{if } x \neq 0$$

and

$$\text{sinc}(0) = 1$$

The function has its maximum value of unity at $x = 0$, and it approaches zero as x approaches infinity the function is oscillatory in nature.

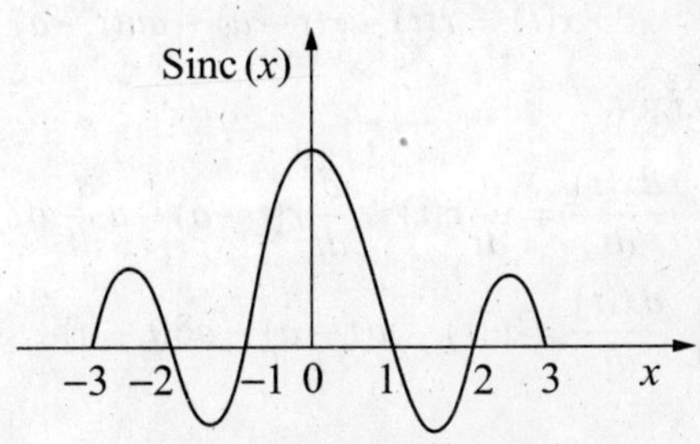

Fig. 1.36. Sinc function

Example 11: Evaluate and sketch the derivatives of the following signals.

(a) $x(t) = u(t) - u(t-a), a > 0$

(b) $x(t) = tu(t) - tu(t-a), a > 0$

(c) $x(t) = \text{sgn}(t)$

SOLUTION:

Since we know that

$$\delta(t) = \frac{d}{dt}u(t) = \frac{d^2 r(t)}{dt^2}$$

(a) $x(t) = u(t) - u(t-a)$

Differentiating both sides w.r.t. t

$$\frac{dx(t)}{dt} = \frac{d}{dt}[u(t) - u(t-a)] = \frac{du(t)}{dt} - \frac{d}{dt}u(t-a)$$

$$\frac{dx(t)}{dt} = \delta(t) - \delta(t-a)$$

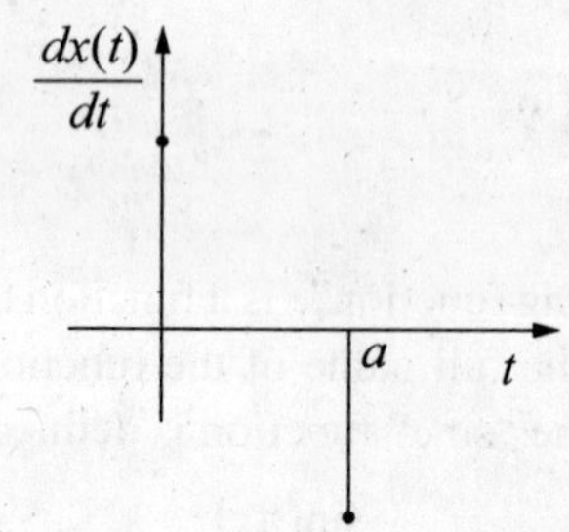

Fig. 1.37. Graphical representation of $\delta(t) - \delta(t-a)$

(b) $x(t) = tu(t) - tu(t-a) = tu(t) - (t-a)u(t-a) - au(t-a)$

We write $tu(t-a)$ as $(t-a)u(t-a) + au(t-a)$

$$x(t) = r(t) - r(t-a) - au(t-a)$$

Differentiating both sides w.r.t. t

$$\frac{dx(t)}{dt} = \frac{d}{dt}r(t) - \frac{d}{dt}r(t-a) - a\frac{d}{dt}u(t-a)$$

$$\frac{dx(t)}{dt} = u(t) - u(t-a) - a\delta(t-a)$$

$$\delta(t) = \begin{cases} 1 & t = 0 \\ 0 & t \neq 0 \end{cases}$$

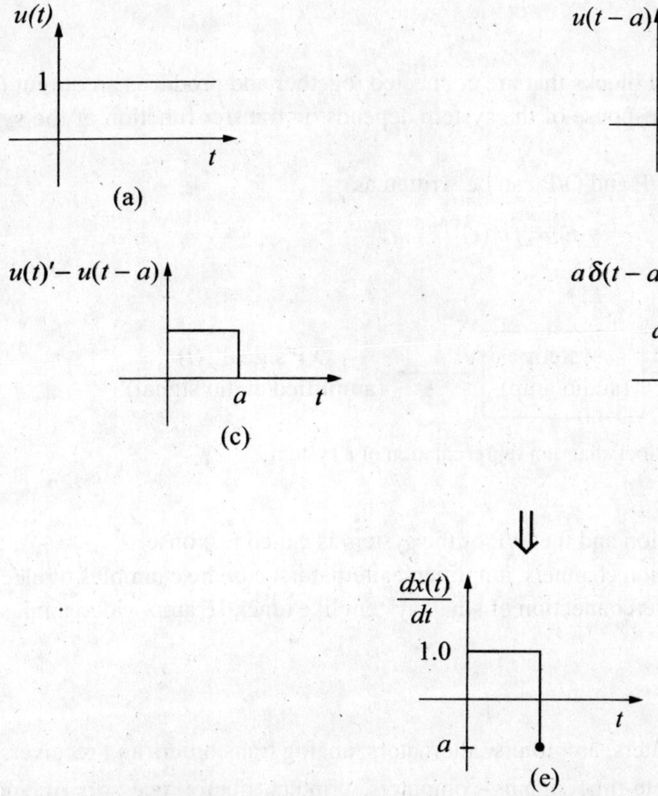

Fig. 1.38. Graphical representation of (a) $u(t)$ (b) $u(t-a)$ (c) $u(t)-u(t-a)$ (d) $a\delta(t-a)$ (e) $u(t)-u(t-a)-a\delta(t-a)$

(c)

$$x(t) = \text{sgn}(t) = \begin{cases} 1 & t > 0 \\ -1 & t < 0 \end{cases}$$

$$x(t) = \text{sgn}(t) = 2u(t) - 1$$

Differentiating both sides w.r.t. t

$$\frac{dx(t)}{dt} = 2\frac{du(t)}{dt} = 2\delta(t)$$

$$\frac{dx(t)}{dt} = 2\delta(t)$$

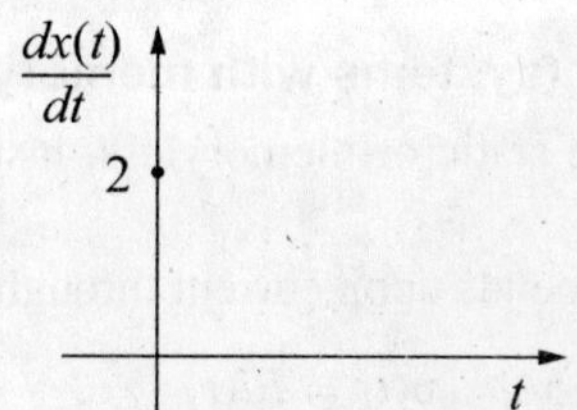

Fig. 1.39. Graphical representation of $2\delta(t)$

1.4 SYSTEM

A system is a set of elements or functional blocks that are connected together and produces an output (O/U) in response to an input (I/P) signal. The response of the system depends on transfer function of the system. I/P signal is $x(t)$ and O/P signal is $y(t)$.

The functional relationship between I/P and O/P can be written as

$$y(t) = f[x(t)]$$

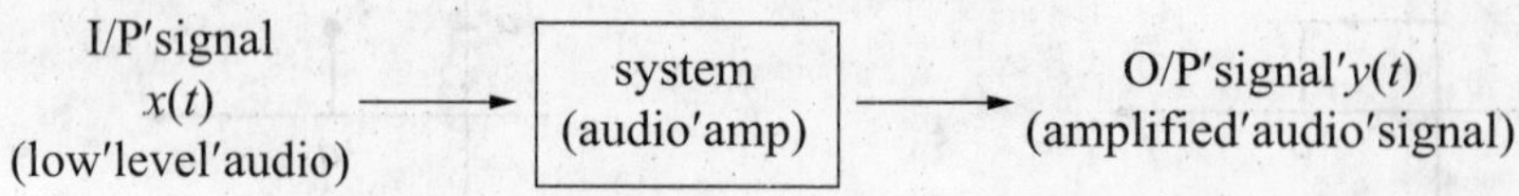

Fig. 1.40. Block diagram representation of a system

The I/P to the system is called excitation and the O/P of the system is called response.

Different types of filters, communication channels, amplifiers, alternators, etc. are examples of electrical system. Big system like a TV set is the interconnection of small system like tuner, IF amp, video amp, sound amp, etc.

1.4.1 Classification of systems

(i) Continuous-time systems–analog filters, amplifiers, alternators, analog transmitters and receivers, etc.

(ii) Discrete-time system handle discrete-time signals–computers, printers, microprocessors, memories, shift registers, etc.

Continuous as well as DTS can be first classified here based on their properties:

Filters: Reject unwanted frequencies present at the input signal.
Amplifier: Increases the strength of the signal present at the input for a car travelling along a highway, the pressure on the accelerator pedal is I/P signal and the speed of the car is O/P signal.

(1) **Dynamicity property:** Static and dynamic system
(2) **Shift invariants:** Time-invariant and time-variant systems
(3) **Linearity property:** Linear and non-linear systems
(4) **Causality property:** Causal and non-causal systems.
(5) **Stability property:** Stable and unstable systems.
(6) **Invertibility property:** Invertible and non-invertible systems.

1.4.2 Static and dynamic systems (systems with memory and without memory)

The continuous time system is said to be static or memoryless, instantaneous if its O/P depends upon the present input only.

e.g., voltage drop in the resistance depends upon current through it

$$\upsilon(t) = Ri(t)$$

Physically these systems contain no energy storage elements. The eqn. relating its O/P signal to its I/P signal contains no derivation, integrals or signal delays.

Another simple memoryless (static) system is identity systems.

$$y(t) = x(t)$$

A dynamic system or system with memory is a system with an O/P signal that at any specified time depends on the value of the I/P signal (present) as well as the past value.

$$\upsilon(t) = \frac{1}{c}\int_{-\infty}^{t} i(t)dt$$

Voltage across the capacitor depends upon the present as well as the past current values.

These systems have one or more energy storage elements.

I/P – O/P relationship of a dynamic CTS is described by its differential equation.

The DTS can also be static or dynamic. If the O/P of DTS depends upon the present input sample only, then it is called a static or memoryless or instantaneous system.

$$y(n) = 10x(n)$$

$$y(n) = 15 \cdot x^2(n) + 10x(n)$$

Dynamic DTS

$$y(n) = x(n) + x(n-1)$$

$$\Rightarrow \quad y(n) = \sum_{k=0}^{4} x(n-k) = x(n) + x(n-1) + x(n-2) + x(n-3) + x(n-4)$$

Example 12: Check whether the given systems are instantaneous or not.

(1) $y(t) = 3x(t) - x^2(t)$

(2) $y(t) = 4x(t) + x(t^{-1})$

(3) $y(t) = k\int_{-\infty}^{t} x(\tau)d\tau$

(4) $y(n) = 2x(n)$

(5) $y(n) = x(n-1)$

(6) $y(n) = y[n-1] + 2x[n]$

(7) $y(n) = \sum_{k=-\infty}^{n} x[k]$

(8) $y(n) = 2x^2(n) + 3x(n)$

1.4.3 Time-invariant and time-variant systems

ACTS is time invariant if the time shift in the I/P signal results in corresponding time shift in the O/P.

Let $y(t) = f[x(t)]$, i.e. $y(t)$ is response for $x(t)$

Then if $x(t)$ is delayed by time t_1, then O/P will also be delayed by the same time, i.e.

$$f[x(t-t_1)] = y(t-t_1)$$

Similarly, if the I/P and O/P characteristics of the DTS do not change with shift of time origin, such systems are called shift invariant or time-invariant system.

Let the system has I/P $x(n)$ and corresponding O/P $y(n)$, i.e.
$y(n) = f[x(n)]$. Then the system is shift invariant or time invariant if and only if

$$f[x(n-k)] = y(n-k)$$

Here k is a constant if the I/P is delayed by k samples, then O/P is also delayed by the same number of samples for time-invariant system.

Time variant:

(1) Change in ambient temperature
(2) Rainfall per month
(3) Noise effect in radio communication channels

Time invariant:

(1) Bus and train arrival and departures
(2) File handling in C language
(3) Printing documents by the printer

Example 13: Check whether the given systems are time invariant or time variant.

(1) $y(t) = \sin x(t)$
(2) $y(t) = tx(t)$
(3) $y(t) = \sin tx(t-2)$
(4) $y(t) = \int_{-\infty}^{t} x(\tau)d\tau$

SOLUTION:

(1) $y(t) = \sin x(t)$ (a)
Consider an I/P obtained by shifting $x(t)$ by t_1

$$x_1(t) = f[x(t)] = \sin x(t-t_1) \quad (1)$$

O/P corresponding to $x_1(t)$ is

$$y_1(t) = y(t-t_1) \quad (2)$$

In (a) replacing t by $(t-t_1)$; $y_2(t) = y(t-t_1) = \sin x(t-t_1)$
Since $y_1(t) = y_2(t)$ so time invariant

(2) $y(t) = tx(t)$
Consider an I/P obtained by shifting $x(t)$ by t_1

$$y_1(t) = y(t,t_1) = f[x(t)] = tx(t-t_1)$$

Now delay the O/P $y(t)$ by t_1

$$y_2(t) = y(t-t_1) = (t-t_1)x(t-t_1)$$

Since $y_1(t) \neq y_2(t)$, so system is time variant.

(3) $y(t) = \sin tx(t-2)$
Let us determine the O/P of the system for delayed input $x(t-t_1)$ i.e.

$$y_1(t) = y(t,t_1) = f[x(t-t_1)] = \sin tx(t-t_1-2) \quad (1)$$

Now delay the O/P by t_1

$$y_2(t) = y(t-t_1) = \sin(t-t_1)x(t-t_1-2) \quad (2)$$

Since $y_1(t) \neq y_2(t)$ system is time variant.

(4)

$$y(t) = \int_{-\infty}^{t} x(\tau)d\tau$$

$$y_1(t) = y(t,t_1) = \int_{-\infty}^{t} x(\tau - t_1)d\tau; \ \tau - t_1 = \lambda$$

$$y(t,t_1) = \int_{-\infty}^{t-t_1} x(\lambda)d\lambda \quad (1)$$

Now delay O/P by t_1; $y_2(t) = y(t-t_1) = \int_{-\infty}^{t-t_1} x(\tau)d\tau$

Since $y_1(t) = y_2(t)$ system is time invariant.

Example 14: Determine whether the following DTS are time invariant or not.

(1) $y(n) = x(n) - x(n-1)$ {time invariant}
(2) $y(n) = x(n)\cos w_0 n$ {time variant}
(3) $y(n) = x^2(n-2)$ {time invariant}

SOLUTION:

(1) $y(n) = x(n) - x(n-1)$

Let us apply the input to this system which is delayed by 'k' samples

Delay input by k samples $y_1(n) = y(n,k) = f[x(n-k)] = x(n-k) - x(n-k-1)$

On delaying output by k samples $y_2(n) = y(n-k) = x(n-k) - x(n-k-1)$ time invariant.

(2) $y(n) = x(n)\cos\omega_0 n$

Let us apply the input to this system which is delayed by 'k' samples

$$y_1(n) = y(n,k) = x(n-k)\cos\omega_0(n) \quad (1)$$

$$y_2(n) = y(n,k) = y(n-k) = x(n-k)\cos\omega_0(n-k) \quad (2)$$

Time variant

(3) $y(n) = x^2(n-2)$

Let us apply the input to this system which is delayed by 'k' samples

$$y_1(n) = y(n,k) = x^2(n-k-2) \quad (1)$$

$$y_2(n) = y(n,k) = y(n-k) = x^2(n-k-2) \quad (2)$$

Time invariant

1.4.4 Linear and non-linear systems

A system is said to be linear if it satisfies the superposition theorem. Consider the two systems as follows.

$$y_1(t) = f[x_1(t)] \text{ is } x_1(t) \text{ is the input and } y_1(t) \text{ is output}$$

$$\text{and } \quad y_2(t) = f[x_2(t)] \text{ is } x_2(t) \text{ is the input and } y_2(t) \text{ is output}$$

Then the CTS is linear if

$$f[a_1x_1(t) + a_2x_2(t)] = a_1y_1(t) + a_2y_2(t)$$

Here a_1 and a_2 are arbitrary constants.

Similarly, the DTS is said to be linear if it satisfies superposition principle.

$$y_1(n) = f[x_1(n)]$$
$$y_2(n) = f[x_2(n)]$$

The DTS is linear if

$$f[a_1x_1(n) + a_2x_2(n_1)] = a_1y_1(n) + a_2y_2(n)$$

A system that obeys the superposition principle is said to be a linear system. Superposition principle states that the response to a weighted sum of input signals be equal to the corresponding weighted sum of the outputs of the system to each of the individual input signals.

Example 15: For a CTS, the I/P and O/P relationship is given by

$$y(t) = tx(t)$$

Check whether the system is linear.

SOLUTION:

Consider two arbitrary inputs $x_1(t)$ and $x_2(t)$, the respective O/P becomes

$$y_1(t) = tx_1(t)$$
$$\text{and} \quad y_2(t) = tx_2(t)$$

Let $x_3(t) = ax_1(t) + bx_2(t)$ (where a and b are constants) is the input to the system, then corresponding O/P may be expressed as

$$y_3(t) = t[ax_1(t) + bx_2(t)] = atx_1(t) + btx_2(t) \quad (1)$$
$$y_3'(t) = ay_1(t) + by_2(t) = atx_1(t) + btx_2(t) \quad (2)$$
$$(1) = (2) \quad \text{Linear}$$

Example 16: $y(t) = x^2(t)$ Linear or not

SOLUTION:

Consider two arbitrary input $x_1(t)$ and $x_2(t)$, the respective O/P becomes

$$y_1(t) = x_1^2(t)$$
$$\text{and} \quad y_2(t) = x_2^2(t)$$

Weighted sum of output $\{y_3{}'(t)\}$

$$y_3{}'(t) = ay_1(t) + by_2(t) = ax_1^2(t) + bx_2^2(t) \quad (1)$$

Let $x_3(t) = ax_1(t) + bx_2(t)$ (where a and b are constants) is the I/P to the system, then corresponding O/P may be expressed as

The output due to weighted sum of input is $\{y_3(t)\}$

$$y_3(t) = x_3^2(t) = [ax_1(t) + bx_2(t)]^2 = a^2x_1^2(t) + b^2x_2^2(t) + 2abx_1(t)x_2(t) \quad (2)$$
$$(1) \neq (2) \text{ Non-linear}$$

Example 17: $y(n) = x[n+1] - x[n-1]$ Linear or not

SOLUTION:

$$y_1(n) = x_1[n+1] - x_1[n-1]$$

and

$$y_2(n) = x_2[n+1] - x_2[n-1]$$

$$y_3{}'[n] = ay_1[n] + by_2[n] = ax_1[n+1] - ax_1[n-1] + bx_2[n+1] - bx_2[n+1] \tag{1}$$

$$\text{Let}\quad x_3[n] = ax_1[n] + bx_2[n]$$

$$y_3[n] = x_3[n+1] - x_3[n-1]$$

$$= ax[n+1] + bx[n+1] - ax_2[n-1] - bx_2[n-1] \tag{2}$$

$$(1) = (2) \text{ Linear}$$

Example 18: Check for

(1) Static/dynamic
(2) Linear/non-linear
(3) Time invariant or time variant

(a) $y(t) = At^2x(t) + Btx(t-2)$

(b) $y(n) = x^2(n) + \dfrac{1}{x^2(n-1)}$

SOLUTION:

(a) (1) Dynamic (have memory)

(2) $y_3{}'(t) = ay_1(t) + by_2(t) = aAt^2x_1(t) + aBtx_1(t-2) + bAt^2x_2(t) + bBtx_2(t-2)$

$y_3(t) = At^2x_3(t) + Btx_3(t-2) = Aat^2x_1(t) + bAt^2x_2(t) + aBtx_1(t-2) + bBtx_2(t-2)$

Linear

Time variant

(b) (1) Dynamic

(2)
$$y_3{}'(n) = ay_1(n) + by_2(n) = ax_1^2(n) + \frac{a}{x_1^2(n+1)} + bx_2^2(n) + \frac{b}{x_2^2(n-1)}$$

$$y_3(n) = x_3^2(n) + \frac{1}{x_3^2(n+1)} = [ax_1(n) + bx_2(n)]^2 + \frac{1}{[ax_1(n) + bx_2(n)]^2}$$

$$= a^2x_1^2(n) + b^2x_2^2(n) + 2abx_1(n)x_2(n) + 1$$

Non-linear

(3)
$$y(n) = x^2(n-k) + \frac{1}{x^2(n-k-1)}$$

Time invariant

1.4.5 Stable and unstable system

When every bounded input produces bounded output then the system is called Bounded Input Bounded Output (BIBO) stable. This criterion is applicable for both the continuous and discrete-time systems.

The input is said to be bounded if there exists some finite number M_x such that

$$\text{CT I/P} \quad |x(t)| \leq M_x < \infty$$
$$\text{DT I/P} \quad |x(n)| \leq M_x < \infty$$

Similarly, the O/P is bounded if there exists some finite number M_y such that

$$\text{CT O/P} \quad |y(t)| \leq M_y < \infty$$
$$\text{DT O/P} \quad |y(n)| \leq M_y < \infty$$

If the system produces unbounded O/P for bounded I/P, then it is unstable.

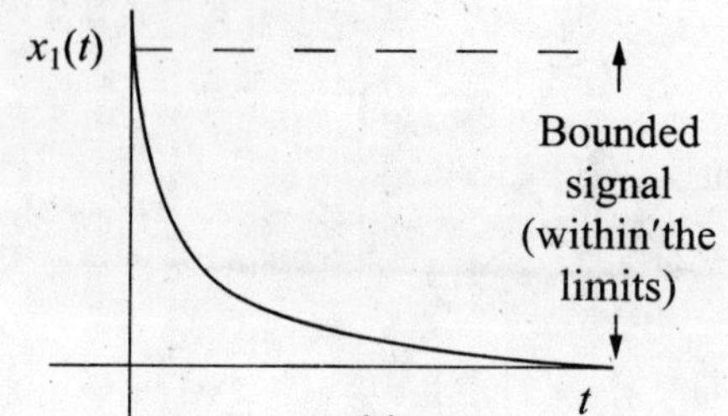

(a)

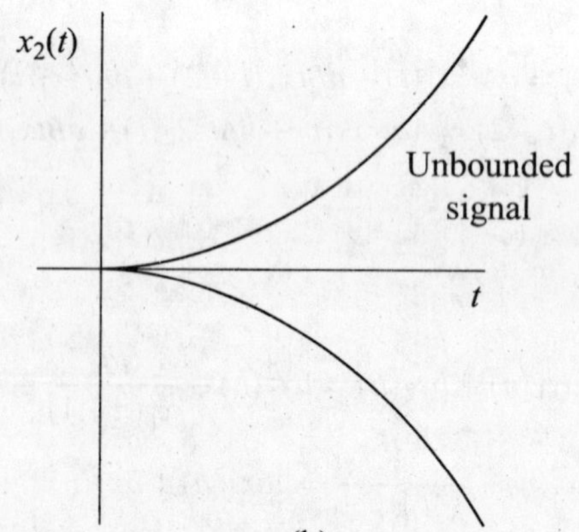

(b)

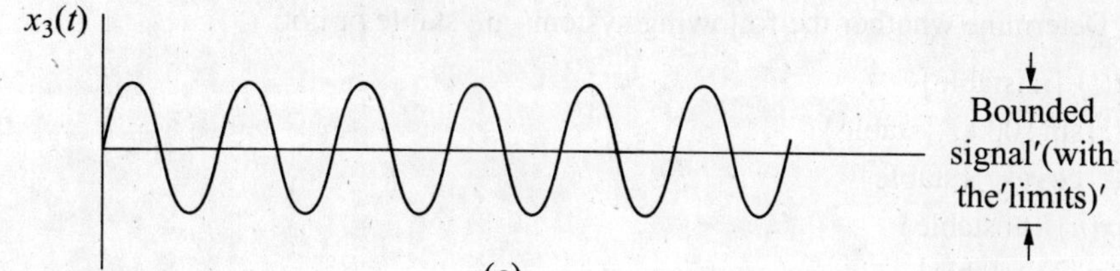

(c)

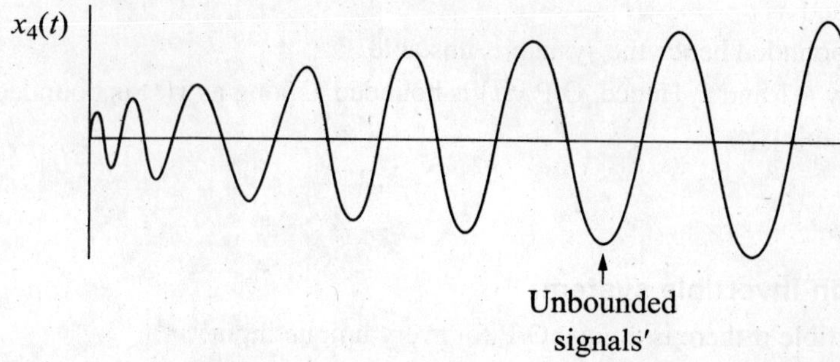

(d)

Fig. 1.41. (a), (c) Bounded signal (b), (d) Unbounded signal

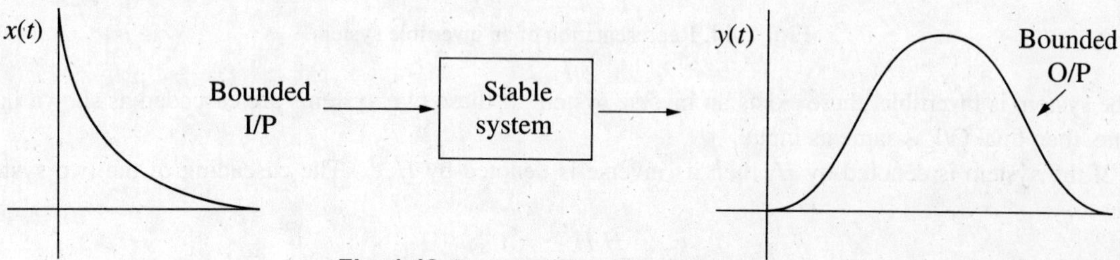

Fig. 1.42. Representation of a stable system

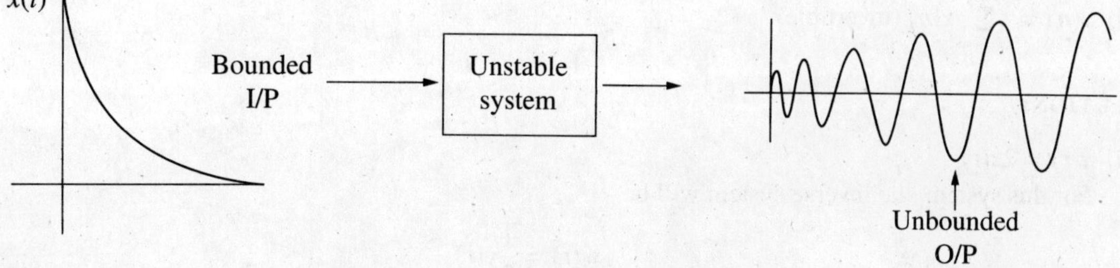

Fig. 1.43. Representation of an unstable system

Example 19: Determine whether the following systems are stable or not.

(a) $y(t) = tx(t)$ {unstable}
(b) $y(t) = x(t)\sin 100\pi\, t$ {stable}
(c) $y(t) = x(t)\cos\omega_0 t$ {stable}
(d) $y(n) = nx(n)$ {unstable}
(e) $y(n) = x(n/2)$ {stable}

SOLUTION:

(a) Let $x(t)$ be bounded and as $t \to \infty$, $y(t) \to \infty$
Hence, the O/P is unbounded hence the system is unstable.
(b) Value of sine in is b/w -1 and 1, Hence, O/P $y(t)$ is bounded as long as $x(t)$ is bounded [stable]
(c) $y(t) = x(t)\cos\omega_0 t$ [stable]
(d) Unstable
(e) Stable

1.4.6 Invertible and non-invertible system

A system is said to be invertible if there is unique O/P for every unique input.

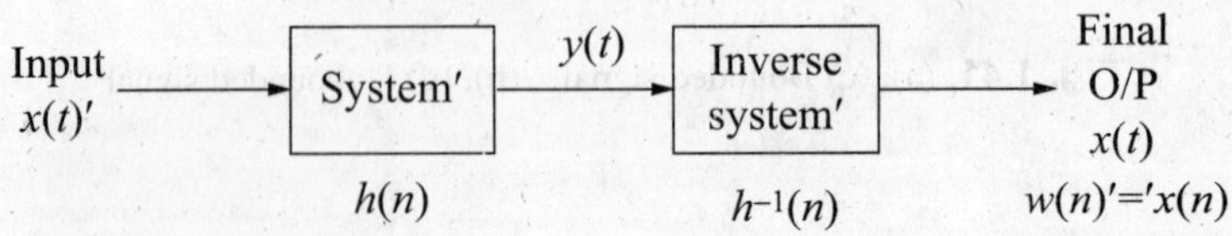

Fig. 1.44. Representation of an invertible system

If the system is invertible, there exists an inverse system, if these two systems are cascaded as shown in the figure, then final O/P is same as input.

If the system is denoted by H, then its inverse is denoted by H^{-1}. The cascading of the two systems gives

$$H\,H^{-1} = 1$$

Example 20: Determine whether the following systems are invertible.

(i) $y(t) = x(t)$ (invertible)
(ii) $y(t) = x^2(t)$ (non-invertible)
(iii) $y(n) = \sum_{k=-\infty}^{n} x(n)$ (invertible)

SOLUTION:

(i) $y(t) = 2x(t)$
For this system, the inverse system will be

$$w(t) = \frac{1}{2}y(t)$$

Hence, if the two systems are cascaded, input is first multiplied by 2.
Then the O/P of the first system is divided by 2. Therefore, original input is obtained back.

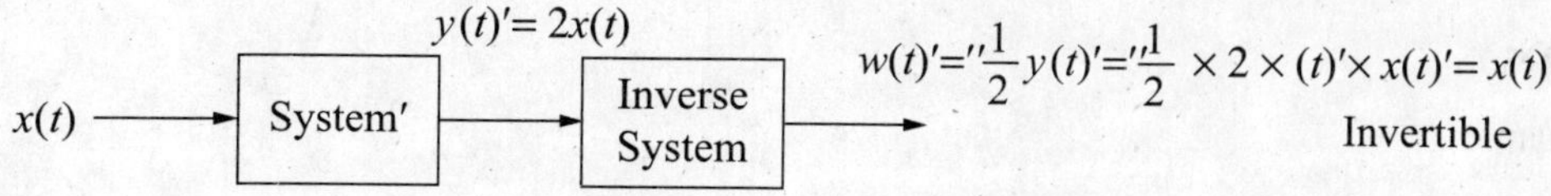

Fig. 1.45. Representation of an invertible system

(ii) $y(t) = x^2(t)$

This system squares the input. Hence, inverse system takes square root.

$$\begin{aligned} w(t) &= \sqrt{y(t)} \\ &= \sqrt{x^2(t)} \\ &= \pm\, x(t) \end{aligned}$$

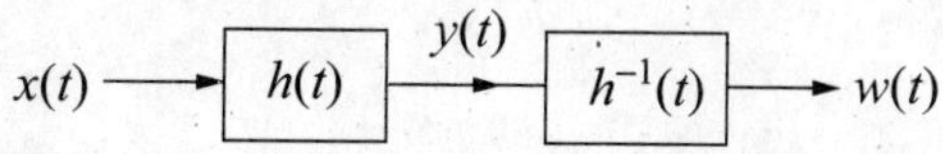

Fig. 1.46. Representation of an invertible system

$h(t)$ is a squarer

$h^{-1}(t)$ is an under router

Thus, two outputs are possible: $x(t)$ or $-x(t)$. This means there is no unique output for unique I/P. Hence, this system is non-invertible.

(iii)

$$y(n) = \sum_{k=-\infty}^{n} x(n)$$

$$\begin{aligned} y(n) &= x(n) + x(n-1) + x(n-2) + \ldots \\ y(n-1) &= x(n-1) + x(n-2) + x(n-3) + \ldots \\ y(n) &= x(n) + y(n-1) \\ x(n) &= y(n) - y(n-1) \end{aligned}$$

Thus, input $x(n)$ can be obtained back. Hence, inverse system exists and its equation is

$$w(n) = y(n) - y(n-1)$$

Example 21: Check whether the following systems are

(1) Static or dynamic

(2) Linear or non-linear

(3) Causal or non-causal

(4) Time invariant or time variant

(a) $\dfrac{dy(t)}{dt} + 10y(t) = x(t)$

(c) $y(n) = \log_{10}|x(n)|$

(b) $y(n) = x(n)\,x(n-1)$

SOLUTION:

(1) dynamic

(2)

$$\frac{dy_1(t)}{dt} + 10y_1(t) = x_1(t) \qquad (1) \qquad\qquad \frac{dy_2(t)}{dt} + 10y_2(t) = x_2(t) \qquad (2)$$

On multiplying (1) by a and (2) by b

$$a\frac{dy_1(t)}{dt} + 10ay_1(t) + b\frac{dy_2(t)}{dt} + 10by_2(t) = ax_1(t) + bx_2(t)$$

$$\frac{d}{dt}[ay_1(t) + by_2(t)] + 10[ay_1(t) + by_2(t)] = ax_1(t) + bx_2(t) \qquad (1)$$

$$\frac{d}{dt}y_3(t) + 10y_3(t) = x_3(t) \qquad (2)$$

Linear

(3) Causal

(4) Stable

(5) This is time invariant since its O/P does not change with time t.

(b) (2) $y(n) = \log_{10}|x(n)|$

(i) Static

(ii) Non-linear

(iii) Causal

(iv) Time invariant

(c) (3) $y(n) = x(n)x(n-1)$

(i) Dynamic

(ii) Non-linear

(iii) Causal

(iv) Time invariant

1.5 ANALOGOUS SYSTEM INTRODUCTION

A physical system is a collection of physical objects connected together to perform an objective. The first step towards analysis of system is formulation of its mathematical model. The mathematical models are formed by application of one or more fundamental laws which relate to the physical nature of the system or its elements, e.g. Kirchhoff's law and Ohm's law is used for electrical networks, mechanical systems are analysed by the application of Newton's law and D'Alember principle, etc. It has been seen that two different physical systems can be described by the same mathematical model thereby originating the idea of analogous systems. The concept of analogous system is very useful in the study of complex systems like electrical, mechanical, hydraulic system, etc. A non-electrical system is studied in terms of its electrical analog because experiments can be easily conducted on electrical systems and analysed. Any change in the parameters can be achieved easily in electric circuit and studied by altering the parameter, so as to accomplish the desired response. Afterwards corresponding analogous mechanical quantities can be changed by analogous amount to design the desired mechanical system.

Systems which are governed by the same type of equations or whose different equations are of identical form (variable could be different) are called **Analogous Systems.**

1.6 MECHANICAL SYSTEM

Analysis of mechanical system makes use of three idealised elements, namely the mass, the spring, and the damper. Equations of mechanical systems are formulated from Newton's laws of motion. The motion of mechanical elements can be translatory, rotational or a combination of both.

1.6.1 Translation systems

The motion along a straight line is called the translatory motion. The variable which describes the translatory motion of mechanical systems are velocity, acceleration and displacement. The elements involved in the translatory motion are mass (M), damper (D) and spring (K). There are three forces which resist translatory motion.

(a) **Inertial Force** – If a force is applied to a mass, as shown in Fig. 1.47(a), then according to Newton's second law of motion the inertia force will be equal to the product of mass M and acceleration 'a' and it is given by

$$f_M = Ma = M\frac{dv(t)}{dt} = M\frac{d^2x(t)}{dt^2}$$

(b) **Damping Force** – If a force is applied to the damper as shown in Fig. 1.47(b), the damping force for viscous friction is proportional to the difference in velocity at two ends of the damper and it is given by

$$f_D = D(v_1 - v_2)$$

$$= D\left[\frac{dx_1(t)}{dt} - \frac{dx_2(t)}{dt}\right]$$

(c) **Spring Force** – If a force is applied to a spring, as shown in Fig. 1.47(c), the spring force (reaction or restoring force) is produced which tries to bring it back to its free length. The spring force is proportional to the displacement (Hooke's law) and it is given by

$$f_k = Kx(t) = K\int_{-\infty}^{t} v(t)dt$$

$$= k\left[\int_{0}^{t} v(t)dt + x(0)\right]$$

Here f_M, f_D, f_K are Newton forces

a is acceleration in m/sec^2

v, v_1, v_2 are velocities in m/sec

x, x_1, x_2 are displacements in m

M is mass in kg

D is the coefficient of viscous friction in N/m/sec

K is spring constant called the stiffness in N/m

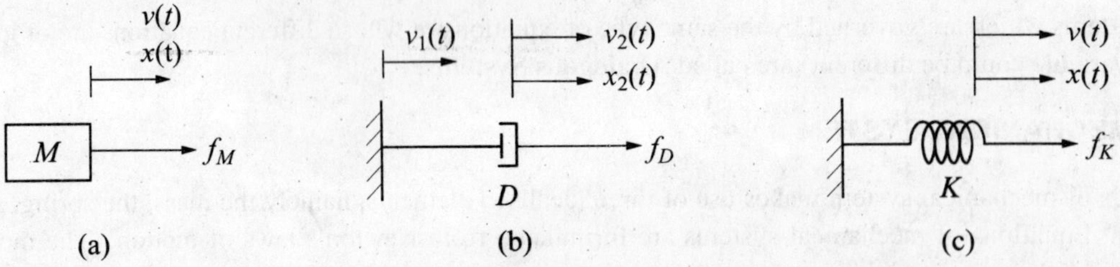

Fig. 1.47.

(a) Mass element
(b) Damper element
(c) Spring element

1.6.2 Rotational system

The rotational motion of a body can be defined as the movement of a body around its fixed axis. Basic elements of rotational motion are mass (J), spring stiffness (K) and rotational friction (f). Variables of interest are the torque (T), angular velocity ($\dot{\theta}$) and angular displacement (θ). Network representation is shown in Fig. 1.48.

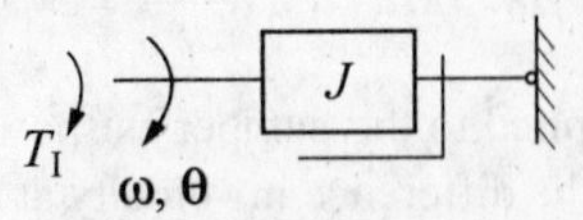

Fig. 1.48.

(a) **Inertial Torque** – It is equal to the moment of inertia (I) times the angular acceleration ($\ddot{\theta}$)

$$T_I = I\alpha$$
$$= I\frac{dw}{dt} = I\frac{d^2\theta}{dt^2}$$

(b) **Damping Torque** – It is equal to the viscous friction coefficient (D_θ) times the angular velocity (ω), i.e.

$$T_D = D_\theta \omega$$
$$= D_\theta \frac{d\theta}{dt}$$

(c) **Spring Torque** – It is equal to the torsional stiffness of the spring (K_θ) times the angular displacement, i.e.

$$T_K = K_\theta \theta = K_\theta \int_{-\infty}^{t} \omega dt$$
$$= K_\theta \left[\int_0^t \omega dt + \theta(0) \right]$$

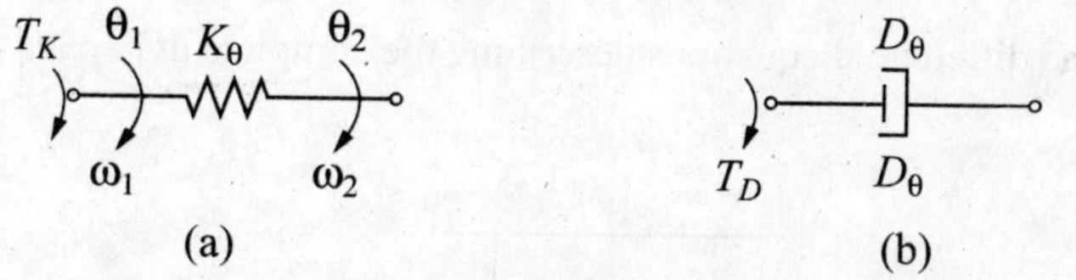

Fig. 1.49.

Here T_I, T_D, T_K are the torques in N-mt

- α – angular acceleration in rad/sec^2
- ω – angular velocity in rad/sec
- θ – angular displacement in rad
- I – moment of inertia in kg mt^2
- D_θ – rotational viscous friction coefficient in N–mt/rad/sec.
- K_θ – tortional stiffness of the spring in N–mt/rad.

1.6.3 Equivalent mechanical system (Node basis)

While drawing analogous networks, it is always better to draw the equivalent mechanical system from the given mechanical system. To draw such a system use the following steps.

Step 1 – Due to applied force, identify the displacements in the mechanical system.

Step 2 – Identify the elements which are under the influence of different displacements.

Step 3 – Represent each displacement by a separate node, using nodal analysis.

Step 4 – Show all the elements in parallel under the respective nodes which are under the influence of respective displacement.

Step 5 – Elements causing same change in displacement will get connected in parallel in between the respective nodes.

Consider the given system

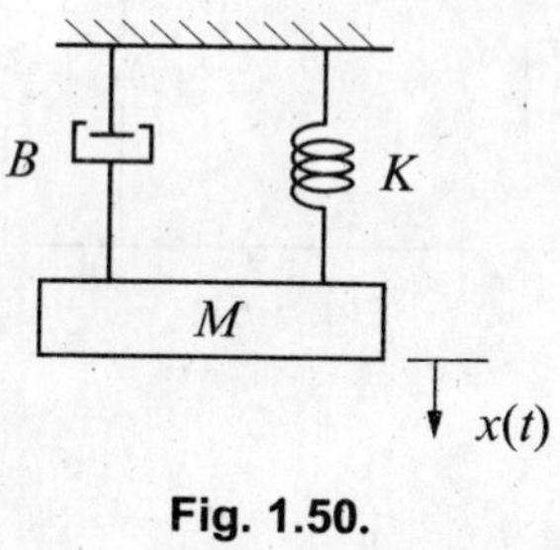

Fig. 1.50.

All elements are under same displacement.

$$M\frac{d^2x}{dt^2} + kx + B\frac{dx}{dt} = 0$$

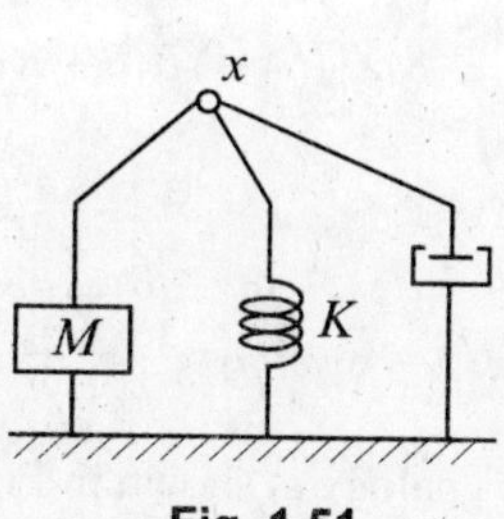

Fig. 1.51.

Example 22: Obtain the differential equations describing the complete dynamics of the mechanical system given below.

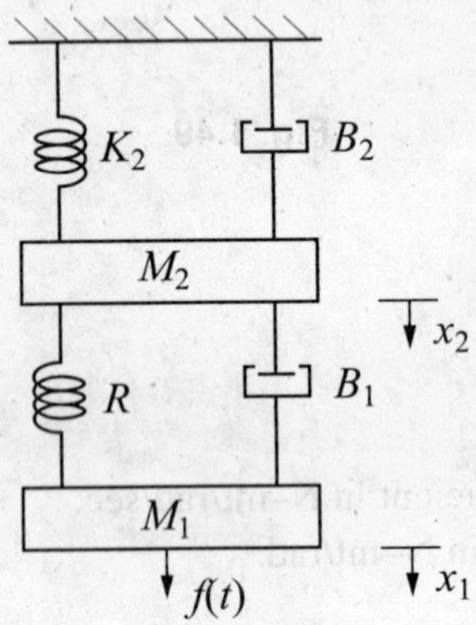

Fig. 1.52.

SOLUTION:

The mechanical network diagram is shown in Fig. 1.53.

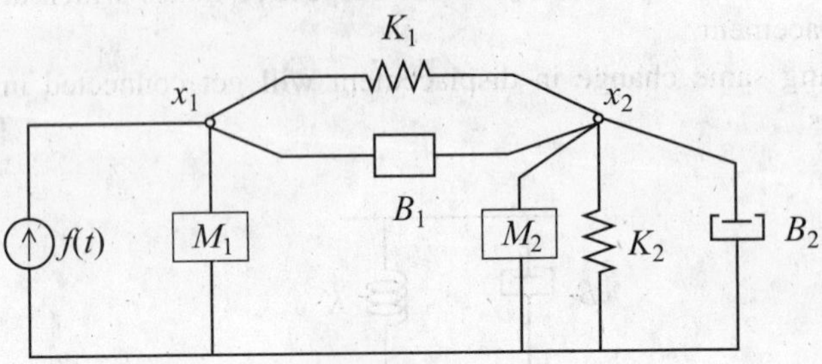

Fig. 1.53.

Node 'x_1'

$$f(t) = M_1\ddot{x}_1 + K_1(x_1 - x_2) + B_1(\dot{x}_1 - \dot{x}_2)$$

or

$$F(S) = M_1S^2X_1(S) + K_1(X_1(S) - X_2(S)) + B_1S(X_1(S) - X_2(S))$$
$$= (S^2M_1 + K_1 + B_1S)X_1(S) - (B_1S + K_1)X_2(S)$$

Node 'x_2'

$$K_1(x_1 - x_2) + B_1(\dot{x}_1 - \dot{x}_2) = M_2\ddot{x}_2 + K_2x_2 + B_2\dot{x}_2$$

$$[M_2S^2 + (B_1 + B_2)S + (K_1 + K_2)]X_2(S) - (B_1S + K_1)X_1(S) = 0$$

The electrical analog based on force voltage analogy is shown in Fig. 1.54.

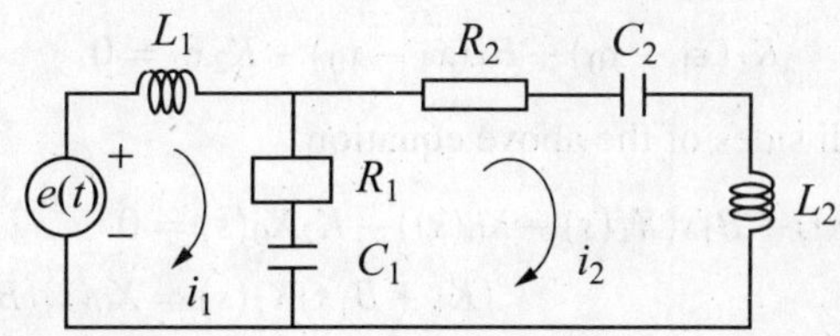

Fig. 1.54.

Analogous quantities $(f-v)$

$$f \to e,\ M \to L,\ B \to R,\ K \to \frac{1}{c},\ x \longrightarrow q,\ \dot{x} \to i.$$

Hence, the electrical analogous equations are

$$E(S) = \left((L_1 S^2 + R_1 S + \frac{1}{C_1} \right) Q_1(S) - \left(R_1 S + \frac{1}{C_1} \right) Q_2(S)$$

and

$$\left(L_2 S^2 + \frac{R_2 S}{C_2} + R_1 S + \frac{1}{C_1} \right) Q_1(S) - \left(R_1 S + \frac{1}{C_1} \right) Q_2(S) = 0$$

Converting the above equations into differential equation form

$$e(t) = L_1 \frac{di_1}{dt} + \frac{1}{C_1} \int i_1 dt + R_1 i_1 - R_1 i_2 \frac{-1}{C_1} \int i_2 dt$$

and

$$L_2 \frac{di_2}{dt} + R_2 i_2 + R_1 i_2 + \frac{1}{C_2} \int i_2 dt$$
$$+ \frac{1}{C_2} \int i_2 dt - R_1 i_1 - \frac{1}{C_1} \int i_1 dt = 0$$

Note: Above equations can be obtained if we apply Kirchhoff's law to the electrical network shown in figure.

Example 23: Obtain the transfer function of the above system shown in Fig. 1.55.

SOLUTION:

According to Newton's law, we can write

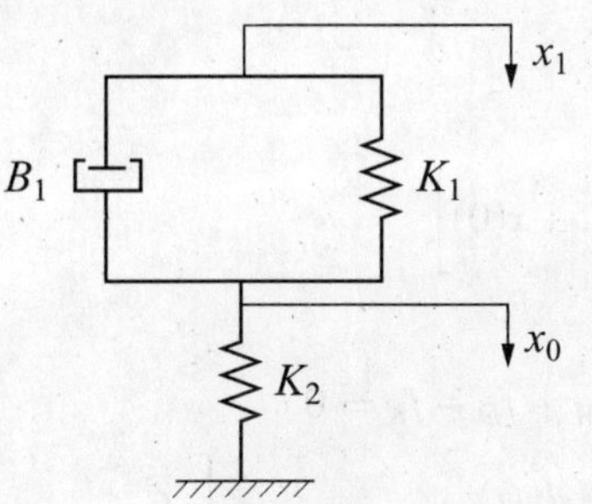

Fig. 1.55.

$$K_1(x_1 - x_0) + B_1(\dot{x}_1 - x_0) - K_2 x_0 = 0$$

Apply Laplace transform on both sides of the above equation

$$K_1(X_1(s) - X_0(s)) + B_1 s(X_1(s) - X_0(s)) - K_2 X_0(s) = 0$$

$$(K_1 + B_1 s)X_1(s) = X_0(s)(B_1 s + K_2 + K_1)$$

$$\frac{X_0(s)}{X_0(s)} = \frac{B_1 s + K}{B_1 s + K_1 + K_2}$$

K_θ – Torsional stiffness of the spring in M-mt/rad

1.7 D'ALEMBERT'S PRINCIPLE

This principle is used for the derivation of equation of motion in mechanical systems. It applies the condition of static equilibrium to solve the problems in dynamics by considering both the externally applied driving forces and the reaction forces of mechanical elements which resist motion. It is actually a slighted modified form of Newton's law of motion and can be stated as:

"For any body, the algebraic sum of externally applied forces and the forces resisting motion in any given direction is zero."

Consider a translation mechanical system shown in Fig. 1.56 consisting of a mass(M), dash pot D and spring K.

Taking the direction of external force (f) is as a reference direction. All the forces in the direction of external force considered position and the forces opposite to the external force taken as –ve.

External force: f

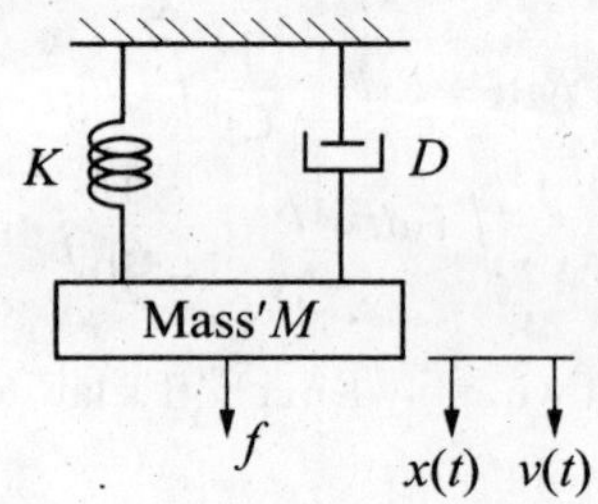

Fig. 1.56.

Resisting forces –

(a) Inertial force $f_M = -M\dfrac{dv(t)}{dt}$

(b) Damping force $f_D = -Dv(t)$

(c) Spring force $f_K = -K\left[\int_0^t v(t)dt + x(0)\right]$

According to D'Alembert's principle –

$$f + f_M + f_D + f_K = 0$$

$$f = M\frac{dv(t)}{dt} + DV(t) + K\left[\int_0^t v(t)dt + x(0)\right] \tag{1}$$

Equation (1) is the dynamic equation for the translation mechanical system as shown in Fig. 1.56. Similarly, we can write equation for rotation mechanical system. D'Alembert's principle for rotational mechanical system can be stated as:

"For any body, the algebraic sum of externally applied torques and the torques resisting rotation about any axis is zero".

External torque → T

Resisting torques are →

(a) Inertial torque $T_I = -I\dfrac{dw(t)}{dt}$

(b) Damping torque $T_D = -Dw(t)$

(c) Spring torque $T_K = -K_\theta\left[\int_0^t w(t)dt + \theta(0)\right]$

By D'Alembert's principle

$$T + T_I + T_D + T_K = 0 \tag{1}$$

$$T = I\frac{dw(t)}{dt} + D_\theta w(t) + K_\theta\left[\int_0^t w(t)dt + \theta(0)\right] \tag{2}$$

Note that eqn. (1) is identical to eqn. (2). Therefore, the translational mechanical system is identical to rotational mechanical system.

1.8 FORCE VOLTAGE ANALOGY – (f–v)

Consider a mechanical system shown in Fig. 1.57.

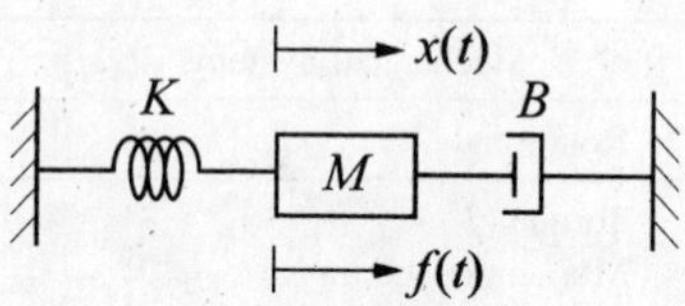

Fig. 1.57.

Mechanical network of the mechanical system under consideration is shown in figure. The nodal equation of the mechanical network at 'x' is

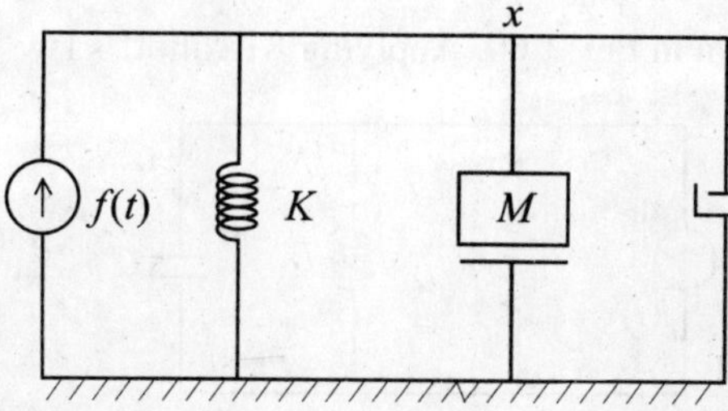

Fig. 1.58.

$$f(t) = f_M + f_B + f_K \tag{1}$$

$$f(t) = M\frac{d^2x}{dt^2} + B\frac{dx}{dt} + Kx \tag{2}$$

Now consider an electrical network shown in Fig. 1.59. The mesh equation by application of Kirchhoff's law is

$$e = Ri(t) + L\frac{di(t)}{dt} + \frac{1}{C}\int i(t)dt \tag{3}$$

Since $i = \dfrac{dq}{dt}$, therefore

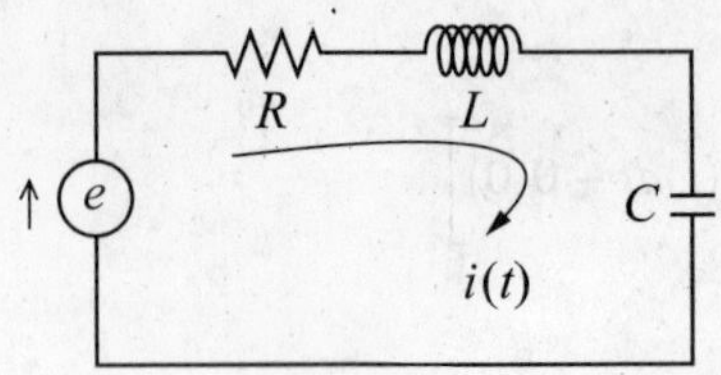

Fig. 1.59.

$$e = \frac{Rdq}{dt} + \frac{Ld^2q}{dt^2} + \frac{q}{C}, \text{ or} \tag{4}$$

Comparing eqns. (2) and (4), we can infer that both these equations are identical in nature and hence analogous. The terms which occupy corresponding positions in these two equations are called analogous quantities. Analogous quantities between mechanical and electrical system based on force-voltage analogy are listed in Table 1.1.

Table 1.1. Analogous quantities (force-voltage analogy)

Mechanical Systems		
Translatory	Rotational	Electrical Systems
Force (f)	Torque (T)	Voltage (e)
Mass (M)	Moment of inertia (I)	Inductance (L)
Viscous friction coefficient (B)	Viscous friction coefficient (f)	Resistance (R)
Spring stiffness (K)	Torsional spring stiffness (K)	Reciprocal of capacitance $(1/C)$
Displacement (x)	Angular displacement (θ)	Charge (q)
Velocity	Angular velocity	Current (i)

1.8.1 Force-current analogy

Consider an electrical network shown in Fig. 1.60. Applying Kirchhoff's law, we get

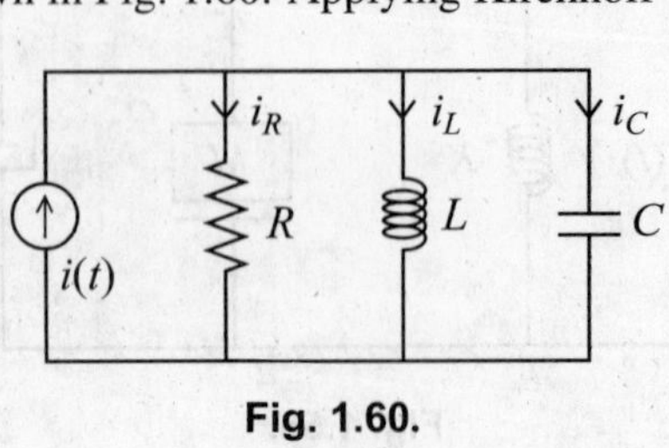

Fig. 1.60.

$$i_R + i_L + i_C = i(t) \tag{5}$$

$$\frac{e}{R} + \frac{1}{L}\int e\,dt = \frac{Cde}{dt} = i(t) \tag{6}$$

But $e = \frac{d\phi}{dt}$, where (ϕ) = flux linkages,

Therefore,

$$\frac{1}{R}\frac{d\phi}{dt} + \frac{\phi}{L} + C\frac{d^2\phi}{dt^2} = i_{(t)} \tag{7}$$

Rearranging, we get

$$i(t) = C\frac{d^2\phi}{dt^2} + \frac{1}{R}\frac{d\phi}{dt} + \frac{\phi}{L} \tag{8}$$

Equations (2) and (8) are identical and hence analogous. The analogous terms based on force-current analogy are listed in Table 1.2.

Table 1.2. Analogous quantities (force-current analogy)

Mechanical Systems		
Translatory	Rotational	Electrical Systems
Force (f)	Torque (T)	Current (i)
Displacement (x)	Angular displacement (θ)	Flux linkages (F)
Velocity $(\dot{x})$	Angular velocity $(\dot{\theta})$	Voltage (e)
Mass (M)	Moment of inertia (I)	Capacitance (C)
Viscous friction coefficient (B)	Viscous friction coefficient (F)	Reciprocal of resistance $(1/R)$
Spring constant (K)	Torsional spring constant (K)	Reciprocal of inductance $(1/L)$

Example 24: Write differential equations for the system shown in Fig. 1.61 and draw its analogous circuit.

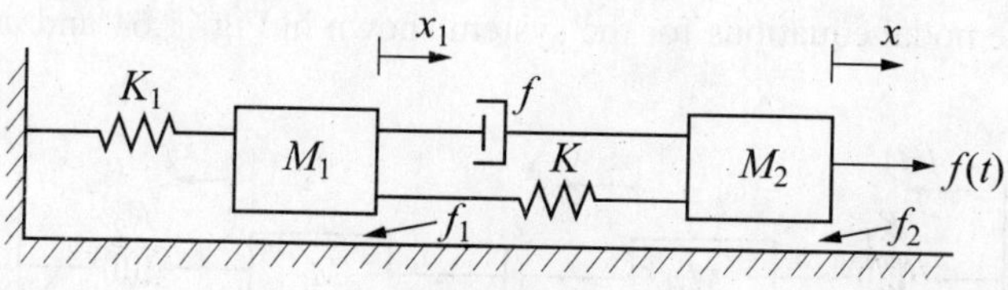

Fig. 1.61.

The network diagram of the above mechanical system is shown in Fig. 1.62.

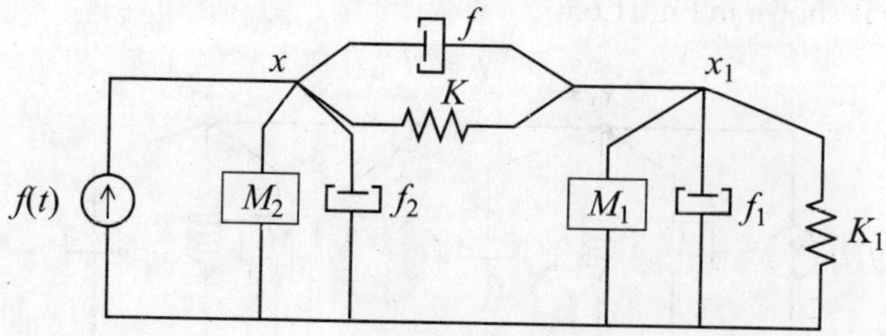

Fig. 1.62.

The nodal equation at node 'x' is

$$f(t) = M_2\frac{d^2x}{dt^2} + f_2\frac{dx}{dt} + K(x - x_1) + f\left(\frac{dx}{dt} - \frac{dx_1}{dt}\right) \quad (1)$$

The nodal equation at node 'x_1' is

$$K_1x_1 + f_1\frac{dx_1}{dt} + M_1\frac{d^2x_1}{dt^2} + f\left(\frac{dx_1}{dt} - \frac{dx}{dt}\right) + K(x_1 - x) = 0 \quad (2)$$

As per $f - v$ analogy

$$M \to L,\ f \to R,\ K \to C,\ x \to \dot{q},\ \frac{dx}{dt} = i,\ f(t) \to v(t)$$

Converting the node eqns. (1) and (2) in its electrical analogous equations, we get

$$v(t) = L_2\frac{di}{dt} + R_2 i + R(i - i_1) + \frac{1}{C}\int(i - i_1)dt \quad (3)$$

$$L_1\frac{di_1}{dt} + R_1 i_1 + \frac{1}{C_1}\int i\,dt = R(i - i_1) + \frac{1}{C}\int(i - i_1)dt \quad (4)$$

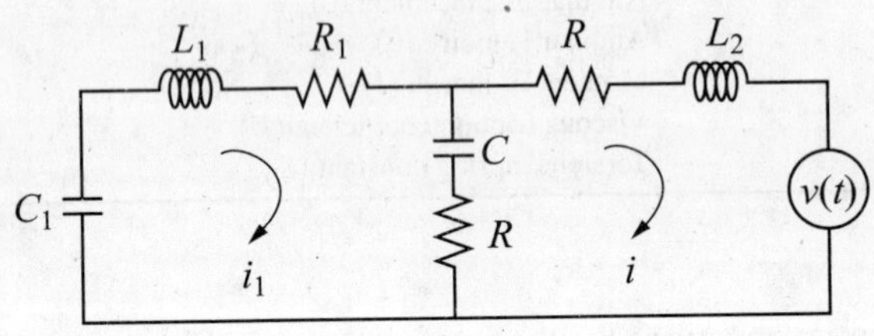

Fig. 1.63. Electrical analog circuit based on $f - v$ analogy

Example 25: Obtain the nodal equations for the system shown in Fig. 1.64 and draw its electrical analog based on $f - i$ analogy.

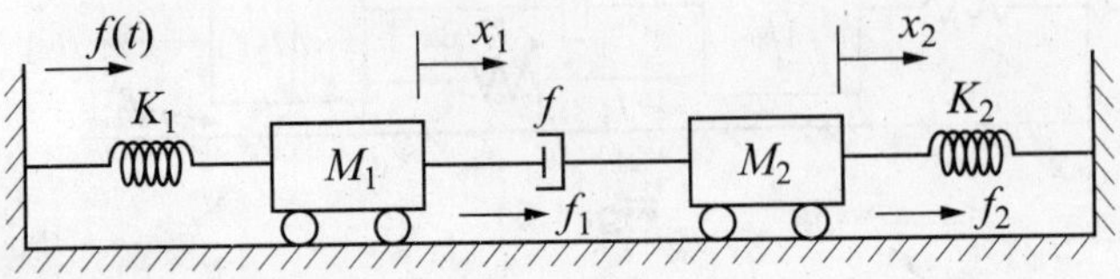

Fig. 1.64.

SOLUTION:

The mechanical n/w is shown in Fig. 1.65.

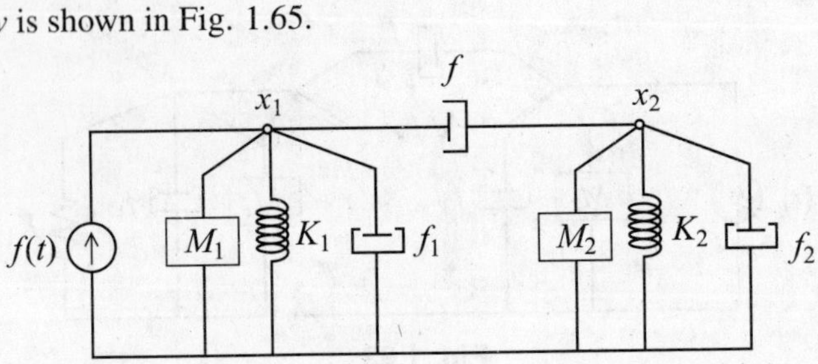

Fig. 1.65.

At node 'x_1'

$$M_1\ddot{x}_1 + f_1\dot{x}_1 + f(\dot{x}_1 - \dot{x}_2) + k_1 x_1 = f(t) \tag{1}$$

At node 'x_2'

$$M_2\ddot{x}_2 + f_2\dot{x}_2 + k_2 x_2 = f(\dot{x}_1 - \dot{x}_2) \tag{2}$$

$f - i$ analogous circuit has similar presentation except its notation as

$$f(t) \to i(t),\ M \to C,\ K \to \frac{1}{L},\ f \to R$$

Hence,

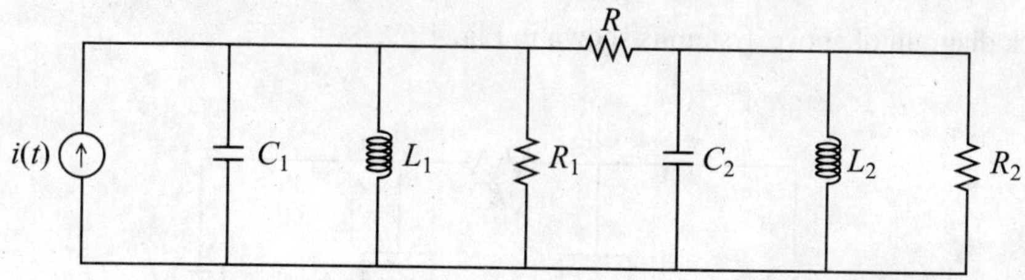

Fig. 1.66.

Electrical analog circuit based on $f - i$ analogy is shown in the above figure.

Example 26: Obtain the transfer function of the system shown in Fig. 1.67.

SOLUTION:

According to Newton's law, we get

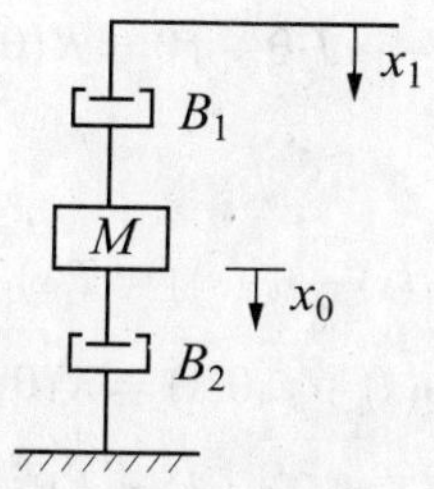

Fig. 1.67.

$$B_1(\dot{x}_1 - \dot{x}_0) = M\ddot{x}_0 + B_2\dot{x}_0$$

$$B_1 S(X_1(S) - X_0(S)) = MS^2 X_0(S) + B_2 S X_0(S)$$

$$B_1 S X_1(S) = X_0(S)[MS^2 + B_2 S + B_1 S]$$

$$\frac{X_0(S)}{X_1(S)} \Rightarrow \left(\frac{B_1}{MS^2 + B_1 + B_2}\right)$$

Example 27: Obtain the transfer function of mechanical network as shown in Fig. 1.68.

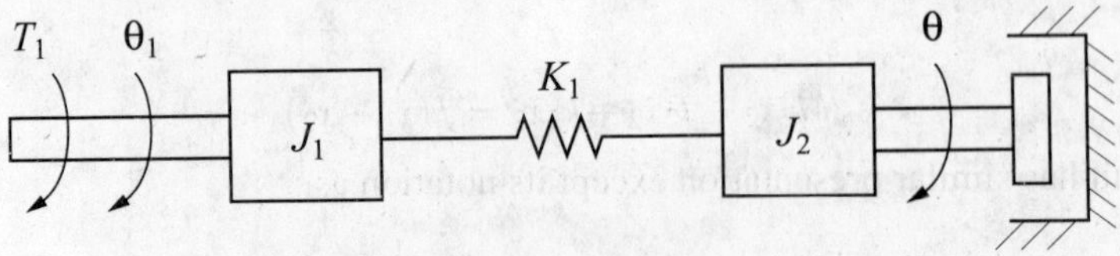

Fig. 1.68.

SOLUTION:

The network diagram of above system is shown in Fig. 1.69.

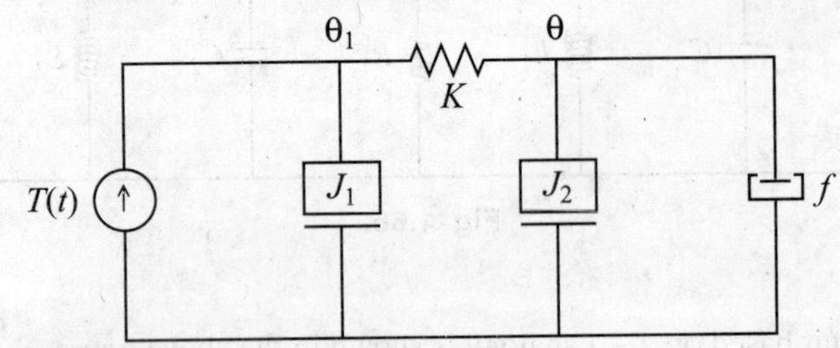

Fig. 1.69.

The nodal equations are

$$\text{Node } \theta_1 \longrightarrow J_1\ddot{\theta}_1 + K(\theta_1 - \theta) = T(t) \tag{1}$$

$$\text{Node } \theta \longrightarrow J_2\ddot{\theta} + f\dot{\theta} = K(\theta_1 - \theta) \tag{2}$$

Laplace transform of eqns. (1) and (2) is

$$J_1S^2\theta_1(s) + K(\theta_1(s) - \theta_2(s)) = T(s) \tag{3}$$

$$J_2S^2\theta(s) + fs\theta(s) = K(\theta_1 - \theta)$$

$$\theta(s)[J_2S^2 + fs + k] = K\theta_1(s)$$

$$\theta_1(S) = \theta(S)\left[\frac{J_2S^2 + fs + k}{k}\right]$$

Substituting the value of $\theta_1(s)$ in eqn. (3).

$$(J_1S^2 + K)\theta(s)\left(\frac{J_2S^2 + fs + K}{K}\right) - K\theta(s) = T(s)$$

$$\text{Transfer function } \frac{\theta(s)}{T(s)} = \frac{K}{(J_1S^2 + K)(J_2S^2 + fs + K) - K^2}$$

Mechanical Coupling Device

1.8.2 (a) Gear trains

Gear train is a mechanical device that transmits energy from one part of the system to another in such a way that force, torque, speed and displacement may be altered. The inertia and friction of the gears are neglected in ideal case. Consider a gear system as shown in the Fig. 1.70.

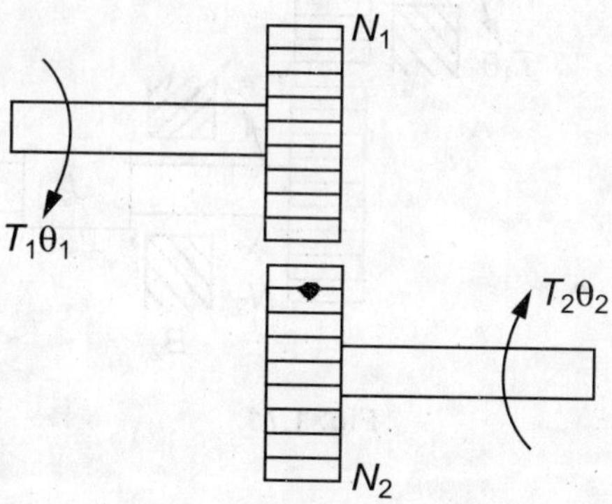

Fig. 1.70. Gear system

The number of teeth on the surface of the gears is proportional to the radii r_1 and r_2 of the gears.

$$\text{i.e.,} \quad r_1 N_2 = r_2 N_1$$

The distance travelled along the surface of each gear is same.

$$\text{i.e.,} \quad \theta_1 r_1 = \theta_2 r_2$$

The work done by one gear is same as the other

$$\text{i.e.,} \quad T_1\theta_1 = T_2\theta_2$$

$$\text{i.e.,} \quad \boxed{\frac{T_1}{T_2} = \frac{\theta_2}{\theta_1}} = \frac{r_1}{r_2} = \frac{N_1}{N_2}$$

(1) The number of teeth N are proportional to the radius r of the gear.

(2) The distance travelled on each gear is same.

(3) Work done = $T\theta$ by each gear is same.

1.8.3 (b) Gear train with inertia and friction

In practice, gear trains have friction and inertia which cannot be neglected. Consider such practical gear arrangement connected to load as shown below.

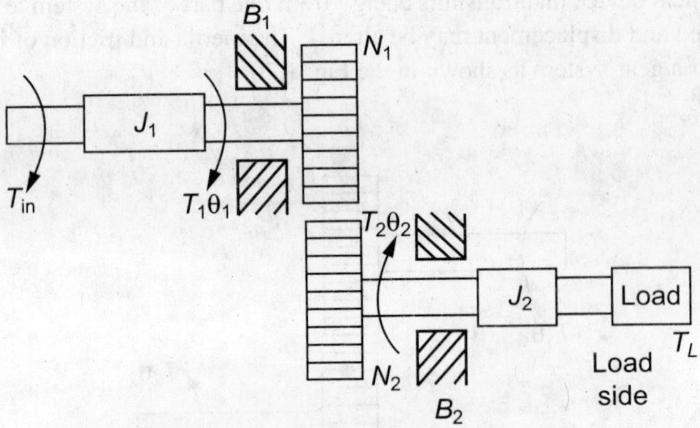

Fig. 1.71.

T_{in} – applied torque

$T_1 T_2$ – torque transmitted to gears

$N_1 N_2$ – no. of teeth

$\theta_1 \theta_2$ – angular displacement

$J_1 J_2$ – inertia of gears

$B_1 B_2$ – friction cofficients

Torque eqn. of side 1 is

$$T_{in} = J_1 \frac{d^2\theta_1(t)}{dt^2} + B_1 \frac{d\theta_1(t)}{dt} + T_1(t) \quad (1)$$

Torque on side 2 is,

$$T_2 = J_2 \frac{d^2\theta_2(t)}{dt^2} + B_2 \frac{d\theta_2(t)}{dt} + T_2(t) \quad (2)$$

Now

$$\frac{T_1}{T_2} = \frac{N_1}{N_2} = \frac{\theta_2}{\theta_1}$$

$$T_2 = \frac{N_2}{N_1} T_1$$

Substituting in eqn. (2)

$$\frac{N_2}{N_1} T_1 = J_2 \frac{d^2\theta_2(t)}{dt^2} + B_2 \frac{d\theta_2(t)}{dt} + T_L(t)$$

$$T_1 = \frac{J_2 N_1}{N_2} \frac{d^2\theta_2(t)}{dt^2} + \frac{B_2 N_1}{N_2} \frac{d\theta_2(t)}{dt} + \frac{N_1}{N_2} T_L(t)$$

Substituting value of T_1 in eqn. (1)

$$T_{in} = J_1 \frac{d^2\theta_2(t)}{dt^2} + B_1 \frac{d\theta_1(t)}{dt} + \frac{J_2 N_1}{N_2} \frac{d^2\theta_2(t)}{dt^2}$$

$$+ \frac{B_2 N_1}{N_2} \frac{d\theta_2(t)}{dt} + \frac{N_1}{N_2} T_2(t)$$

Substituting $\theta_2 = \frac{N_1}{N_2}\theta_1$

$$\therefore \quad T_{in} = J_1 \frac{d^2\theta_1}{dt^2} + B_1 \frac{d\theta_1}{dt} + \frac{N_1}{N_2} J_2 \frac{N_1}{N_2} \frac{d^2\theta}{dt}$$

$$+ \frac{N_1}{N_2} B_2 \frac{N_1}{N_2} \frac{d\theta_1}{dt} + \frac{N_1}{N_2} T_L$$

$$T_{in} = \left[J_1 + \left(\frac{N_1}{N_2}\right)^2 J_2 \right] \frac{d^2\theta_1}{dt^2} + \left[B_1 + \left(\frac{N_1}{N_2}\right)^2 B_2 \right] \frac{d\theta_1}{dt} + \frac{N_1}{N_2} T_L$$

$\therefore J_{le}$ = equivalent inertia referred to primary side

$$J_{le} = J_1 + \left(\frac{N_1}{N_2}\right)^2 J_2$$

B_{le}= equivalent friction referred to primary side

$$B_{le} = B_1 + \left(\frac{N_1}{N_2}\right)^2 B_2$$

$$\therefore \quad T_{in} = J_{le} \frac{d^2\theta_1}{dt^2} + B_{le} \frac{d\theta_1}{dt} + \frac{N_1}{N_2} T_L$$

Similarly, the equation can be written referred to load side also, where applied torque gets transferred to load as $\left(\frac{N_2}{N_1}\right) T$.

$$\left(\frac{N_2}{N_1}\right) T_{in} = J_{2e} \frac{d^2\theta_2}{dt^2} + B_{2e} \frac{d\theta_2}{dt} + T_L$$

where

$$J_{2e} = J_2 + \left(\frac{N_2}{N_1}\right)^2 J_l$$

$$B_{2e} = B_2 + \left(\frac{N_2}{N_1}\right)^2 B_1$$

1.8.4 Belt or chain drives

Belt or chain drives perform the same function as that of gear train. Assuming that there is no slippage between the belt and pulleys, we can write

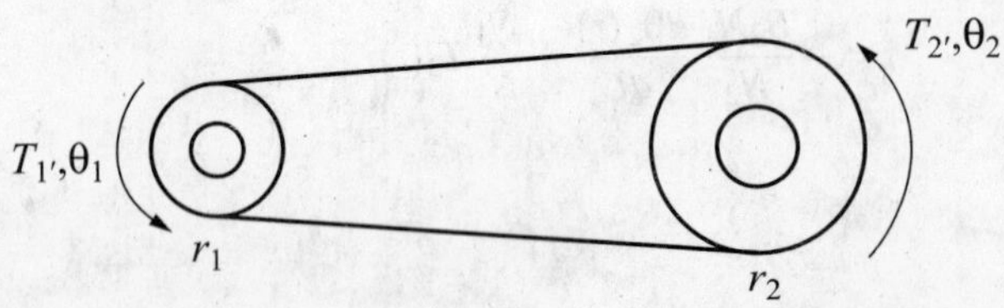

Fig. 1.72.

Then $\dfrac{T_1}{T_2} = \dfrac{r_1}{r_2} = \dfrac{\theta_2}{\theta_1}$

1.8.5 Mechanical levers

The level system is shown in Fig. 1.73. This transmits translational motion and forces, similar to gear trains.

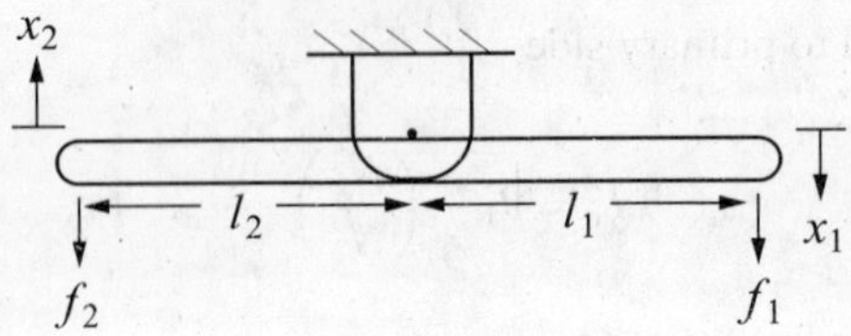

Fig. 1.73.

By law of moment

$$f_1 l_1 = f_2 l_2$$

By work done $\quad f_1 x_1 = f_2 x_2$

$$\text{Hence,} \quad \frac{f_1}{f_2} = \frac{x_2}{x_1} = \frac{l_2}{l_1}$$

Example 28: Obtain differential equations $x(t)$ and $f(t)$ for the system shown in Fig. 1.74. Draw its analogous circuit.

SOLUTION:

The network diagram is shown in figure $f^1(t) \cdot l_2 = f(t) \cdot l_1$

$$f^1(t) = \frac{l_1}{l_2} f(t)$$

and

$$f^1(t) = M\ddot{x} + B\dot{x} + Kx$$

$$\frac{l_1}{l_2} f(t) = M\ddot{x} + B\dot{x} + Kx$$

$$f(t) = \frac{l_2}{l_1}[M\ddot{x} + B\dot{x} + Kx]$$

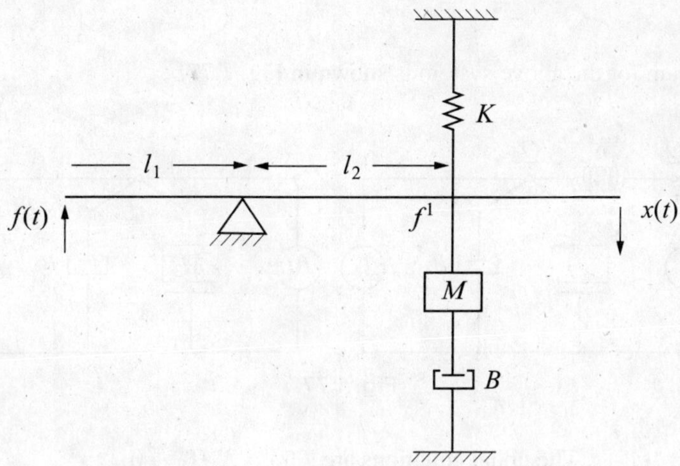

Fig. 1.74.

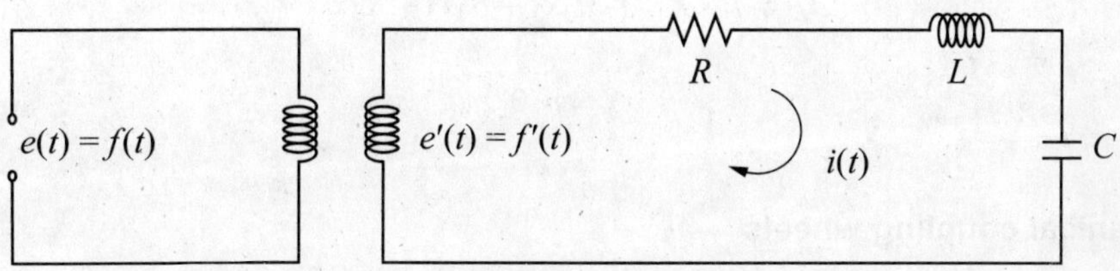

Fig. 1.75.

Using $f - v$ analogy, the analogous circuit is given below.

$$M \rightarrow L, B \rightarrow R, K = \frac{1}{C}$$

Example 29: Obtain the nodal equations for the system shown here.

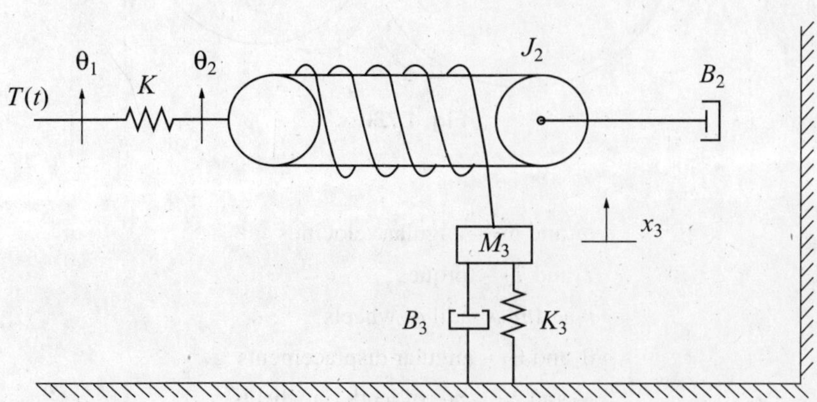

Fig. 1.76.

SOLUTION:

The network diagram for the above system is shown in Fig. 1.77.

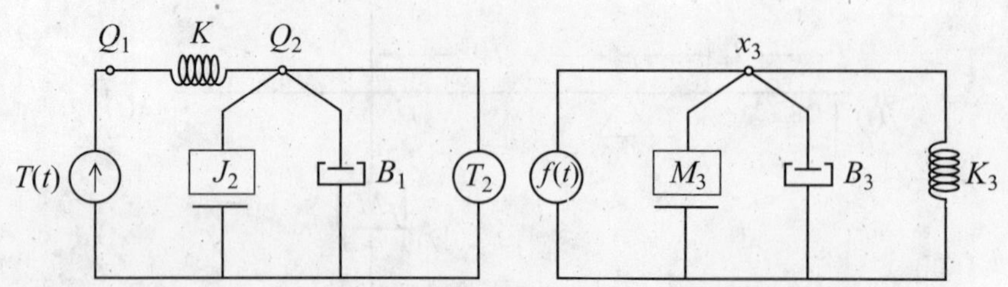

Fig. 1.77.

The nodal equations are $T(t) = K_1(\theta_1 - \theta_2)$ (1)

$$J_2\ddot{\theta}_2 + B_2\dot{\theta}_2 + K_1(\theta_1 - \theta_2) + T_L = 0 \quad (2)$$

$$T_L = f \times r \quad (3)$$

$$T_L = f \times r = r[M_3\ddot{x}_3 + B_3\dot{x}_3 + K_3x_3] \quad (4)$$

and

$$x_3 = r \cdot \theta_2 \quad (5)$$

1.8.6 Mechanical coupling wheels

Here the wheels are mechanically coupled. At the points A and B, wheels experience equal and opposite forces and the contact will be same.

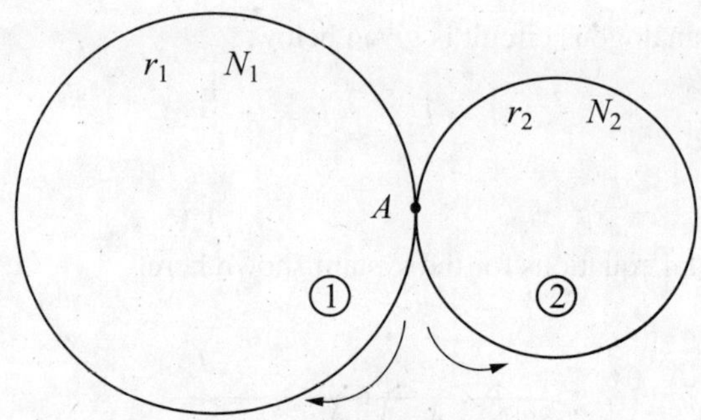

Fig. 1.78.

Let

ω_1 and ω_2 – angular velocities

T_1 and T_2 – torques

r_1 and r_2 – radii of wheels

θ_1 and θ_2 – angular displacements

N_1 and N_2 – no. of teeth on wheels

In ideal case, work done by wheel 1 will be equal to the work done by wheel 2

Work done by wheel 1 = $T_1\theta_1$

Work done by wheel 2 = $T_2\theta_2$

$$\therefore T_1\theta_1 = T_2\theta_2$$

$$\frac{T_1}{T_2} = \frac{\theta_2}{\theta_1} \tag{1}$$

Linear distance travelled along the surface of wheel (1) = $\theta_1 r_1$

Linear distance travelled along the surface of wheel (2) = $\theta_2 r_2$

$$\theta_1 r_1 = \theta_2 r_2$$

$$\frac{\theta_1}{\theta_2} = \frac{r_2}{r_1} \tag{2}$$

Combining eqns. (1) and (2)

$$\frac{T_1}{T_2} = \frac{\theta_1}{\theta_2} = \frac{r_1}{r_2}$$

$$\frac{T_1}{T_2} = \frac{r_1}{r_2} \tag{3}$$

No. of teeth on wheels ∝ radius of wheels

$$N \propto r$$

$$\frac{N_1}{N_2} = \frac{r_1}{r_2}$$

$$\frac{T_1}{T_2} = \frac{N_1}{N_2} \tag{4}$$

and

$$\theta_1 = \theta_2 \frac{r_2}{r_1}$$

$$\dot{\theta}_1 = \dot{\theta}_2 \frac{r_2}{r_1}$$

and

$$\dot{\theta}_2 = \omega_2$$

so

$$\omega_1 = \omega_2 \frac{r_2}{r_1}$$

$$\frac{\omega_1}{\omega_2} = \frac{r_1}{r_2} \tag{5}$$

From all the above equations

$$\frac{T_1}{T_2} = \frac{\theta_2}{\theta_1} = \frac{r_1}{r_2} = \frac{N_1}{N_2} = \frac{\omega_2}{\omega_1}$$

If we consider $r_1 : r_2$ as the turn ratio of an ideal transformer then angular velocity will be analogous to current and torque will be analogous to voltage. This is $f - v$ analogy.

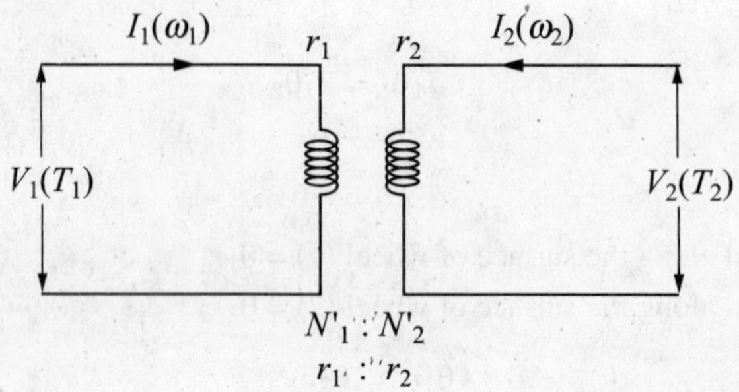

Fig. 1.79.

In case of $f - i$ analogy

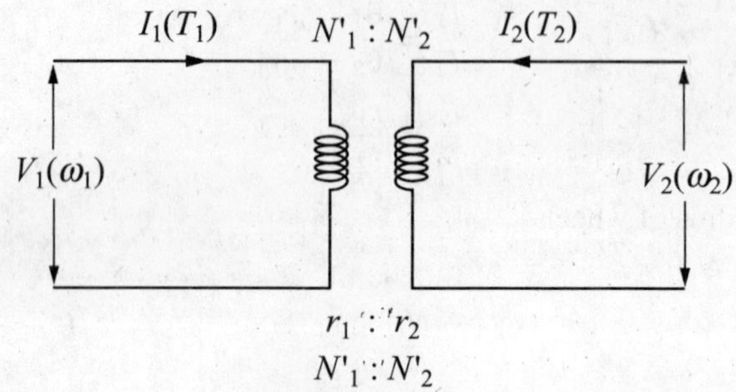

Fig. 1.80.

$N_1' : N_2'$ are the turn ratio of the transformer.

Example 30: (a) Find the ratio $r_1 : r_2$

(b) If $\omega_1 = 10$ rad/sec. Calculate ω_2

(c) If torque for wheel 1 is 5 N-mt. What will be the torque on wheel 2?

(d) If the displacement of wheel 1 is 20° then what is the displacement of wheel 2?

(e) Draw its electrical $f - v$ analogy.

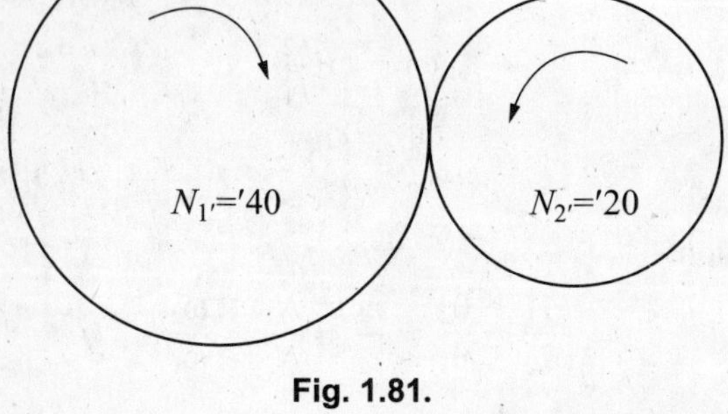

Fig. 1.81.

SOLUTION:

It is known that

Linear velocity $V_1 = V_2$

Work done $\omega_1 = \omega_2$

$T_1\theta_1 = T_2\theta_2$

Linear distance travelled by wheel 1 and wheel 2 is $\theta_1 r_1 = \theta_2 r_2$

(a) Radius and no. of teeth

$$\frac{r_1}{r_2} = \frac{N_1}{N_2} = \frac{40}{20} = 2.$$

(b)

$$\theta_1 r_1 = \theta_2 r_2$$

$$\text{So } \dot{\theta}_1 r_1 = \dot{\theta}_2 r_2$$

$$\omega_1 r_1 = \omega_2 r_2$$

$$\frac{\omega_1}{\omega_2} = \frac{r_2}{r_1} = \frac{1}{2}$$

$$\omega_2 = \omega_1 \times 2 = 10 \times 2 = 20 \text{ rad/sec.}$$

(c)

$$\frac{T_1}{T_2} = \frac{\theta_2}{\theta_1} = \frac{r_1}{r_2} = 2$$

$$T_2 = T_2 = \frac{T_1}{2} = \frac{S}{2}$$

$$T_2 = 2.5 \text{ N-mt.}$$

(d)

$$\frac{\theta_1}{\theta_2} = \frac{r_2}{r_1} = \frac{N_2}{N_1} = \frac{1}{2}$$

$$\frac{20}{\theta_2} = \frac{1}{2} \qquad \theta_2 = 40° \text{ rad} \qquad \textbf{Ans.}$$

(e) If $r_1 : r_2$ is the turn ratio of an ideal transfer, then angular velocity will be analogous to current and torque will be analogous to voltage. This is $f - v$ analogy.

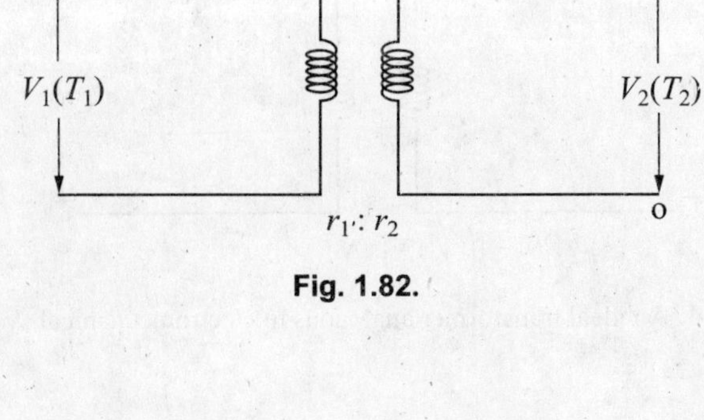

Fig. 1.82.

1.9 ELECTROMECHANICAL SYSTEM

A system in which both electrical and mechanical systems (elements) occur in combination is called electromechanical system—microphones, telephone receivers, loudspeakers, and electrical machines, etc.

Consider a conductor of length l lies perpendicular in uniform magnetic field with flux density B, the current I through conductor produces an upward force on it by Fleming's left-hand rule, as shown in Fig. 1.83.

$$\boxed{f = BIL} \quad \boxed{\frac{I}{F} = \frac{1}{Bl}} \tag{1}$$

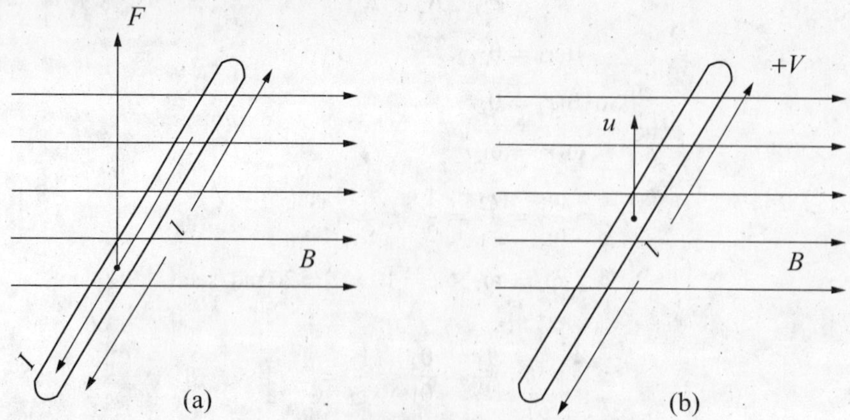

Fig. 1.83. Conductor placed in a uniform magnetic field.

(a) Current I in conductor produces force (Motoring Action)

(b) Motion of conductor with velocity induces voltage V (Generating Action)

In Fig. 1.83(b) the conductor moving with upward velocity u will have induced an emf V with polarity as indicated by Fleming's right-hand rule as

$$\boxed{V = Blu} \quad \boxed{\frac{v}{u} = Bl} \tag{2}$$

The eqns. (1) and (2) are analogous to an ideal transformer with turn ratio $N_1 : N_2 :: Bl : 1$. It carries V and I in the primary side as electrical quantities and u and f are mechanical quantities in the secondary side as shown in Fig. 1.84.

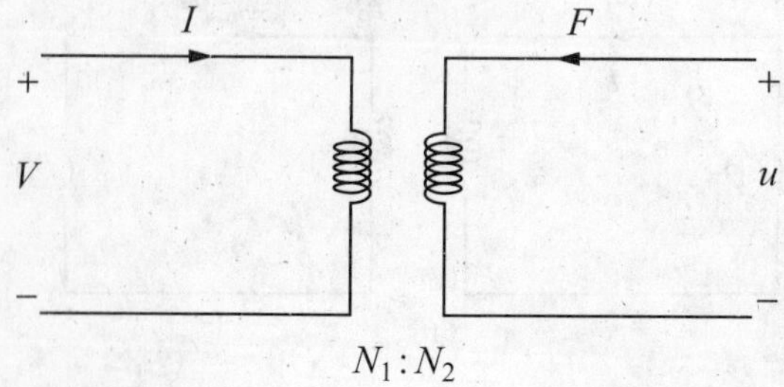

Fig. 1.84. An ideal transformer analogous to electromechanical system.

Since f and u are analogous to i and V respectively, therefore, this is termed force current $(f - I)$ analogy.

In electromechanical system, if one aims to solve the system for the electrical quantities (i.e. V and I), we refer the transformer to primary side, i.e. all the mechanical quantities on the secondary side are referred to primary side.

1.10 LIQUID LEVEL SYSTEM

For the liquid level system of Fig. 1.85, we consider the flow through a short pipe into the tank, q_i is inflow rate of the liquid in mt^3/sec, q_0 is outflow rate of liquid in mt^3/sec and h_0 is the height of liquid in mt.

The resistance R for the liquid flow can be defined as

$$R = \frac{\text{Change in level difference}}{\text{Change in rate of flow}}$$

For laminar flow, $R = \dfrac{\Delta h}{\Delta q}$ is constant and analogous to electrical resistance.

Under steady conditions, we have $q_i = q_0$ and h_0 is the steady liquid level in tank.

Let Δq_i is the small increase in the liquid inflow rate which causes increase of Δh_0 in liquid level in the tank. The resultant increase of liquid outflow rate Δq_0 is given by

$$\Delta q_0 = \frac{\Delta h}{R}$$

Liquid flow rate balanced equation is given by (rate of liquid inflow – rate of liquid outflow) = rate of liquid storage in tank.

$$\Delta q_i - \frac{\Delta h}{r} = c\frac{d}{dt}(\Delta h)$$

where C is the cross-sectional area of the tank and analogous to electrical capacitance.

$$RC\frac{d}{dt}(\Delta h) + \Delta h = R\Delta q_i \qquad (1)$$

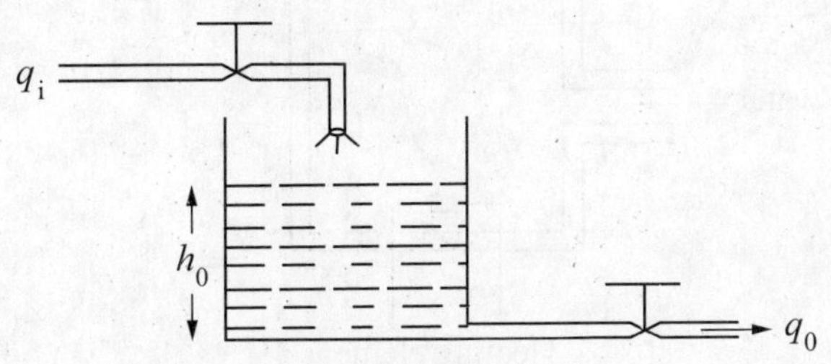

Fig. 1.85. A liquid level system.

Equation (1) is mathematical equation of Fig. 1.85.

Take Laplace on eqn. (1)

$$RCS\,\Delta H(s) + \Delta H(s) = R\Delta Q_i(s)$$

$$\frac{\Delta H(S)}{\Delta Q_i(S)} = \frac{R}{RSC+1}$$

The electrical analogous circuit is shown in Fig. 1.86.

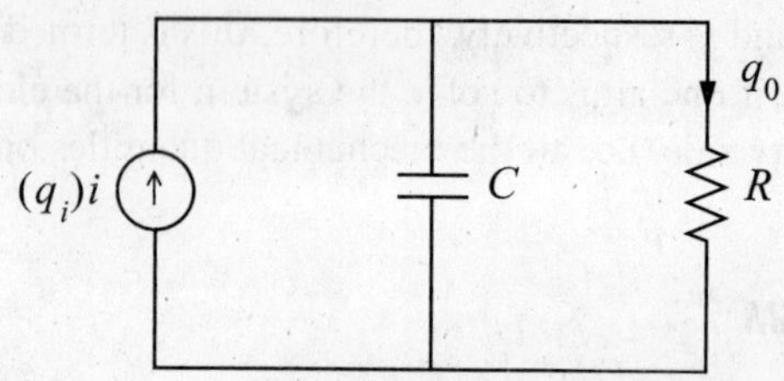

Fig. 1.86. Electrical analog of the liquid level system of Fig. 1.85.

Table 1.3. Electrical analogous quantities of liquid level system

Liquid level system	Electrical system
Liquid flow, mt^3	Charge (Coulomb)
Liquid flow rate, mt^3/sec.	Current (Amp)
Height, mt	Voltage (Volts)
Liquid resistance, mt/mt^3/sec	Resistance (Ohm)
Liquid capacitance, mt^3/mt	Capacitance (Farad)

1.11 THERMAL SYSTEM

In thermal system, there is transfer of heat from one substance to another substance. The thermal system can be analyzed in terms of resistance and capacitance. Let us consider a thermal system as shown in Fig. 1.87.

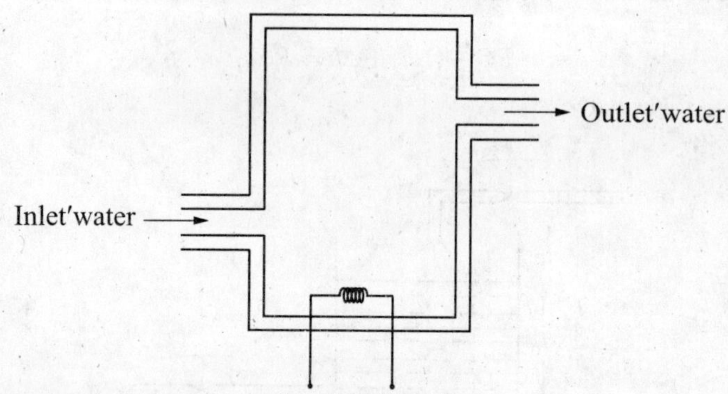

Fig. 1.87.

1.12 HEAT TRANSFER SYSTEM

Suppose there is no heat stored in the insulation and water is kept at uniform temperature.

Let θ_i = temperature of inlet water (°C)

θ_0 = temperature of outlet water (°C)

$\theta =$ temperature of surroundings

$q =$ rate of heat flow to heating element (J/sec)

$q_i =$ rate of heat flow to the water

$q_t =$ rate of heat flow through tank insulation

$C =$ thermal capacitance $(J/°\text{C})$

$R =$ thermal resistance $(°\text{C}/J\text{-}S^{-1})$

Rate of heat flow of the water in tank

$$q_i = C\frac{d\theta_0}{dt} \tag{1}$$

The rate of heat flow from water to surroundings through insulation

$$q_t = \frac{\theta_0 - \theta}{R}$$

$$q = q_i + q_t$$

$$q = C\frac{d\theta_0}{dt} + \frac{\theta_0 - \theta}{R} \tag{2}$$

Neglect the term $\dfrac{\theta}{R}$

$$q = C\frac{d\theta_0}{dt} + \frac{\theta_0}{R} \tag{3}$$

Take Laplace on both sides of eqn. (3)

$$Q(S) = CS\theta_0(S) + \frac{\theta_0(S)}{R}$$

$$\text{Transfer function} = \frac{\theta_0(S)}{Q(S)} = \frac{R}{1+CSR}$$

$$= \frac{R}{CR\left(S + \dfrac{1}{RC}\right)}$$

∴ Time constant of thermal system is RC.

1.12.1 Thermometer

Consider Fig. 1.49. A thermometer is immersed in a tub containing water.

Let $\theta_i =$ temperature of water in tub.

$\theta_0 =$ temperature indicated by thermometer

then, rate of heat flow to the thermometer

$$\frac{dq}{dt} = \frac{\theta_i - \theta_\circ}{R}$$

The indicated temperature rises at the rate

$$\frac{d\theta_\circ}{dt} = \frac{1}{C}\frac{dq}{dt} = \frac{1}{C}\left(\frac{\theta_i - \theta_\circ}{R}\right)$$

Take the Laplace transform on both sides

$$S\theta_o(S) = \frac{1}{RC}[\theta_i(S) - \theta_o(S)]$$

$$\theta_o(S)\left[S + \frac{1}{RC}\right] = \frac{1}{RC}\theta_i(S)$$

$$\frac{\theta_o(S)}{\theta_i(S)} = \frac{1/RC}{S + \dfrac{1}{RC}}$$

Consider $\theta_i(t)$ = unit step input = $u(t)$

$$\theta_i(S) = \frac{1}{S}$$

$$\theta_o(S) = \frac{1}{S(1+SRC)}$$

Take inverse Laplace $\theta_o(t) = 1 - e^{-t/RC}$

∴ Time constant = RC

Example 31: A thermometer has a time constant of 15.33 sec. It is quickly taken from a temperature 0 °C to a water bath having a temp. 100 °C. What temperature will be indicated after 60 sec?

SOLUTION:

The thermometer is subjected to a step input of 100 °C

$$\theta_i(S) = \frac{100}{S}$$

It is known that

$$\frac{\theta_o(S)}{\theta_i(S)} = \frac{1}{1+SRC}$$

$$\theta_o(s) = \frac{100}{S(1+SRC)}$$

$$= \frac{100}{S(1+ST)} \quad T \rightarrow \text{time constant}$$

$$= \frac{100}{(1+15.33\,S)S}$$

$$= 100(1 - e^{-t/T})$$

$$\theta_o(t) = 100\left(1 - e^{\frac{-60}{15.33}}\right)$$

$$\theta_o(t) = 98^\circ\text{C}$$

Temperature indicated by thermometer after 60 sec = 98 °C

1.13 ANALYSIS OF FIRST ORDER AND SECOND ORDER LINEAR SYSTEM BY CLASSICAL METHOD

The continuous-time system is described by differential equations. To analyze a system, it is necessary to be familiar with the mathematical concept of differential equations and the solution techniques.

The order of system is same as the order of the differential equation that represents the highest derivative involved in that differential equation. It also gives information about the number of energy storage elements present in the system.

As the differential equation contains no partial derivatives, it is considered ordinary. Here, the differential equations that are formed for the system analysis will be linear, ordinary, differential equations (LODE) with constant coefficients.

1.13.1 First order system

(i) First order homogeneous differential equation

Consider a differential equation

$$\frac{dy(t)}{dt} + py(t) = 0$$

Here p is constant

$$\frac{dy(t)}{dt} = -py(t)$$

$$\int \frac{dy(t)}{y(t)} = \int -p\,dt$$

On integrating

$$\int \frac{dy(t)}{y(t)} = \int -p\,dt$$

In $[y(t)] = -Pt + C$

Take $C = \ln K$

$$\ln[y(t)] = -pt + \ln K.$$

$$\ln\left[\frac{y(t)}{K}\right] = -pt$$

$$\frac{y(t)}{K} = e^{-pt} \Rightarrow y(t) = Ke^{-pt}$$

Here K is constant.

This is the particular solution.

(ii) First order non-homogeneous differential equation.

Consider a differential equation

$$\frac{dy(t)}{dt} + py(t) = Q$$

where p is constant and Q may be a function of time.

In the above equation multiply e^{pt} on both sides of equation

$$e^{pt}\frac{dy(t)}{dt} + pe^{pt}y(t) = Qe^{pt}$$

Since $d(a \cdot b) = a\,db + b\,da$, so LHS of above equation will be

$$\frac{d(e^{pt}y(t))}{dt} = Qe^{pt}$$

Then the general solution of the above second order system will be

$$y(t) = K_1 e^{p_1 t} + K_2 e^{p_2 t}$$

Here p_1, p_2 may be real or complex and K_1, K_2 are constants.
If $p_1 = p_2$, then the solution of second order differential equation will be

$$y(t) = k_1 e^{p_1 t} + k_2 t e^{p_1 t}$$

Integrate the above equation on both sides

$$\int d\left(e^{pt} y(t)\right) dt = \int Q e^{pt} dt$$

$$e^{pt} y(t) = \int Q e^{pt} dt + C$$

or

$$y(t) = e^{-pt} \int Q e^{pt} dt + C e^{-pt}$$

where C is a constant.

1.13.2 Second order differential equation

Consider a second order differential equation

$$a\frac{d^2 y(t)}{dt^2} + b\frac{dy(t)}{dt} + cy(t) = 0$$

Suppose e^{pt} is the solution of above equation, then it should satisfy the above equation.

$$ap^2 e^{pt} + be^{pt} p + ce^{pt} = 0$$

$$e^{pt}(ap^2 + bp + c) = 0$$

$$p = \frac{-b \pm \sqrt{b^2 - 4ac}}{2a}$$

and

$$p_1 = \frac{-b + \sqrt{b^2 - 4ac}}{2a}, \quad p_2 = \frac{-b - \sqrt{b^2 - 4ac}}{2a}$$

(A) OBJECTIVE TYPE QUESTIONS

Q1: Periodic and random signals are power signals whereas non-periodic and deterministic signals are energy signals.
(a) True (b) False

Q2: A power signal has infinite energy whereas an energy signal has zero average power.
(a) True (b) False

Q3: A complex-valued signal $x(t)$ has a real part $x_1(t)$ and an imaginary part $x_2(t)$ shown in figure below 1.1 P & 1.2 P. The signal $x(t)$ is
(a) conjugate symmetric (b) asymmetric
(c) symmetric (d) antisymmetric

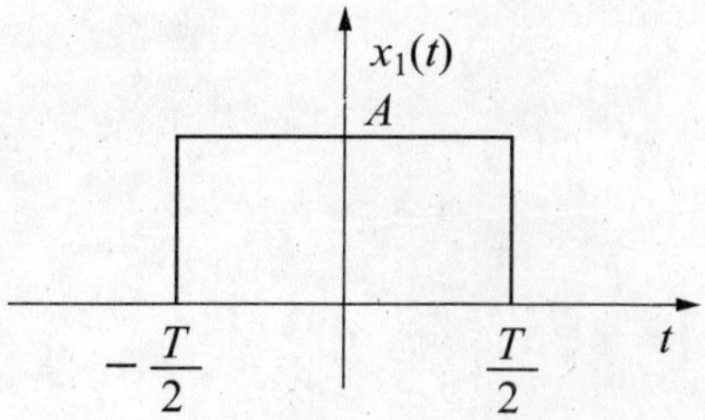

Fig. 1.1 P.

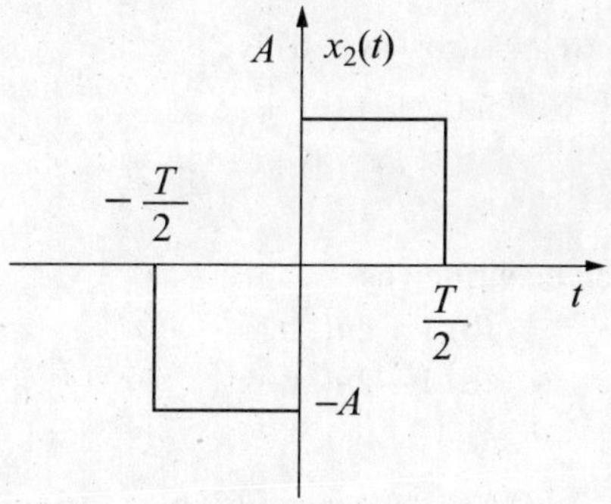

Fig. 1.2 P.

Q4: Whether $x(t) = 3e^{j5\pi t}$ is

(a) periodic and energy signal (b) non-periodic and energy signal

(c) periodic and power signal (d) non-periodic and power signal

Q5: Whether $x(t) = 3\sin^2 t$ is

(a) non-periodic (b) periodic

Q6: Determine whether $x(t) = \pi^2 u(t)$ is

(a) energy signal (b) power signal (c) neither energy nor power signal

Q7: Determine whether $x(t) = e^{-3t}$ is

(a) even (b) odd (c) neither even nor odd

Q8: Unit step function can be written as

(a) $1 - u(-t)$ (b) $1 + u(-t)$

(c) $1 - u(-t)$ (d) $1 + u(-t)$

Q9: Value of $\int_{-\infty}^{\infty} g(z)\delta(t-z)dt$ given as

(a) $g(t)$ (b) $g(-t)$ (c) $-\delta(t)$

Q10: Derivative of $u(t)$ is

(a) $\delta(-t)$ (b) 1

(c) $-\delta(-t)$ (d) $\delta(t)$

Q11: Power of $x(t) = A\cos(wt+\theta)$ is

(a) $-\frac{A^2}{2}$ (b) $\frac{A^2}{2}$

(c) $1+\frac{A^2}{2}$ (d) $1-\frac{A^2}{2}$

Q12: Sgn (t) can be written as

(a) $1-2u(t)$ (b) $1+2u(-t)$

(c) $1-2u(-t)$ (d) $1+2u(t)$

Answers

(1) a	(2) a	(3) a	(4) c	(5) b	(6) c
(7) c	(8) a	(9) a	(10) d	(11) b	(12) c

(B) OBJECTIVE TYPE QUESTIONS

Q1: For a linear system, which point is correct

(a) $f[x(t+T)] = f(x(t)) + f[x(T)]$ (b) $f[x(t-T)] = f[x(t)] - f[x(T)]$

(c) $f(x_1+x_2) = f(x_1) + f(x_2)$ (d) all are correct

Q2: Memory in a discrete-time system is analog of

(a) memory in a continuous-time system

(b) energy storage in a continuous-time system

(c) sampled memory of a continuous-time LTI system

(d) none of these

Q3: A continuous-time system with impulse response $h(t) = e^{at}u(t)$ is stable and when

(a) stable for all a and for $a = 0$

(b) stable for $a = 10$ and for $a < 0$

(c) stable for $a < 10$ and for all a

(d) none of these

Q4: Find the linear system from the following:

(a) $y(n) = x(n)x(n-1)$ (b) $y(n) = \frac{1}{2}nx(n)$

(c) $y(n) = [x(n)]^2x(n-1)$ (d) $y(n) = [x(n)]^2x(n+1)$

Q5: Find the linear system from the following:

(a) $y(n) = \frac{1}{2}n\{x(n)\}^2$ (b) $y(n) = \frac{1}{4}[x(n)+x(n-1)+x(n-2)]$

(c) $y(n) = x(n+1)x(n+2)$ (d) $y(n) = x^2(n+1)x(n-2)$

Q6: A system having impulse response $h(t)$ will be BIBO stable if

(a) $\int_{-\infty}^{\infty} |h(t)|dt < \infty$ (b) $\int_{-\infty}^{\infty} |h(t)|dt > \infty$

(c) $\int_{-\infty}^{\infty} |h(t)|dt > \infty$ (d) $\int_{-\infty}^{\infty} |h(t)|dt = 0$

Q7: The system $y(t) = Ax(t) + B$ is

(a) linear system (b) non-linear system

(c) partly (a) and partly (b) (d) all of these

Q8: The system $\frac{d^2y(t)}{dt^2} + 4\frac{dy(t)}{dt} + 2y(t) = x(t)$ is

(a) linear system (b) non-linear system

(c) partly (a) and partly (b) (d) none of these

Q9: The system $y(t) = x(t) + 4x(t+3)$ is

(a) causal system (b) non-causal system

(c) none of these

Q10: The system $y(t) = x(t) + \frac{1}{2}x(t-1)$ is

(a) causal system (b) non-causal system

(c) none of these

Answers

1 (c) 2 (c) 3 (a) 4 (b) 5 (b) 6 (a) 7 (b) 8 (a) 9 (b) 10 (a)

(C) OBJECTIVE TYPE QUESTIONS

Q1: Consider a simple mass spring system as given in Fig. 1.50. The transfer function

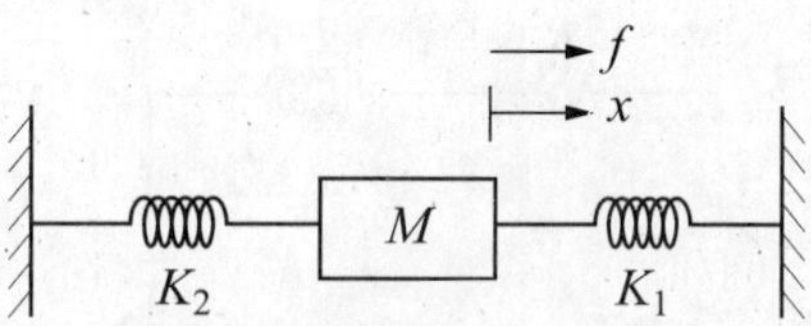

Fig. 1.3 P.

$\frac{X(s)}{F(s)}$ of the given system will be

(a) $\frac{1}{MS^2+FS+K_1+K_2}$ (b) $\frac{1}{MS^2+FS+\frac{K_1K_2}{K_1+K_2}}$

(c) $\frac{1}{MS^2+FS+K_1+K_2}$ (d) $\frac{K_2}{MS^2+S+K_1}$

Q2: The order of the differential equation is

$$\frac{d^2y(t)}{dt^2} + 5\frac{dy(t)}{dt} + 6y(t) = \underline{u}(t)$$

(a) 1 (b) 2

(c) 0 (d) none

Q3: Match the following for $f - v$ analogy

(i) F (1) R

(ii) K (2) L

(iii) M (3) $\frac{1}{C}$

(iv) B (4) V

Q4: In $f-i$ analogy, voltage is represented by

(a) velocity (b) mass

(c) spring constant (d) force

Q5: In rotational system, mass is represented by

(a) friction (b) tortional spring constant

(c) inertia (d) torque

Q6: Which of the following does not represents a mechanical system?

(a) M (c) R

(b) K (d) F

Q7: Which is correctly matched

(a) $F \rightarrow q$

(b) $M \rightarrow L$

(c) $B \rightarrow 1/C$

(d) $K \rightarrow R$

Q8: The order of the system is

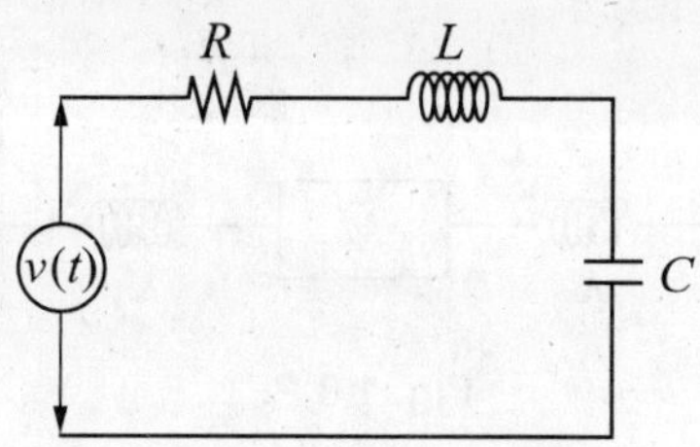

Fig. 1.4 P.

(a) 0 (c) 3

(b) 2 (d) None of these

Q9: In friction wheels, which relation is true?

(a) Torque and radius (b) Angular velocity and radius

(c) Work done $= \dfrac{\text{Torque}}{\text{Angular displacement}}$

Q10: For mechanically coupled wheels which relation is not ture?

(a) Work done is not equal (b) Linear distance travelled by the wheels is same

(c) Work done $= T\theta$ (d) None of these

Answers

(1) b (2) b (3) b (4) a (5) c (6) c (7) b (8) b (8) b (9) a (10) a

QUESTION BANK

Q1: Classify signals according to signal characteristics.

Q2: Sketch the signal $x(t) = e^{-t}$ for an interval $0 \le t \le 2$. Sample the signal with a sample period $T = 0.2$ second and sketch the discrete signal.

Q3: Check whether the following signals are periodic or not and if periodic then find the fundamental period:

(a) $x(n) = \cos\left(\frac{n\pi}{2}\right)\cos\left(\frac{n\pi}{4}\right)$ (b) $x(n) = Re\{e^{jn\pi/12}\} + Im\{e^{jn\pi/18}\}$

(c) $x(n) = \sin(\pi + 0.2n)$ (d) $x(n) = e^{jn\pi/16}\cos\left(\frac{n\pi}{17}\right)$ (e) $x(n) = \cos\left(\frac{n}{10}\right)\cos\left(\frac{n\pi}{10}\right)$

Q4: Signal $x(t)$ is shown in the figure below.

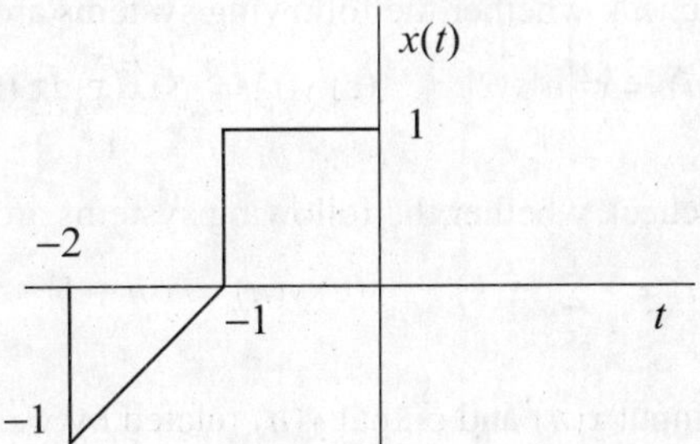

Fig. 1.5 P.

Sketch and label carefully each of the following signals (i) $x(2t+1)$ (ii) $x(4-t/2)$.

Q5: For the following system, check whether the following systems are linear.

(a) $y(t) = t^2 x(t-1)$ (b) $y(n) = x^2[-n]$

(c) $y(n) = x[n+1] - x[n-1]$ (d) $\frac{dy(t)}{dt} + 3ty(t) = t^2 x(t)$ (e) $y(n) = x^2[n]$

Q6: For the signal $x(t)$ sketch the following signals,

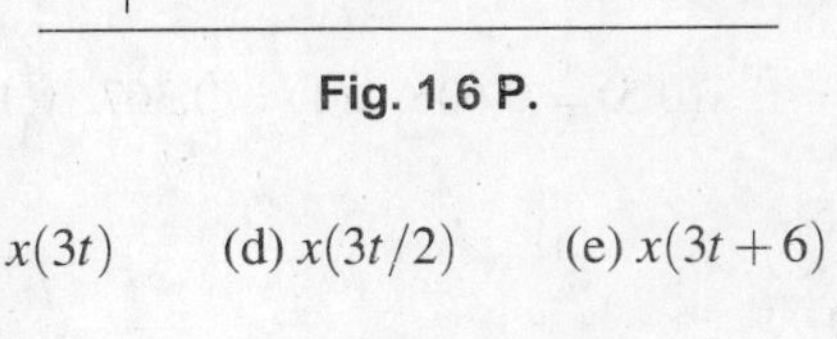

Fig. 1.6 P.

(a) $x(-t)$ (b) $x(t+6)$ (c) $x(3t)$ (d) $x(3t/2)$ (e) $x(3t+6)$ (f) $x(6-t)$

Q7: A continuous-time linear system s with input $x(t)$ and output $y(t)$ yields the following input-output pairs:

$x(t) = e^{j2t} \rightarrow y(t) = e^{j3t}$
$x(t) = e^{-j2t} \rightarrow y(t) = e^{-j3t}$

(i) if $x_1(t) = \cos(2t)$ determine the corresponding output $y_1(t)$ for system s.
(ii) if $x_2(t) = \cos(2(t-1/2))$ determine the corresponding output $y_2(t)$ for system s.

Q8: Consider a causal LTI system with $x(n)$ as input and output $y(n)$ are related by difference equation.

$$y[n] = \frac{1}{4}y[n-1] + x[n] \quad \text{Determine } y[n] \text{ if } x[n] = \delta[n-1]$$

Q9: For the following system, check whether the following systems are stable or not,
(a) $y(t) = x(t)\cos w_0 t$ (b) $y(n) = nx[n]$ (c) $y(n) = x(n/2)$

Q10: For the following system, check whether the following systems are time invariant or not,
(a) $y(t) = \sin tx(t-2)$ (b) $y(n) = x^2[n-2]$ (c) $y(t) = \int_{-\infty}^{t} x(\tau)d\tau$ (a) $y(t) = tx(t) + 3$

Q11: For the following system, check whether the following systems are causal or not,
(a) $y(t) = x(t)x(t-1)$ (b) $y(n) = \sum_{k=-\infty}^{n+1} x(k)$ (c) $y(n) = x[n+1] - x[n-1]$

Q12: Consider a system S with input $x(n)$ and output $y(n)$ related by $y(n) = x(n)[g(n) + g(n-1)]$
(a) if $g(n) = 1$ for all n, show that S is time invariant
(b) if $g(n) = n$, show that S is not time invariant
(c) if $g(n) = 1 + (-1)^n$, show that S is time invariant

Q13:

(a) Determine the energy of the signal

$$x(t) = \cos(10\pi t)u(t)u(2-t)$$

(b) Find the unspecified constants, denoted by C_1, C_2 and C_3 in the following expression:

$$10\delta(t) + 1\dot{\delta}(t) + (2 + C_2)\ddot{\delta}(t) = (3 + C_3)\delta(t) + 5\dot{\delta}(t) + 6\ddot{\delta}(t)$$

where "$\cdot$" and "$\cdot\cdot$" over a symbol denote the first and second order time derivatives respectively.

SOLUTIONS OF QUESTION BANK

Q1: See Section 1.2

Q2:

$$x(t) = e^{-t}$$

$$x(0) = e^{-0} = 1,\ x(0.5) = 0.606,\ x(1) = 0.367,\ x(1.5) = 0.223$$

$$x(2) = 0.135$$

To get the signal $x(t)$, mark the above points and draw a smooth curve joining all points as shown in Fig. 1.
Now $x(nT) = x(t)\big|_{t=hT} = x(t)\big|_{t=0.2h} = x(0.2h) = e^{-0.2h}$

$$x(0) = 1$$

$$x(1) = e^{-0.2} = 0.81 \quad x(6) = e^{-1.2} = 0.30$$

$$x(2) = e^{-0.4} = 0.67 \quad x(7) = e^{-1.4} = 0.25$$

$$x(3) = e^{-0.6} = 0.55 \quad x(8) = e^{-1.6} = 0.20$$

$$x(4) = e^{-0.8} = 0.45 \quad x(9) = e^{-1.8} = 0.17$$

$$x(5) = e^{-1} = 0.37 \quad x(10) = e^{-2} = 0.14$$

so $x(n) = \{1, .81, .67, .55, .45, .37, .30, .25, .20, .17, .14\}$

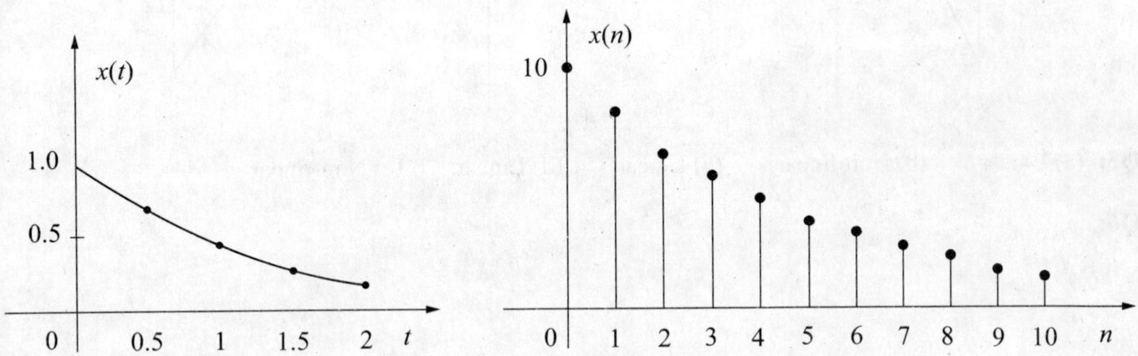

Q3:

(a) Periodic, fundamental period $N = 8$

$$Re\{e^{jn\pi}/12\} = \cos\frac{n\pi}{12}$$

(b)

$$\text{Im}\{e^{jn\pi}/18\} = \sin\frac{n\pi}{18}$$

$$N_1 = \frac{2\pi m}{w_{01}} = \frac{2\pi m}{\frac{\pi}{12}} = 24\ m \quad \text{for } m = 1 N_1 = 24$$

$$N_2 = \frac{2\pi m}{w_{02}} = \frac{2\pi m}{\frac{\pi}{13}} = 36\ m \quad \text{for } m = 1, N_2 = 36$$

$$\frac{N_1}{N_2} = \frac{24}{36} = \frac{2}{3} \Rightarrow N = 3N_1 = 2N_2$$

Since ratio of two integers, hence periodic.

So fundamental period $N = 3 \times 24 = 2 \times 36 = 72$

(c) $x(n) = \sin(\pi - 0.2n)$

$$N = \frac{2\pi m}{0.2} = 10\pi m$$

For any value of m, N is not an integer, so non-periodic

(d) Periodic, fundamental period $N = 544$

(e) Non-periodic

Q4:

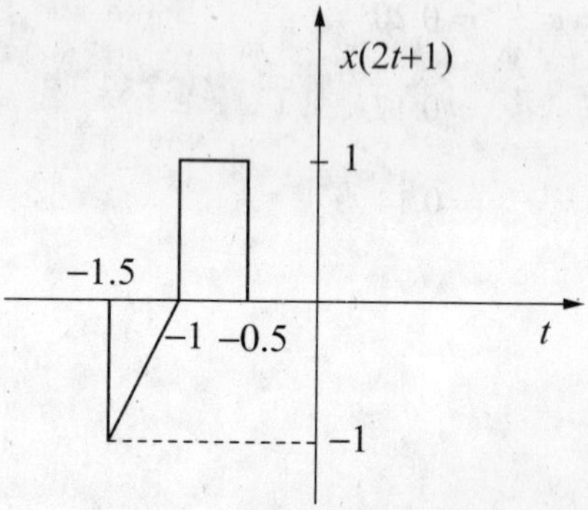

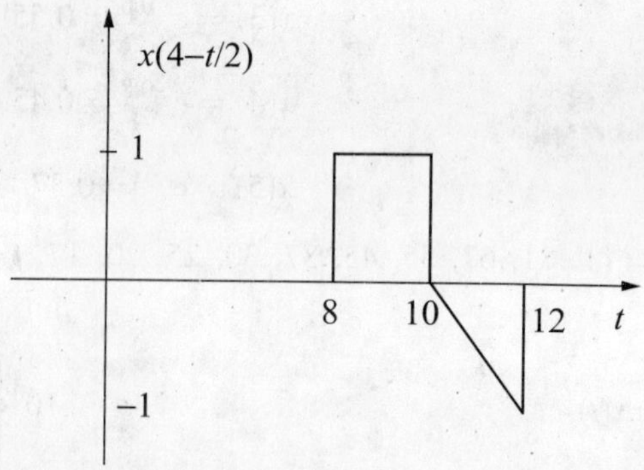

Q5: (a) Linear (b) Non-linear (c) Linear (d) Linear (e) Non-linear

Q6:

(a)

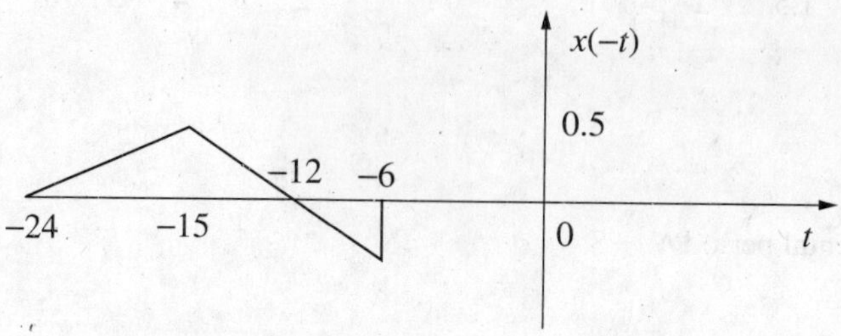

(f)

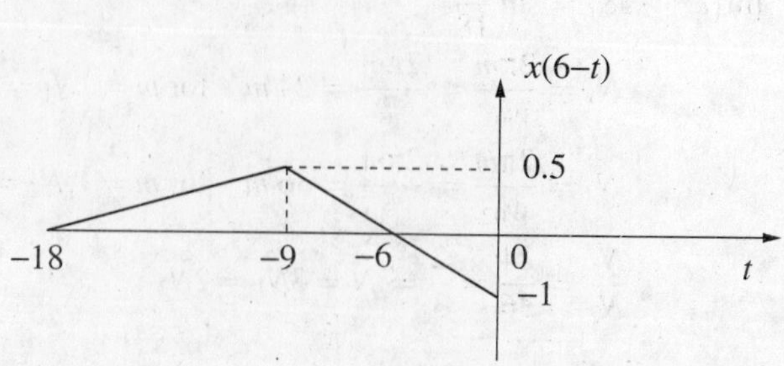

Q7:

(i)
$$x_1(t) = \cos 2t = \frac{e^{j2t} + e^{-j2t}}{2} = \frac{1}{2}\left[e^{j2t} + e^{-j2t}\right]$$
$$e^{j2t} \to e^{j3t},\ e^{-j2t} \to e^{-j3t}$$
$$\text{So } x_1(t) \to y_1(t) = \frac{1}{2}\left[e^{j3t} + e^{-j3t}\right] = \cos 3t$$

(ii)
$$x_2(t) = \cos\left(2\left(t - \frac{1}{2}\right)\right) = \cos\,(2t-1) = \frac{e^{j(2t-1)} + e^{-j(2t-1)}}{2}$$
$$= \frac{e^{j2t}\cdot e^{-j} + e^{-j2t}e^{j}}{2} = \frac{1}{2}\left[e^{j2t}e^{-j} + e^{-j2t}e^{j}\right]$$
$$\text{so } x_2(t) \to y_2(t) = \frac{1}{2}\left[e^{j3t}e^{-j} + e^{-j3t}e^{j}\right] = \frac{1}{2}\left[e^{j(3t-1)} + e^{-j(3t-1)}\right]$$
$$y_2(t) = \cos\,(3t-1)$$

Q8:
$$y(n) = \frac{1}{4}y(n-1) + x(n)$$

and
$$x(n) = \delta(n-1)$$

Since system is causal $y(n) = 0$ for $n < 1$
$$y(1) = \frac{1}{4}y(0) + x(1) = 0 + 1 = 1$$
$$y(2) = \frac{1}{4}y(1) + x(2) = \frac{1}{4} + 0 = \frac{1}{4}$$
$$y(3) = \frac{1}{4}y(2) + x(3) = \frac{1}{4} \times \frac{1}{4} + 0 = \left(\frac{1}{4}\right)^2$$
$$y(4) = \frac{1}{4}y(3) + x(4) = \frac{1}{4} \times \left(\frac{1}{4}\right)^2 + 0 = \left(\frac{1}{4}\right)^3$$
$$y(n) = \left(\frac{1}{4}\right)^{n-1}$$

Q9: (a) Stable (b) Unstable (c) Stable

Q10: (a) Time variant (b) Time invariant (c) Time invariant (d) Time variant

Q11: (a) Causal (b) Non-causal (c) Non-causal

Q12: $y(n) = x(n)\,[g(n) + g(n-1)]$

(a)
$$g(n) = 1 \Rightarrow y(n) = x(n)[1+1] = 2x(n)$$
$$y_1(n) = 2x(n-k) \text{ shift in input } K \text{ samples}$$
$$y_2(n) = 2x(n-k) \text{ shift in output } y_2(n) = y(n-k)$$

Since $y_1(n) = y_2(n)$,
Time invariant

(b)
$$g(n) = n \Rightarrow y(n) = x(n)[n + (n-1)] = (2n-1)x(n)$$
$$y_1(n) = (2n-1)x(n-k)$$
$$y_2(n) = [2(n-k)-1]\,x(n-k)$$

Time variant

(c) if $g(n) = 1 + (-1)^n$
$$y(n) = x(n)\left[1 + (-1)^n + 1 + (-1)^{n-1} = x(n)\left[2 - (-1)^{n-1} + (-1)^{n-1}\right] = 2x(n)\right.$$

Time invariant

Q13:

(a) $x(t) = \cos 10\pi t u(t)u(2-t)$

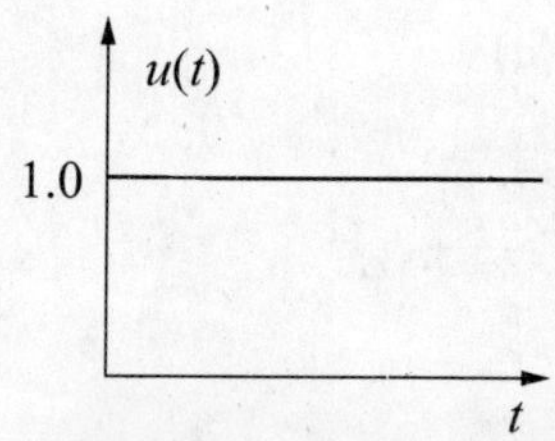

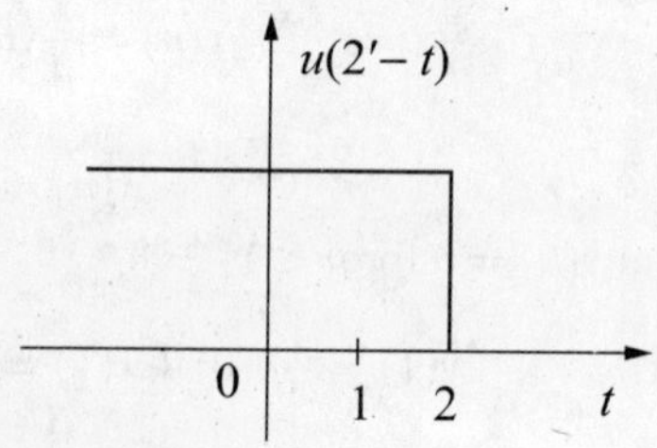

$$E = \int_{-\infty}^{\infty} (x(t))^2 dt = \int_0^2 \cos^2 2\pi t\,dt = \frac{1}{2}\left[\int_0^2 (1 + \cos 4\pi t)dt\right] = 1 \text{ joule}$$

(b) $10\delta(t) + (1\dot{\delta}(t) + (2 + C_2)\ddot{\delta}(t) = (3 + C_3)\delta(t) + 5\dot{\delta}(t) + 6\ddot{\delta}(t)$
On comparing L.H.S. and R.H.S., we get

$$10 = 3 + C_3$$
$$C_3 = 7 \quad (1)$$
$$C_1 = 5 \quad (2)$$
$$2 + C_2 = 6$$
$$C_2 = 4 \quad (3)$$

So unspecified constants are

$$C_1 = 5$$
$$C_2 = 4$$
$$C_3 = 7$$

QUESTION BANK WITH ANSWERS

Q1: Obtain the nodal equation for the system shown in Fig. 1.7 P and draw its analogous electrical network.

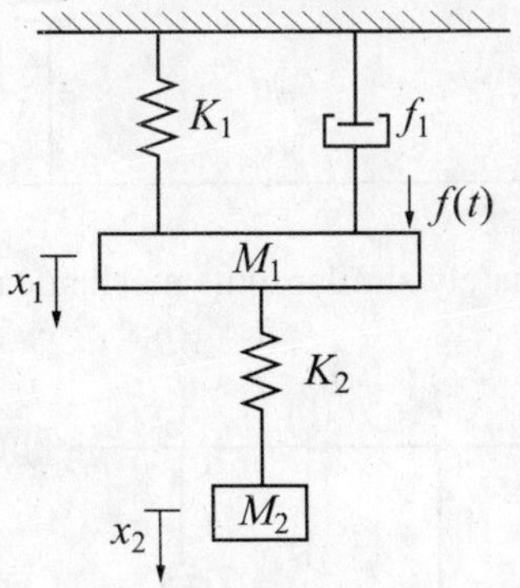

Fig. 1.7 P.

SOLUTION:

The mechanical diagram of the given system is shown below.

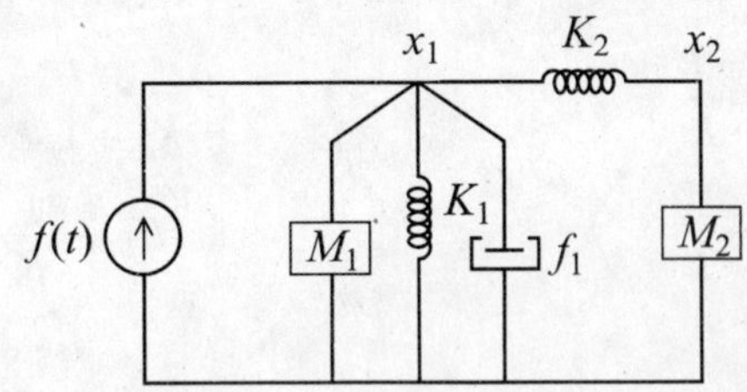

Node 'x_1'

$$f(t) = M_1\ddot{x}_1 + K_1x_1 + f_1\dot{x}_1 + K_2(x_1 - x_2)$$

$$f(s) = M_1S^2\, x_1(s) + k_1\, x_1(s) + F_1(s)s\, x_1(s) + K_2(x_1(s) - x_2(s))$$

Node 'x_2'

$$K_2(-x_1 + x_2) + M_2\ddot{x}_2 = 0$$

or

$$M_2S^2\, x_2(s) + K_2(\, x_2(s) - x_1(s)) = 0$$

$$K_2\, x_1(s) = K_2\, x_2(s) + S^2M_2\, x_2(s)$$

Electrical analog based on $f - v$ analogy is shown below.

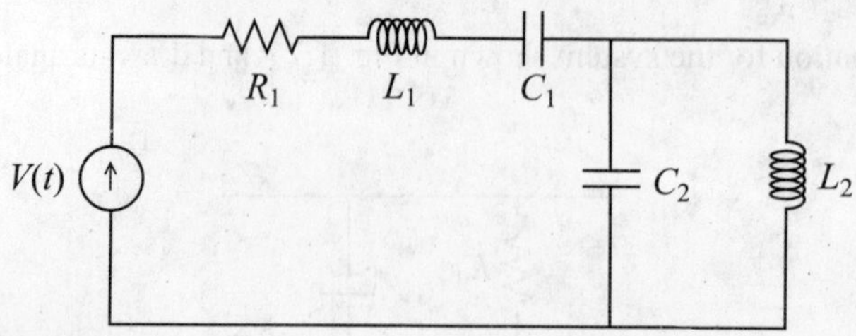

Basically $f - i$ analogy is approximately similar to its mechanical network diagram.

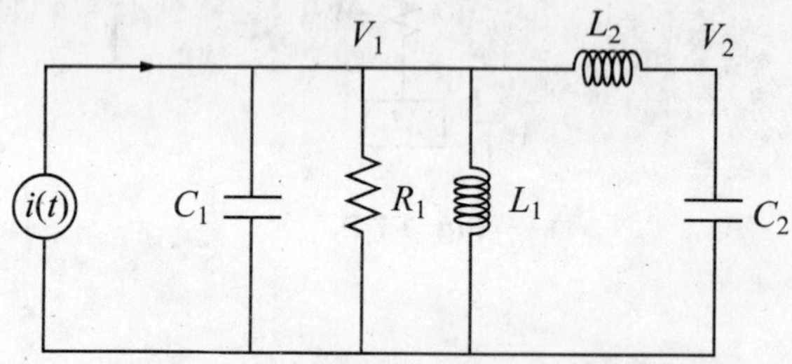

$f - v \Rightarrow$	$f - i \Rightarrow M_1 \rightarrow C_1$
$f \rightarrow v$	$x \rightarrow \theta$
$M \rightarrow L$	$f \rightarrow i$
$B \rightarrow R$	$K \rightarrow 1/L$
$K \rightarrow \dfrac{1}{C}$	$B_1 \rightarrow \dfrac{1}{R}$
$x \rightarrow q$	
$\dot{x} \rightarrow i$	

Q2: Obtain the performance equations for the system shown in Fig. 1.8 P in Laplace transform.

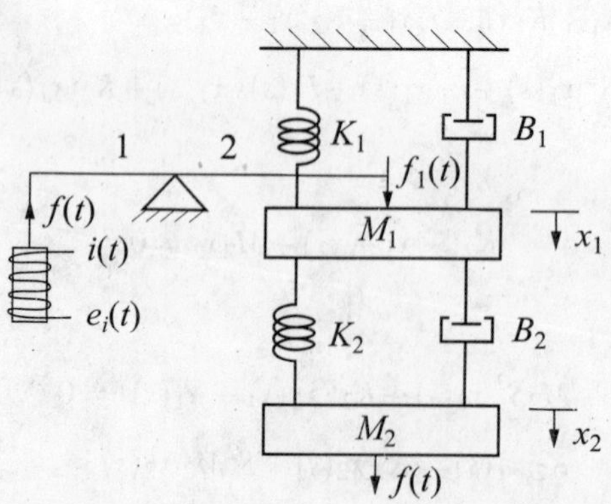

Fig. 1.8 P.

SOLUTION:

The equations in time domain are

$$e_i(t) = Ri(t) + L\frac{di(t)}{dt}$$
$$E_i(s) = [R + LS]I(s) \tag{1}$$

and

$$f(t)\cdot 1 = f_1(t)\times 2$$
$$f(t) = 2f_1(t)$$

and at M_1

$$f_1(t) = M_1\ddot{x}_1 + K_1x_1 + B_1\dot{x}_1 + K_2(x_1 - x_2) + B_2(\dot{x}_1 - \dot{x}_2)$$
$$F_1(s) = M_1S^2x_1(s) + K_1x_1(s) + B_1Sx_1(s) + K_2(x_1(s) - x_2(s)) + B_2(Sx_1(s) - Sx_2(s))$$
$$F_1(s) = (M_1S^2 + K_1 + B_1S + K_2 + B_2S)x_1(s) - x_2(s)(K_2 + SB_2) \tag{2}$$

and at M_2

$$M_2\ddot{x}_2 + K_2(x_2 - x_1) + B_2(\dot{x}_2 - \dot{x}_1) = 0$$
$$M_2S^2\,x_2(s) + K_2(x_2(s) - x_1(s)) + B_2S(x_2(s) - x_1(s)) = 0$$
$$(M_1S^2 + K_2 + SB_2)\,x_2(s) = x_1(s)[K_2 + B_2S] \tag{3}$$

Q3: Obtain differential equation for the dynamics of the system shown in Fig. 1.9 P. Draw its $f-i$ and $f-v$ analogy.

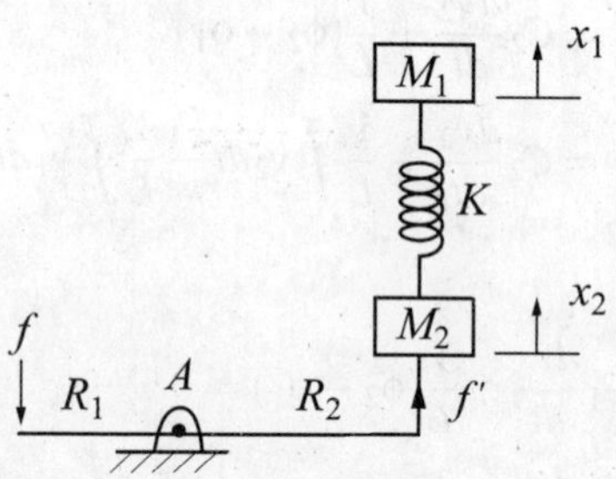

Fig. 1.9 P.

SOLUTION:

At $A \rightarrow$

Torques should be balanced so

$$f.R_1 = f'R_2$$

and at $M_2 \rightarrow$

$$f' = M_2\ddot{x}_2 + K_2(x_2 - x_1)$$

At $M_1 \rightarrow$

$$M_1\ddot{x}_1 + K(x_1 - x_2) = 0$$

Here the mechanical diagram of figure will be

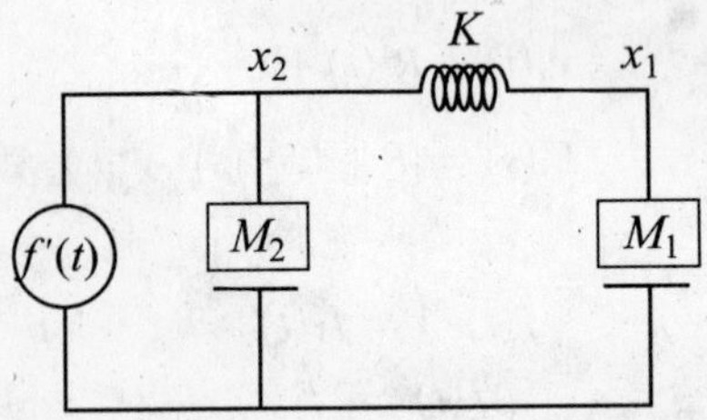

$f-i$ Analogous circuit equations $\Rightarrow$

$$f \rightarrow i$$
$$M \rightarrow C$$
$$K \rightarrow \frac{1}{L}$$
$$B \rightarrow \frac{1}{R}$$
$$x \rightarrow \phi = \int \dot{v} dt$$
$$\dot{x} = v$$

Equation (1) can be written as

$$i' = C_2 \frac{dv_2}{dt} + \frac{1}{L}(\phi_2 - \phi_1)$$
$$i' = C_2 \frac{dv_2}{dt} + \frac{1}{L}\int v_2 dt - \frac{1}{L}\int v_1 dt \tag{1}$$

Equation (2) can be written as

$$C_1 \frac{dv_1}{dt} + \frac{1}{L}(\phi_2 - \phi_1) = 0$$
$$C_1 \frac{dv_1}{dt} + \frac{1}{L}\int v_2 dt - \frac{1}{L}\int v_1 dt = 0 \tag{2}$$

From eqns. (1) and (2)

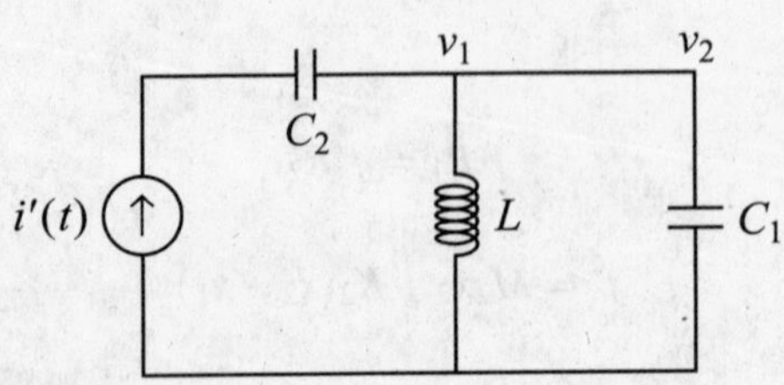

$f-v$ and $f-i$ analogous circuit of the above mechanical network will be given by these equations.

$f-v$ analogous equation $\Rightarrow$.

$$f^1 = M_2\ddot{x}_2 + K_2(x_2 - x_1) \tag{3}$$

As

$$M \to L$$
$$K \to 1/C$$
$$f \to v$$
$$B \to R$$
$$x \to q\int i\,dt$$
$$\ddot{x} \to \frac{di}{dt}$$

So electrical form $(f-v)$ of eqn. (1) is

$$\begin{aligned} V' &= L_2\frac{di_2}{dt} + \frac{1}{C}(q_2 - q_1) \\ &= L_2\frac{di_2}{dt} + \frac{1}{C}\int i_2 dt - \frac{1}{C}\int i_1 dt \end{aligned} \tag{4}$$

Similarly,

$$\begin{aligned} &M_1\ddot{x}_1 + K(x_1 - x_2) = 0 \\ &C_1\frac{di_1}{dt} + \frac{1}{C}\int i_1 dt - \frac{1}{C}\int i_2 dt = 0 \end{aligned} \tag{5}$$

From eqns. (4) and (5) {using KVL analysis}

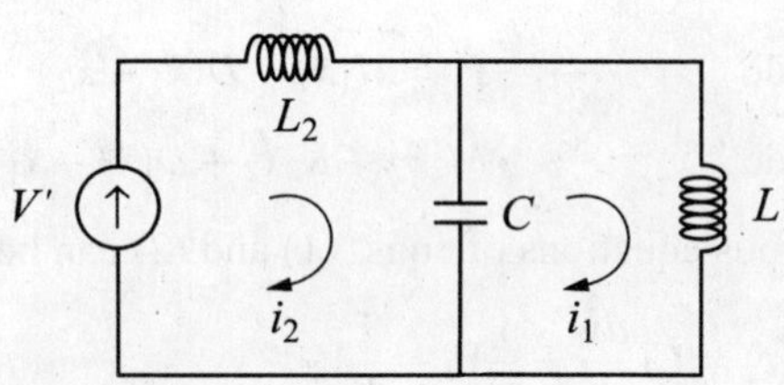

Q4: Draw the electrical analogous circuit of mechanical system shown in Fig. 1.10 P using current analogy.

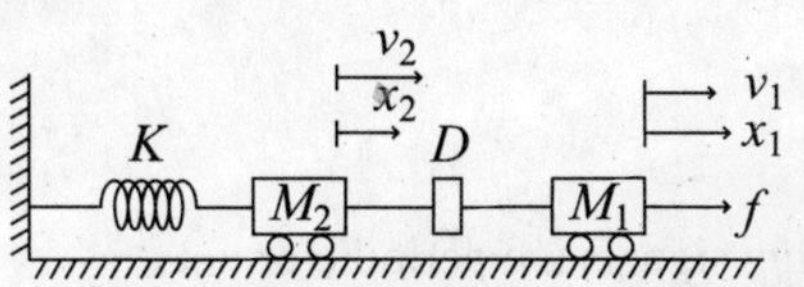

Fig. 1.10 P.

SOLUTION:

The required mechanical equivalent circuit is given below.

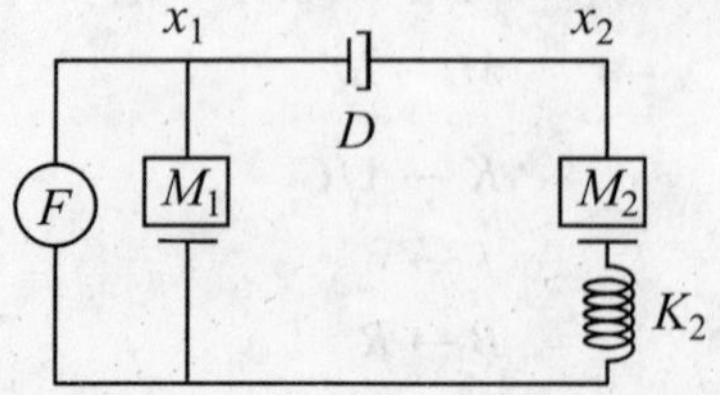

(a)

$f - i$ Analogy —

$f \to i, m \to c,\ k \to \frac{1}{L},\ D \Rightarrow \frac{1}{R}\ x \to \phi\ x \Rightarrow \frac{d\phi}{dt} = V$

The electrical $(f - i)$ analogous circuit is shown here

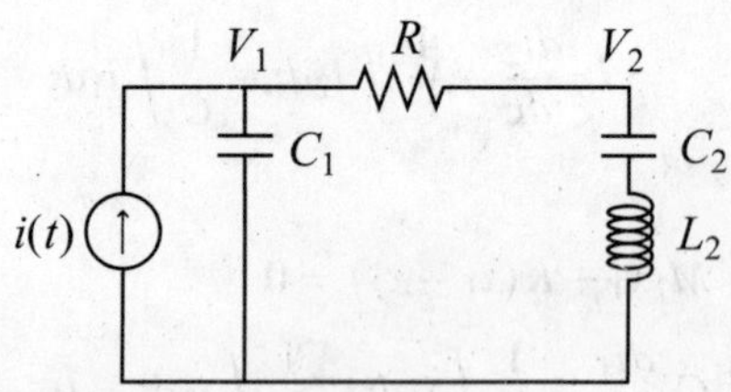

(b)

Equation corresponding to Fig. (a)

$$\text{At node } x_1 \quad \to \quad F = M_1\ddot{x}_1 + D(\dot{x}_1 - \dot{x}_2) \tag{1}$$

$$\text{At node } x_2 \quad \to \quad M_2\ddot{x}_2 + K_2x_2 + D(\dot{x}_2 - \dot{x}_1) = 0 \tag{2}$$

Equivalent electrical $(f - i)$ analogous equations of eqns. (1) and (2) can be written as

$$i(t) = C_1\frac{dv}{dt} + \frac{1}{R}V_1 - \frac{1}{R}V_2 \tag{3}$$

$$\frac{C_2dV_2}{dt} + \frac{1}{R_2}\phi(V_2 - V_1) + \frac{1}{L_2}\int V_2dt = 0 \tag{4}$$

Equations (3) and (4) will be same if we apply the node analysis on the electrical equivalent network of mechanical network.

UNSOLVED PROBLEMS

Q1: Determine whether the following signals are periodic or not:

(a) $x(t) = \sin 15\pi t$ (b) $x(t) = \cos 20\pi t + \cos 20t$

(c) $x(n) = \sin\left(\frac{6\pi n}{7}\right)$ (d) $x(n) = \sin\left(\frac{n}{s}\right)$ {Ans. (a) P (b) NP (c) P (d) NP}

Q2: Assume $x_1(t)$ and $x_2(t)$ are periodic signals with periods T_1 and T_2 respectively. Under what conditions is the sum $x(t) = x_1(t) + x_2(t)$ be periodic and what will be the period of $x(t)$ if it is periodic.

{Ans. $T = mT_1 = nT_2$}

Q3: Whether the signal $x(t) = u(t+a) - u(t-a)$ is a power signal or an energy signal?.

{Ans. $E = 2a$, energy signal}

Q4: Determine whether the given signal is a power signal or an energy signal or neither.

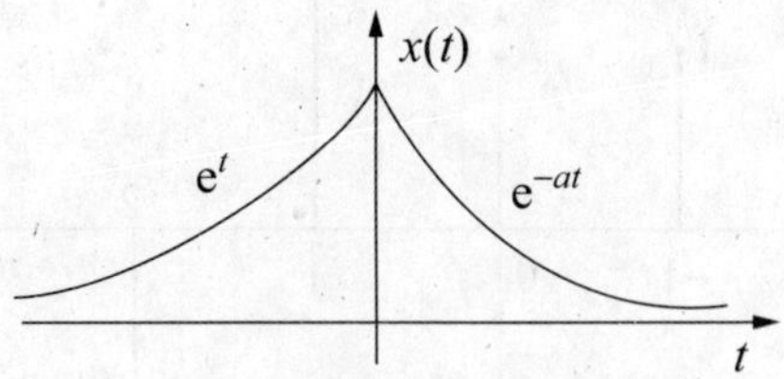

Fig. 1.11 P.

{Ans. Energy signal, $E = \frac{1}{a}$ }

Q5: Prove that $x(t) = A$ is a power signal.

Q6: A discrete-time sequence is given as

$$x(n) = \left\{ \begin{matrix} 1, 1, 1, 1, \frac{1}{2}, \frac{1}{2} \\ \uparrow \end{matrix} \right\}$$

(a) Sketch $x(n)$

(b) Determine $x(n)u(2-n)$ and sketch it.

{Ans. $x(n)u(2-n) = \{1, 1, 1, 1, 0, 0\}$

Q7: For the signal $x(t)$ of figure, sketch the following signals:

(a) $x(-t)$ (b) $x(t+2)$ (c) $x(2t)$ (d) $x(t/2)$ (e) $x(2-t)$ (f) $x(2t+4)$

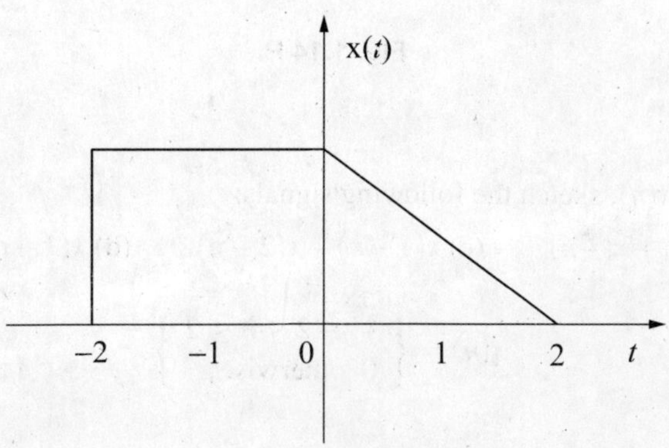

Fig. 1.12 P.

Q8: For the given signal $x[n]$, sketch the following signals:

(a) $x(-n)$ (b) $x(n+2)$ (c) $x(2-n)$ (d) $x(-n)u(n)$

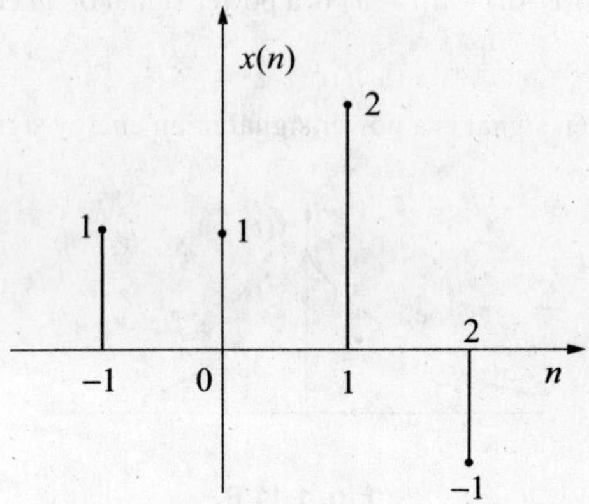

Fig. 1.13 P.

Q9: For the given signal $x(t)$, sketch the following signals:

(a) $x(-t)$ (b) $x(t)+x(-t)$ (c) $x(1-t)+x(2-t)$ (d) $x(1-t)x(2-t)$

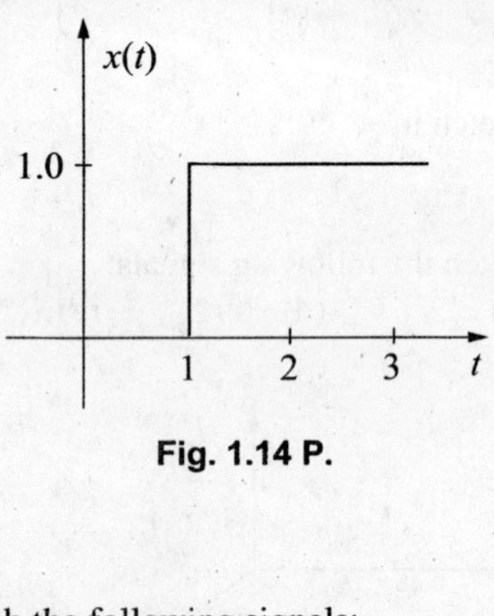

Fig. 1.14 P.

Q10: For the given signal $x(n)$, sketch the following signals:

(a) $x(-n)$ (b) $x(n)+x(-n)$ (c) $x(1-n)+x(2-n)$ (d) $x(1-n)x(2-n)$

$$x(n) = \begin{Bmatrix} 1 & -2 \le n \le 2 \\ 0 & \text{otherwise} \end{Bmatrix}$$

UNSOLVED PROBLEMS

Q1: Draw the electric analog circuit and find $\dfrac{X_0(s)}{X_i(s)}$.

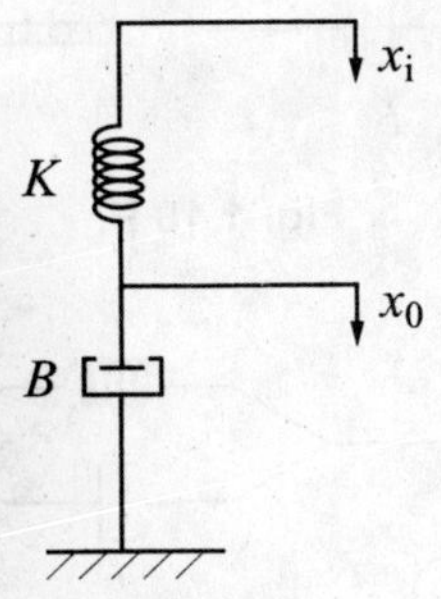

Fig. 1.15 P.

Q2: Draw its mechanical equivalent circuit and find $x_2(s)/F(s)$.

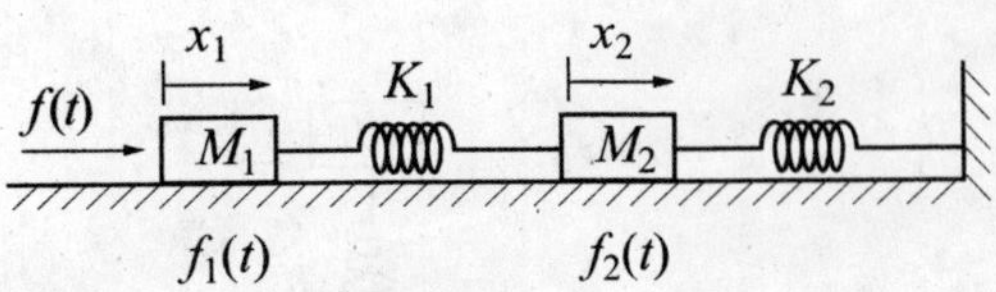

Fig. 1.16 P.

Ans. $\dfrac{K_1}{(S^2M_2+Sf_2+K_1+K_2)(S^2M_1+Sf_2+K_1)-K_1^2}$

Q3: Obtain the nodal equations for the system shown in figure and draw its analogous electrical network.

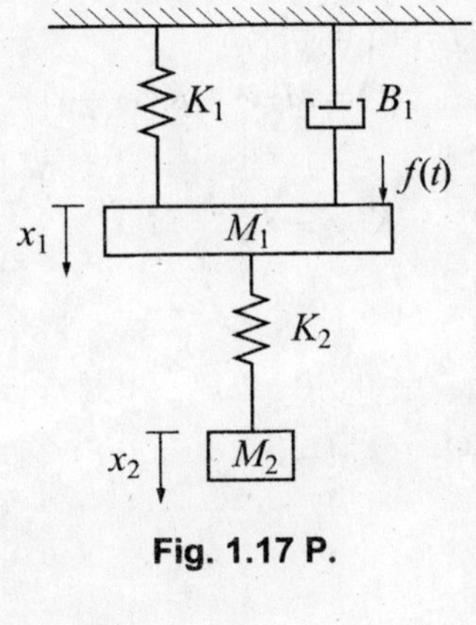

Fig. 1.17 P.

Q4: Obtain differential equations for the dynamics of the system shown in figure.

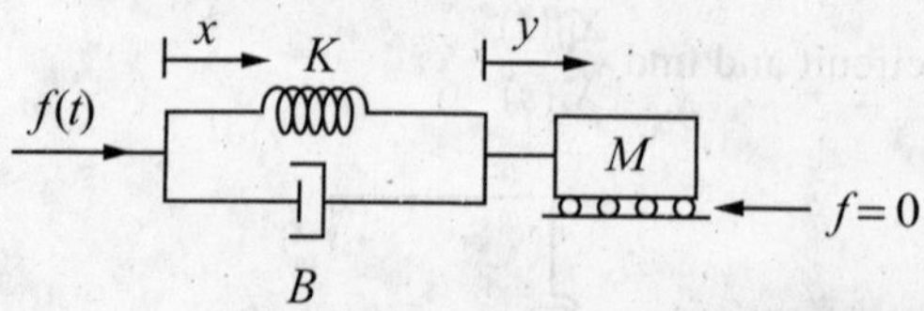

Fig. 1.18 P.

Hint →

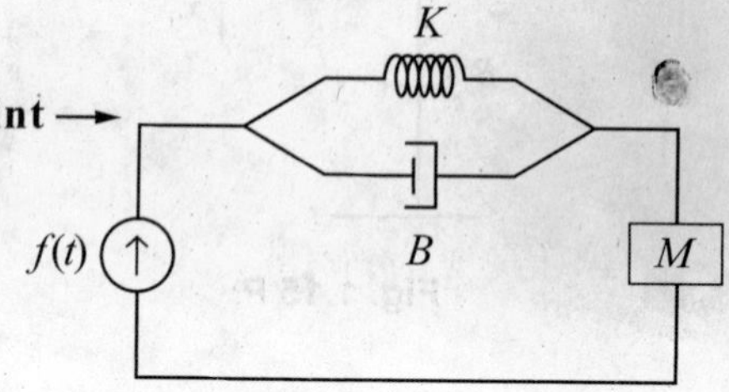

Q5: Draw the mechanical network diagram of the system shown in figure and write the equations of performance.

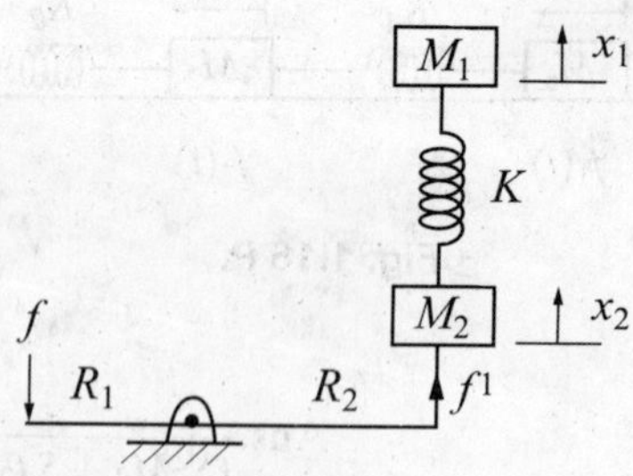

Fig. 1.19 P.

Hint:

$$f \times R_1 = f' \times R_2$$

$$f' = M_2\ddot{x}_2 + K(x_2 - x_1)$$

and

$$K(x_2 - x_1) = M_1\ddot{x}_1$$

Q6: For the gear system shown in figure.

(a) If $\theta_1 = 3$ rad (clockwise), calculate $\theta_4,\ \theta_2,\ \theta_3$

(b) If $\omega_1 = 15$ rad/sec. calculate $\omega_3,\ \omega_4$

(c) If $T_1 = 10$ N-mt, calculate $T_2,\ T_3$

(d) If $\ddot{\theta}_1 = 4$ rad /sec^2, calculate $\ddot{\theta}_3$

(e) Find $r_1 : r_3$ and $r_1 : r_4$

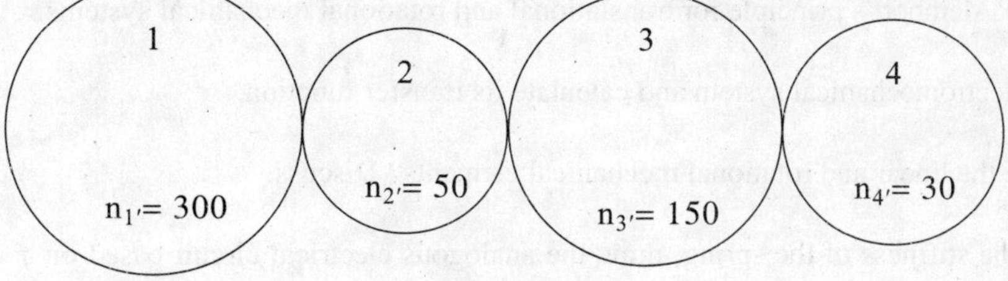

Fig. 1.20 P.

Hint →

$$T \propto r$$

$$n \propto r$$

$$\omega \propto \frac{1}{r}$$

$$\ddot{\theta} \propto \frac{1}{n}$$

Q7: Draw the $f - v$ and $f - i$ analogy circuit for the mechanical system shown in figure.

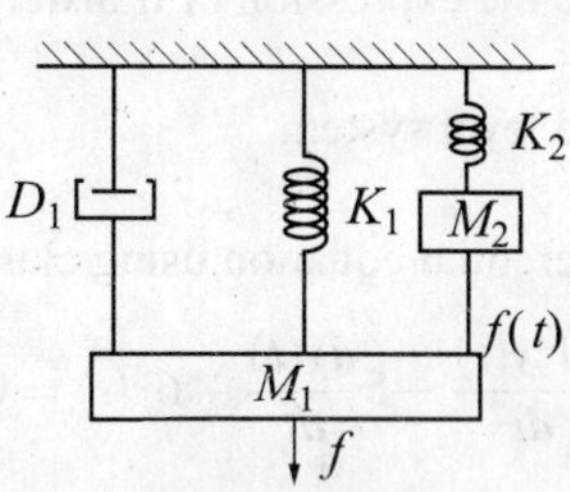

Fig. 1.21 P.

Q8:

(a) Draw the equivalent mechanical network.
(b) Write the differential equation.
(c) Draw analogous electrical network.

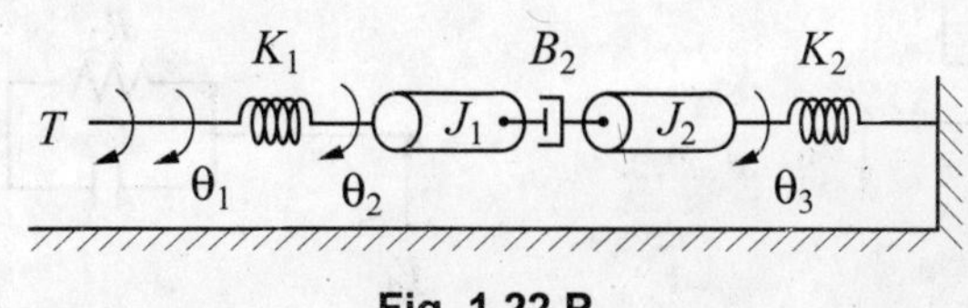

Fig. 1.22 P.

Hint → at B_2 angular displacement is $\theta_2 - \theta_3$

Q9: Explain D'Alembert's principle for translational and rotational mechanical systems.

Q10: Define electromechanical system and calculate its transfer function.

Q11: What are the linear and rotational mechanical elements? Discuss.

Q12: If K is the stiffness of the spring, draw the analogous electrical circuit based on $f-v$ analogy and determine $\frac{X_2(s)}{X_1(s)}$

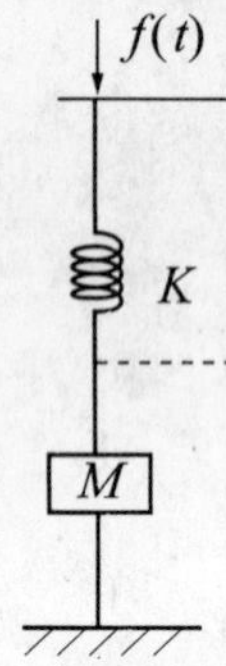

Fig. 1.23 P.

Q13: Define $f-v$ analogy and $f-i$ analogy.

Q14: Explain thermal system and derive the expression of transfer function for the heat transfer system.

Q15: Find the transfer function of liquid level system.

Q16: Find the solution of the given differential equation using classical method.

$$\frac{d^2y(t)}{dt^2}+5\frac{dy(t)}{dt}+6y(t)=0$$

Q17: Show the system in Fig. 1.24 P and Fig. 1.25 P are analogous systems.

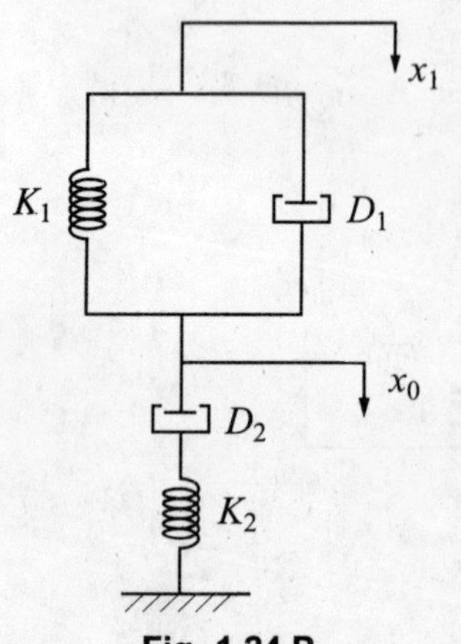

Fig. 1.24 P.

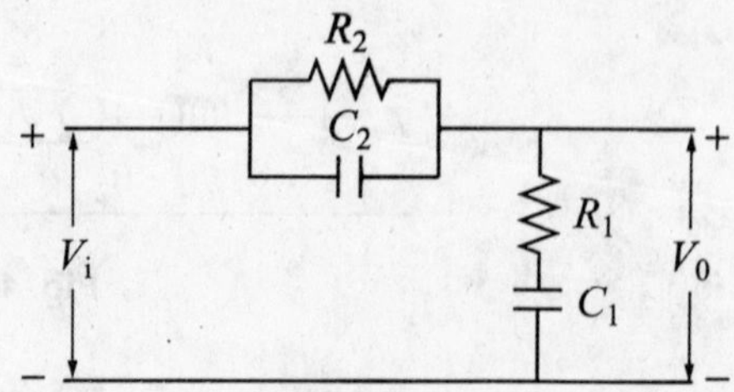

Fig. 1.25 P.

Q18: Draw the mechanical network for the system in figure and draw its analogous system.

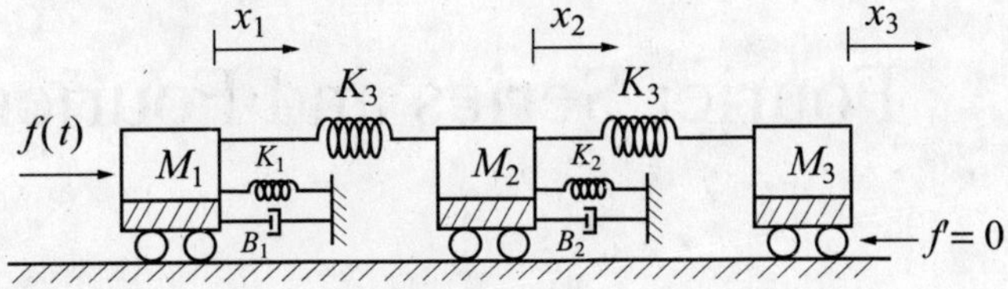

Fig. 1.26 P.

2

Fourier Series and Fourier Transform

2.1 INTRODUCTION

Fourier series is used to get frequency spectrum of a time-domain signal, when signal is a periodic function of time. We have seen that the sum of two sinusoids is periodic provided their frequencies are integer multiple of a fundamental frequency, w_0.

2.2 TRIGONOMETRIC FOURIER SERIES

Consider a signal $x(t)$, a sum of sine and cosine function whose frequencies are integral multiple of w_0

$$x(t) = a_0 + a_1 \cos(w_0 t) + a_2 \cos(2w_0 t) + \cdots$$

$$b_1 \sin(w_0 t) + b_2 \sin(2w_0 t) + \cdots$$

$$x(t) = a_0 + \sum_{n=1}^{\infty} (a_n \cos(nw_0 t) + b_n \sin(nw_0 t)) \tag{1}$$

$a_0, a_1, \ldots, b_1, b_2, \ldots$ are constants and w_0 is the fundamental frequency.

Evaluation of Fourier Coefficients

To evaluate a_0 we shall integrate both sides of eqn. (1) over one period (t_0, t_0+T) of $x(t)$ at an arbitrary time t_0

$$\int_{t_0}^{t_0+T} x(t)dt = \int_{t_0}^{t_0+T} a_0 dt + \sum_{n=1}^{\infty} a_n \int_{t_0}^{t_0+T} \cos(nw_0 t)dt + \sum_{n=1}^{\infty} b_n \int_{t_0}^{t_0+T} \sin(nw_0 t)dt$$

Since $\int_{t_0}^{t_0+T} \cos(nw_0 dt) = 0$

$$\int_{t_0}^{t_0+T} \sin(nw_0 dt) = 0$$

$$a_0 = \frac{1}{T} \int_{t_0}^{t_0+T} x(t)dt \tag{2}$$

To evaluate a_n and b_n, we use the following result:

$$\int_{t_0}^{t_0+T} \cos(nw_0 t)\cos(mw_0 t)dt = \begin{cases} 0 & m \neq n \\ T/2 & m = n \neq 0 \end{cases}$$

For $m \neq n$

$$\frac{1}{2}\int_{t_0}^{t_0+T} 2\cos(nw_0t)\cos(mw_0t)dt$$

$$= \frac{1}{2}\int_{t_0}^{t_0+T} \{\cos(m+n)w_0t + \cos(m-n)w_0t\}dt = 0$$

$$\int_{t_0}^{t_0+T} \cos nw_0 t dt = 0$$

$$= \frac{1}{nw_0}\left[\sin nw_0 t\right]_{t_0}^{t_0+T}$$

$$= \frac{1}{nw_0}[\sin w_0(t_0+T) - \sin w_0]$$

$$= \frac{1}{nw_0}[\sin(nw_0t_0 + 2\pi n)] - \sin nw_0 = 0$$

For $m = n \neq 0$

$$\frac{1}{2}\int_{t_0}^{t_0+T} 2\cos^2 nw_0 t dt = \frac{1}{2}\int_{t_0}^{t_0+T} (1+\cos 2nw_0t)dt = \frac{T}{2}$$

Multiply eqn. (1) by $\cos(mw_0t)$ and integrate over one period.

$$\int_{t_0}^{t_0+T} x(t)\cos(mw_0t)dt = a_0\int_{t_0}^{t_0+T}\cos(mw_0t)dt + \sum_{n=1}^{\infty} a_n \int_{t_0}^{t_0+T}\cos(nw_0t)\cos(mw_0t)dt + \sum_{n=1}^{\infty} b_n \int_{t_0}^{t_0+T}\sin(nw_0t)\cos(mw_0t)dt$$

1st and 3rd terms are zero

$$\sum_{n=1}^{\infty} a_n \int_{t_0}^{t_0+T}\cos(nw_0t)\cos(mw_0t)dt = \int_{t_0}^{t_0+T} x(t)\cos(mw_0t)dt$$

$$a_n = \frac{2}{T}\int_{t_0}^{t_0+T} x(t)\cos(nw_0t)dt \qquad (3)$$

Multiply eqn. (1) by sin (mw_0t) and integrate over one period

$$\int_{t_0}^{t_0+T} x(t)\sin(mw_0t)dt = a_0\int_{t_0}^{t_0+T}\sin(mw_0t)dt + \sum_{n=1}^{\infty} a_n\int_{t_0}^{t_0+T}\cos(nw_0t)\sin(mw_0t)dt +$$

$$\sum_{n=1}^{\infty} b_n\int_{t_0}^{t_0+T}\sin(mw_0t)\sin(nw_0t)dt$$

$$b_n = \frac{2}{T}\int_{t_0}^{t_0+T} x(t)\sin(nw_0t)dt \qquad (4)$$

Example 1:

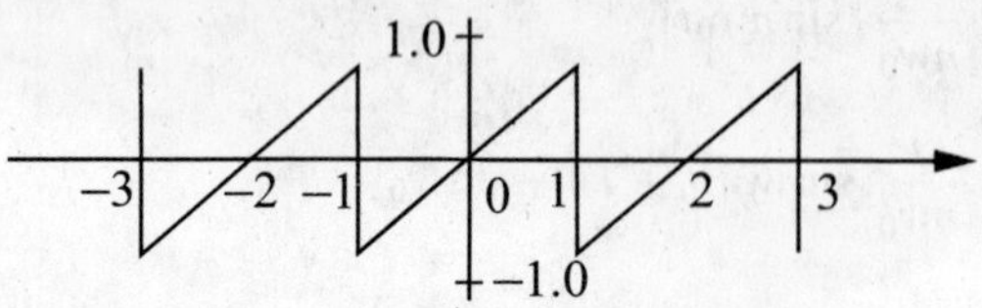

Fig. 2.1.

$$T \to -1 \text{ to } 1 \qquad T = 2 \qquad w_0 = \pi \qquad x(t) = t, -1 < t < 1$$

$$a_0 = \frac{1}{2}\int_{-1}^{1} t\,dt = \frac{1}{4}(1-1) = 0$$

$$a_n = 0$$

$$b_n = \int_{-1}^{1} t\sin\pi nt\,dt = \left[\frac{-t\cos\pi nt}{n\pi} - \frac{\cos\pi nt}{n\pi}\right]_{-1}^{1}$$

$$= \frac{-1}{n\pi}[t\cos\pi nt + \cos\pi nt]_{-1}^{1} = -\frac{1}{n\pi}[2\cos\pi + \cos\pi - \cos\pi]$$

$$b_n = \frac{-2}{n\pi}\cos n\pi = \frac{2}{\pi}\left[\frac{-(-1)^n}{n}\right]$$

$$\begin{matrix} b_1 & b_2 & b_3 & b_4 & b_5 & b_6 \\ \frac{2}{\pi} & \frac{-2}{2\pi} & \frac{2}{3\pi} & \frac{-2}{4\pi} & \frac{2}{5\pi} & \frac{-2\cdots}{6\pi} \end{matrix}$$

$$x(t) = \sum_{n=1}^{\infty}\frac{2}{\pi}\left[\frac{-(-1)^n}{n}\right]\sin n\pi t$$

$$= \frac{2}{\pi}\left[\sin\pi t - \frac{1}{2}\sin 2\pi t + \frac{1}{3}\sin 3\pi t - \frac{1}{4}\sin 4\pi t + \cdots\right]$$

Example 2:

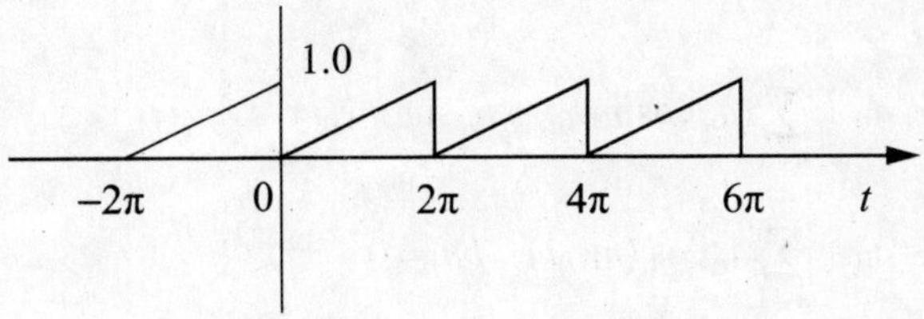

Fig. 2.2.

$$x(t) = \frac{t}{2\pi} \qquad T = 2\pi \qquad w_0 = \frac{2\pi}{T} = 1$$

$$a_0 = \frac{1}{T}\int_0^{2\pi} x(t)dt = \frac{1}{4\pi^2}\left[\frac{1}{2}t^2\right]_0^{2\pi} = \frac{1}{2}$$

$$a_n = \frac{2}{4\pi^2}\int_0^{2\pi} t\cos ntdt = \frac{1}{2\pi^2}\left[\frac{t\sin t}{n} + \frac{\sin nt}{n}\right]_0^{2\pi}$$

$$= \frac{1}{2\pi^2}\left[\frac{2\pi\sin 2n\pi}{n} + \frac{\sin 2n\pi}{n}\right] = 0$$

$$b_n = \frac{2}{4\pi^2}\int_0^{2\pi} t\sin ntdt = \frac{-1}{2\pi^2}\left[\frac{t\cos nt}{n} + \frac{\cos nt}{n}\right]_0^{2\pi}$$

$$= \frac{-1}{2\pi^2}\left[\frac{2\pi\cos 2n\pi}{n} + \frac{\cos 2n\pi}{n} - \frac{1}{n}\right]$$

$$b_n = \frac{-1}{n\pi}$$

$$x(t) = \frac{1}{2} + \sum_{n=1}^{\infty}\left(\frac{-1}{n\pi}\right)\sin nt = \frac{1}{2} + \frac{1}{\pi}\sum_{n=1}^{\infty}\frac{1}{n}\cos(nt + \pi/2)$$

$$= \frac{1}{2} - \frac{1}{\pi}\left[\sin t + \frac{\sin 2t}{2} + \frac{\sin 3t}{3} + \cdots\right]$$

Example 3:

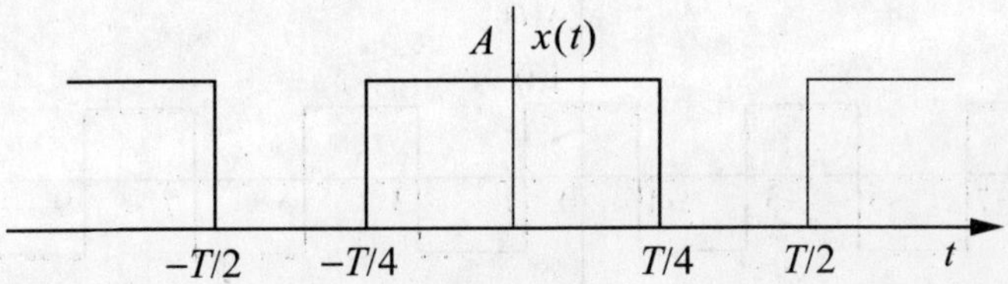

Fig. 2.3. Rectangular waveform

Figure shows a periodic rectangular waveform which is symmetrical to the vertical axis. Obtain its F.S. representation.

$$x(t) = a_0 + \sum_{n=1}^{\infty} (a_n \cos nw_0t + b_n \sin nw_0t)$$

$$x(t) = a_0 + \sum_{n=1}^{\infty} a_n \cos (nw_0t) \quad bn = 0$$

$$x(t) = 0 \quad \text{for } \frac{-T}{2} < t < \frac{-T}{4}$$

$$+A \quad \text{for } \frac{-T}{4} < t < \frac{T}{4}$$

$$0 \quad \text{for } \frac{T}{4} < t < \frac{T}{2}$$

$$a_0 = \frac{1}{T} \int_{-T/4}^{T/4} A dt = \frac{A}{2}$$

$$a_n = \frac{2}{T} \int_{-T/4}^{T/4} A \cos (nw_0t) dt = \frac{2A}{Tnw_0} \left[\sin nw_0 \frac{T}{4} + \sin nw_0 \frac{T}{4} \right]$$

$$a_n = \frac{4A}{2\pi n} \sin \left(\frac{n\pi}{2} \right) = \frac{2A}{\pi n} \sin \left(\frac{n\pi}{2} \right) \quad w_0 = \frac{2\pi}{T}$$

$$a_1 = \frac{4A}{2\pi} = \frac{2A}{\pi}$$

$$a_2 = 0$$

$$a_3 = \frac{2A}{3\pi} \sin \frac{3\pi}{2} = \frac{2A}{3\pi}(-1) = \frac{-2A}{3\pi}$$

$$x(t) = \frac{A}{2} + \frac{2A}{\pi} \left(\cos w_0t - \frac{1}{3} \cos 3w_0t + \frac{1}{5} \cos 5w_0t + \cdots \right)$$

Example 4: Find the trigonometric Fourier series for the periodic signal $x(t)$.

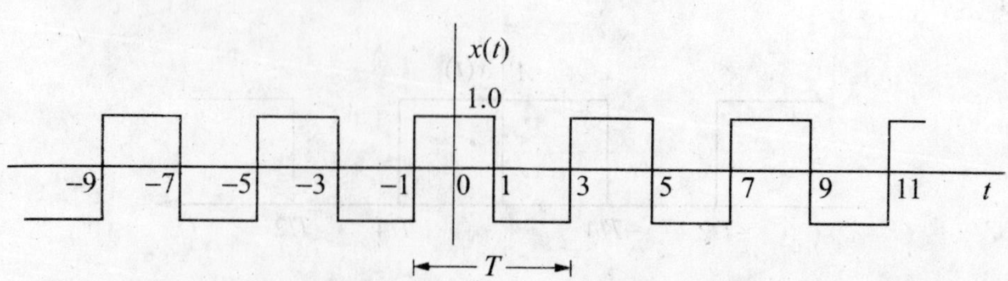

Fig. 2.4.

SOLUTION:

$$b_n = 0 \quad x(t) = \begin{cases} 1 & -1 < t < 1 \\ -1 & 1 < t < 3 \end{cases}$$

$$a_0 = \frac{1}{T}\int_{-1}^{3} x(t)dt = \frac{1}{T}\left[\int_{-1}^{1} dt + \int_{t}^{3}(-1)dt\right] \qquad T = 4$$

$$= \frac{1}{T}[2-2] = 0 \qquad \therefore \quad w_0 = \frac{2\pi}{T} = \frac{2\pi}{4} = \frac{\pi}{2}$$

$$a_n = \frac{2}{T}\left[\int_{-1}^{1} \cos\,(nw_0t)dt + \int_{1}^{3} \cos\,(nw_0t)dt\right]$$

$$= \frac{2}{2\pi n}\left\{\left[2\sin\frac{\pi n}{2}\right] - \left[\sin\frac{3n\pi}{2} - \sin\frac{n\pi}{2}\right]\right\}$$

$$= \frac{1}{n\pi}\left[3\sin\frac{n\pi}{2} - \sin\frac{3n\pi}{2}\right] \qquad \sin\frac{3n\pi}{2} = \sin\left(\pi + \frac{n\pi}{2}\right) = -\sin\frac{n\pi}{2}$$

$$a_n = \frac{4}{n\pi}\sin\left(\frac{n\pi}{2}\right)$$

$$a_n = \begin{cases} 0 & n = \text{even} \\ \dfrac{4}{n\pi} & n = 1,5,9,13 \\ \dfrac{-4}{n\pi} & n = 3,7,11,15 \end{cases}$$

$$x(t) = \frac{4}{\pi}\cos\left(\frac{\pi}{2}t\right) - \frac{4}{3\pi}\cos\left(\frac{3\pi}{2}t\right) + \frac{4}{5\pi}\cos\left(\frac{5\pi}{2}t\right) - \frac{4}{7\pi}\cos\left(\frac{7\pi}{2}t\right) + \cdots$$

$$x(t) = \frac{4}{\pi}\left[\cos\left(\frac{\pi}{2}t\right) - \frac{1}{3}\cos\left(\frac{3\pi}{2}t\right) + \frac{1}{5}\cos\left(\frac{5\pi}{2}t\right)\cdots\right]$$

Example 5: Find the F.S.C. for the continuous-time periodic signal

$$x(t) = 1.5 \qquad 0 \le t < 1$$
$$= -1.5 \qquad 1 \le t < 2$$

with fundamental freq. $w_0 = \pi$

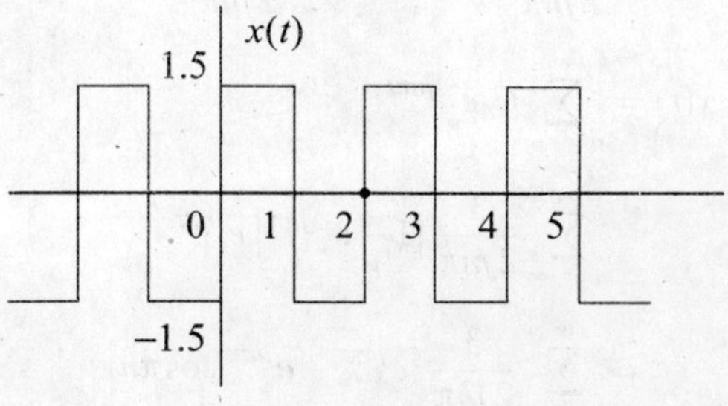

Fig. 2.5.

SOLUTION:

$$T = \frac{2\pi}{w_0} = 2,\ w_0 = \pi$$

$$a_0 = a_n = 0$$

$$b_n = \int_0^1 1.5 \sin n\pi t\,dt - \int_1^2 1.5 \sin n\pi t\,dt$$

$$= \frac{1.5}{n\pi}\Big\{[-\cos n\pi + 1] + [\cos 2n\pi - \cos n\pi]\Big\}$$

$$b_n = \frac{3}{n\pi}[1 - \cos n\pi]$$

$$x(t) = \frac{3}{\pi}\left[2\sin \pi t + \frac{2}{3}\sin 3\pi t + \frac{2}{5}\sin 5\pi t + \cdots\right]$$

$$\frac{6}{\pi}\left[\sin \pi t + \frac{1}{3}\sin 3\pi t + \frac{1}{5}\sin 5\pi t + \cdots\right]$$

$$C_0 = \frac{1}{2}\left[\int_0^1 1.5 dt - 1.5\int_1^2 dt\right] = 0$$

OR

By using complex exponential Fourier series

$$C_n = \frac{1}{2}\left[\int_0^1 1.5 e^{-jn\pi t}dt - 1.5\int_1^2 e^{-jn\pi t}dt\right]$$

$$C_n = \frac{3}{-4jn\pi}\left[e^{-jn\pi t}\Big|_0^1 - e^{-jn\pi t}\Big|_1^2\right]$$

$$= \frac{-3}{4jn\pi}\left[e^{-jn\pi} - 1 - e^{-j2n\pi} + e^{-jn\pi}\right]$$

$$= \frac{3}{2jn\pi}\left[1 - e^{-jn\pi}\right] = \frac{3}{2jn\pi}[1 - \cos n\pi]$$

$$x(t) = \sum_{n=-\infty}^{\infty} C_n e^{-jn\pi t}$$

$$\sum_{n=-\infty}^{\infty} \frac{3}{2jn\pi}\left[1 - e^{-jn\pi}\right] e^{jn\pi t}$$

$$= \sum_{n=-\infty}^{\infty} \frac{3}{2jn\pi}\left[e^{jn\pi t} - e^{jn\pi t}\cos \pi n\right]$$

Example 6: Find cosine F.S. of an half wave rectified sine function.

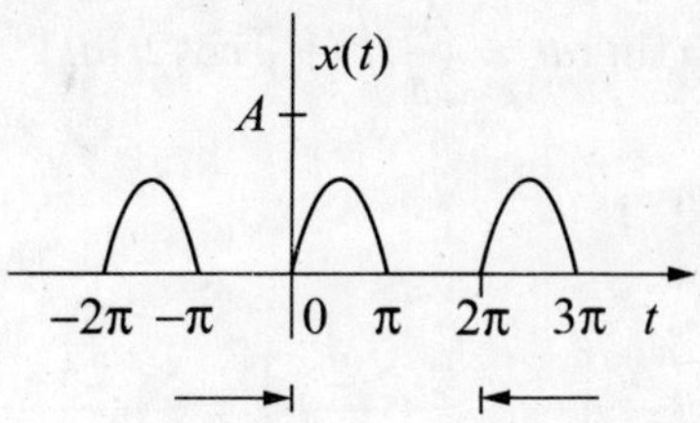

Fig. 2.6.

SOLUTION:

$$x(t) = A\sin w_0 t \qquad 0 \le t \le \pi$$
$$= 0 \qquad \pi \le t \le 2\pi$$

$$T = 2\pi, \quad w_0 = \frac{2\pi}{2\pi} = 1$$

$$a_0 = \frac{1}{T}\int_0^{\pi} A\sin t\,dt = \frac{A}{2\pi}(-\cos)\Big|_0^{\pi} = \frac{A}{\pi}$$

$$a_n = \frac{2}{T}\int_0^{\pi} \cos w_0 t A\sin t\,dt = \frac{A}{\pi}\int_0^{\pi} \sin t\cos nt\,dt$$

$$= \frac{A}{2\pi}\int_0^{\pi}\left[\sin(1+n)t + \sin(1-n)t\right]dt$$

$$= \frac{A}{2\pi}\left[-\frac{\cos(1+n)t}{1+n} - \frac{\cos(1-n)t}{(1-n)}\right]_0^{\pi}$$

$$= \frac{A}{2\pi}\left[-\frac{\cos(n+1)\pi}{n+1} + \frac{1}{n+1} - \frac{\cos(1-n)\pi}{(1-n)} + \frac{1}{1-n}\right]$$

$$b_n = \frac{A}{\pi}\int_0^{\pi} \sin t\sin nt\,dt$$

$$= \frac{A}{2\pi}\left[\int_0^{\pi} \cos(1-n)t - \cos(1+n)t\right]dt$$

$$= \frac{A}{2\pi}\left[\frac{\sin(1-n)t}{1-n}\Big|_0^{\pi} - \frac{\sin(1+n)}{1+n}\Big|_0^{\pi}\right]$$

$$= \frac{A}{2\pi}\left[\frac{\sin(1-n)\pi}{1-n} - \frac{\sin(1+n)\pi}{1+n}\right]$$

$$b_n = 0$$

for $n = 1$

$$= \frac{A}{2\pi}\int_0^{\pi} \sin t \sin t dt = \frac{A}{2\pi}\int_0^{\pi}(1-\cos 2t)dt$$

$$= \frac{A}{2\pi}[\pi] = \frac{A}{2}$$

When n is even

$$= \frac{A}{2\pi}\left[\frac{2}{n+1} - \frac{2}{1-n}\right] = \frac{2A}{\pi(1-n^2)}$$

Example 7:

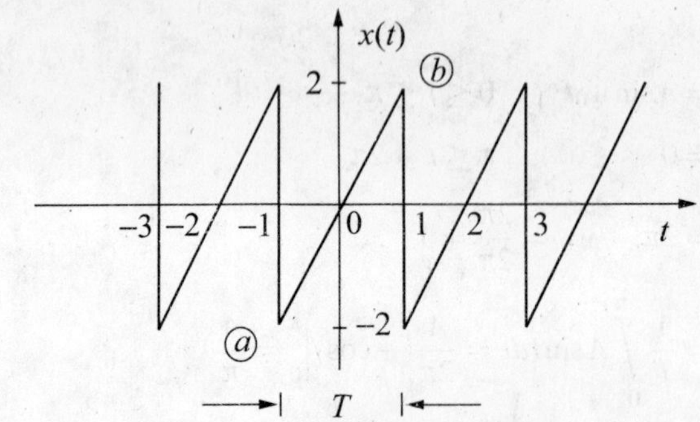

Fig. 2.7.

SOLUTION:

$$T = 2 \quad w_0 = \frac{2\pi}{T} = \pi$$

$$x(t) = \begin{cases} 2t & -1 < t < 1 \\ 0 & \end{cases}$$

Point (a) $(-1,-2)$
Point (b) $(1,2)$

$$y-(-2) = \frac{2-(-2)}{1-(-1)}(x-(-1))$$

$$y+2 = \frac{4}{2}(x+1)$$

$$y+2 = 2x+2$$

$$y = 2x$$

$$x(t) = 2t$$

Since function is an odd function

$$a_n = 0,\ a_0 = \frac{1}{T}\int_{-1}^{1} 2t dt = \frac{1}{2}\times 0 = 0$$

$$b_n = \frac{2}{T}\int_{-1}^{1} t\sin(n\pi t)dt = \frac{2}{T}\left[\frac{-t\cos n\pi t}{n\pi}\Bigg|_{-1}^{1} + \frac{1}{n^2\pi^2}\cos n\pi t\Bigg|_{-1}^{1}\right]$$

$$\frac{2}{2n\pi}[-\cos n\pi - \cos n\pi] = \frac{-2}{n\pi}\cos n\pi$$

$$b_1 = \frac{2}{\pi},\ b_2 = \frac{-1}{\pi},\ b_3 = \frac{+2}{3\pi}$$

$$x(t) = \frac{2}{\pi}\left[\sin(\pi t) - \frac{1}{2}\sin(2\pi t) + \frac{1}{3}\sin(3\pi t)\cdots\right]$$

Exponential Fourier Series

$$x(t) = a_0 + \sum_{n=1}^{\infty}(a_n \cos nw_0 t + b_n \sin nw_0 t)$$

We know that $\cos nw_0 t = \dfrac{e^{jnw_0 t} + e^{-jnw_0 t}}{2},\ \sin nw_0 t = \dfrac{e^{jnw_0 t} - e^{-jnw_0 t}}{2j}$

$$x(t) = a_0 + \sum_{n=1}^{\infty}\frac{1}{2}(a_n - jb_n)e^{jnw_0 t} + \frac{1}{2}(a_n + jb_n)e^{-jnw_0 t}$$

$$= C_0 + \sum_{n=1}^{\infty}\frac{1}{2}\left[C_n e^{jnw_0 t} + C_{-n}e^{-jnw_0 t}\right]$$

$$C_0 = a_0$$

$$C_n = \frac{1}{2}(a_n - jb_n)$$

$$C_n = \frac{1}{2}(a_n + jb_n)$$

$$= C_0 + \sum_{n=1}^{\infty} C_n e^{-jnw_0 t} + \sum_{n=-\infty}^{-1} C_n e^{jnw_0 t}$$

$$x(t) = \sum_{n=-\infty}^{\infty} C_n e^{jnw_0 t}$$

$$C_n = \frac{1}{2}[a_n - jb_n] = \frac{1}{2}\left[\frac{2}{T}\int_{t_0}^{t_0+T} x(t)\cos nw_0 t\,dt - j\frac{2}{T}\int_{t_0}^{t_0+T} x(t)\sin nw_0 t\,dt\right]$$

$$= \frac{1}{T}\left[\int_{t_0}^{t_0+T} x(t)\left[\cos nw_0 t - j\sin w_0 t\right]\right]$$

$$= \frac{1}{T}\int_{t_0}^{t_0+T} x(t)e^{-jnw_0 t}\,dt$$

2.3 CONVERGENCE OF FOURIER SERIES – DIRICHLET CONDITIONS

Existence of Fourier Series: The conditions under which a periodic signal can be represented by an F.S. are known as Dirichlet conditions. F.P. → Fundamental Period

(1) The function $x(t)$ has only a finite number of maxima and minima, if any within the F.P.
(2) The function $x(t)$ has only a finite number of discontinuities, if any within the F.P.
(3) The function $x(t)$ is absolutely integrable over one period, that is

$$\int_0^T |x(t)|dt < \infty$$

2.4 PROPERTIES OF CONTINUOUS FOURIER SERIES

(1) Linearity: If $x_1(t)$ and $x_2(t)$ are two periodic signals with period T with F.S.C. C_n and D_n then F.C. of linear combination of $x_1(t)$ and $x_2(t)$ are given by

$$\text{FS}[Ax_1(t)+Bx_2(t)] = AC_n + BD_n$$

Proof: If $z(t) = Ax_1(t)+Bx_2(t)$

$$a_n = \frac{1}{T}\int_{t_0}^{t_0+T} [Ax_1(t)+Bx_2(t)]\,e^{-jnw_0t} = \frac{A}{T}\int_T x_1(t)e^{-jnw_0t}dt + \frac{B}{T}\int_T x_2(t)e^{-jnw_0t}dt$$

$$a_n = AC_n + BD_n$$

(2) Time shifting: If the F.S.C. of $x(t)$ are C_n then the F.C. of the shifted signal $x(t-t_0)$ are

$$\text{FS}[x(t-t_0)] = e^{-jnw_0\,t_0}C_n$$

Let $t - t_0 = \tau$

$$dt = d\tau$$

$$B_n = \frac{1}{T}\int_T x(t-t_0)e^{-jnw_0t}dt$$

$$= \frac{1}{T}\int_T x(\tau)e^{-jnw_0(t_0+\tau)}d\tau = \frac{1}{T}\int_T x(\tau)e^{-jnw_0\tau}d\tau\cdot e^{-jnw_0t_0}$$

$$B_n = e^{-jnw_0t_0}\cdot C_n$$

(3) Time reversal: $\text{FS}[x(-t)] = C_{-n}$

$$B_n = \frac{1}{T}\int_T x(-t)e^{-jnw_0t}dt = \frac{1}{T}\int_T x(-t)e^{-j(-n)w_0T}dt$$

$$-t = \tau$$

$$dt = -d\tau$$

$$= \frac{1}{T}\int_{-T} x(\tau)e^{-j(-n)w_0\tau}d\tau = C_{-n}$$

(4) Time scaling: $\text{FS}[x(at)] = C_n$ with fundamental frequency aw_0

$$B_n = \frac{1}{T/a}\int\limits_{T/a} x(at)e^{-jnw_0t}dt \quad at = \tau \quad adt = d\tau \quad w_0 = \frac{2\pi}{T} = w_0^1 = aw_0$$

$$= \frac{a}{aT}\int\limits_{aT} x(\tau)e^{-jn\left(\frac{w_0}{a}\right)^{\tau}}d\tau = \frac{1}{T'}\int\limits_{T'} x(\tau)e^{-jnw_0{}'\tau}d\tau$$

$$B_n = C_n$$

(5) Multiplication: Let $x_1(t)$ and $x_2(t)$ are both periodic with period T

$$x_1(t) \overset{\text{F.S.}}{\longleftrightarrow} C_n \quad x_2(t) \overset{\text{F.S.}}{\longleftrightarrow} D_n$$

$$\text{FS}\left[x_1(t)x_2(t)\right] = \sum_{l=-\infty}^{\infty} C_n D_{n-l}$$

(6) Conjugation: If $\text{FS}[x(t)] = C_n$ then $\text{FS}[x^*(t)] = C^*_{-n}$. Therefore,

(1) if $x(t)$ is real and even, then C_n is real and even

(2) if $x(t)$ is real and odd, then C_n is imaginary and odd and $a_0 = 0$

(7) Differentiation: If $\text{FS}[x(t)] = C_n$

$$\text{then} \quad \text{FS}\left[\tfrac{d}{dt}x(t)\right] = jnw_0C_n$$

(8) Integration: If $\text{FS}[x(t)] = C_n$

$$\text{then FS}\left[\int_{-\infty}^{t} x(\tau)d\tau\right] = \tfrac{1}{jnw_0}C_n$$

(9) Parseval's relation for periodic signal:

$$\text{Average power} = \frac{1}{T}\int\limits_{T} |x(t)|^2 dt = \sum_{n=-\infty}^{\infty} |C_n|^2$$

(10) Frequency shifting:

$$\text{FS}\left[e^{jlw_0t}x(t)\right] = C_{n-l}$$

$$C_n = \frac{1}{T}\int\limits_{T} e^{jlw_0t}x(t)e^{-jnw_0t}$$

$$= \frac{1}{T}\int\limits_{T} x(t)e^{-j(n-l)w_0t}dt$$

$$= C_{n-l}$$

Example 8: Compute the exponential series of the following signal.

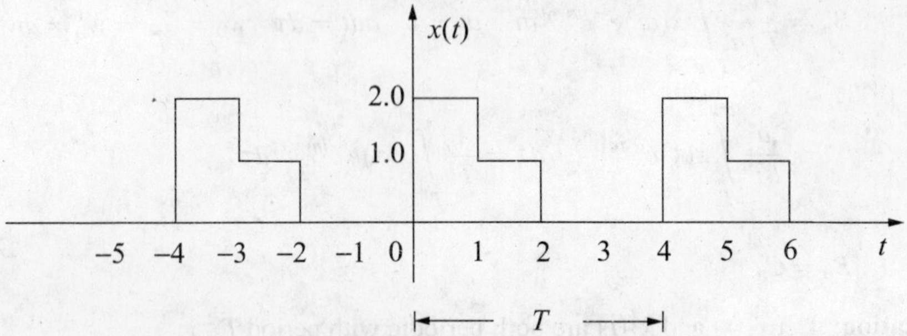

Fig. 2.8.

SOLUTION:

$$T = 4 \quad w_0 = \frac{\pi}{2}$$

$$C_0 = \frac{1}{T}\int_0^T x(t)dt = \frac{1}{4}\left[\int_0^1 2dt + \int_1^2 dt\right] = \frac{3}{4}$$

$$C_n = \frac{1}{4}\left[\int_0^1 2e^{-jn\frac{\pi t}{2}}dt + \int_1^2 e^{-jn\frac{\pi t}{2}}dt\right]$$

$$= \frac{1}{4}\left\{\frac{-4}{jn\pi}\left[e^{-jn\frac{\pi}{2}} - 1\right] - \frac{2}{jn\pi}\left[e^{-jn\pi} - e^{-jn\frac{\pi}{2}}\right]\right\}$$

$$= \frac{-1}{2jn\pi}\left[2e^{\frac{-jn\pi}{2}} - 2 + e^{-jn\pi} - e^{-jn\frac{\pi}{2}}\right] = \frac{-1}{2jn\pi}\left[e^{-jn\frac{\pi}{2}} + e^{-jn\frac{\pi}{2}} - 2\right]$$

$$= -\frac{1}{jn\pi}\left[1 - \frac{1}{2}(-1)^n - \frac{1}{2}e^{-jn\frac{\pi}{2}}\right] \quad x(t) = \frac{3}{4} + \sum_{n=-\infty}^{\infty}\frac{1}{jn\pi}\left[e^{jn\frac{\pi}{2}} - \frac{1}{2}(-1)^n e^{jn\frac{\pi}{2}} - \frac{1}{2}\right]$$

Example 9:

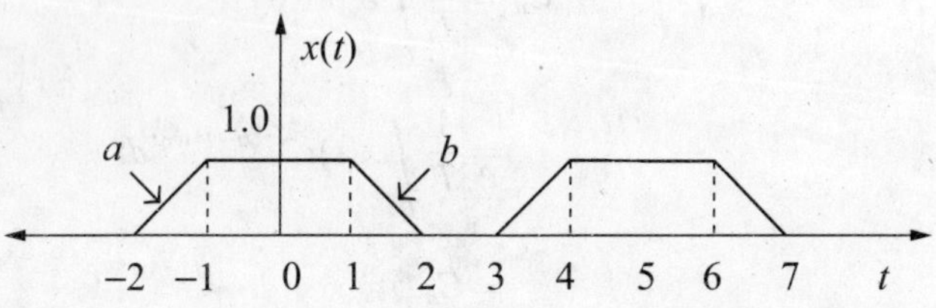

Fig. 2.9.

SOLUTION:

$$T=5 \quad w_0=\frac{2\pi}{5}$$

$$x(t)=\begin{cases} t+2 & -2<t<-1 \\ 1.0 & -1<t<1 \\ 2-t & 1<t<2 \end{cases}$$

(a)
$$(-2,0)(-1,1)$$

$$(y-1)=\frac{-1}{-1}(x+1)$$

$$y=t+2$$

(b) $(1,1)(2,0)$

$$y-0=\frac{1}{-1}(x-2)$$

$$y=-x+2=-t+2$$

$$C_0=\frac{1}{5}\left[\int_{-2}^{-1}(t+2)dt+\int_{-1}^{1}dt+\int_{1}^{2}(2-t)dt\right]$$

$$C_0=\frac{3}{5}$$

$$C_n=\frac{1}{5}\left[\underbrace{\int_{-2}^{-1}(t+2)e^{-j\frac{2n\pi}{5}}dt}_{A}+\underbrace{\int_{-1}^{1}e^{-j\frac{2n\pi}{5}}dt}_{B}+\underbrace{\int_{1}^{2}(2-t)e^{-j\frac{2n\pi}{5}}dt}_{C}\right]$$

$$A=\int_{-2}^{-1}e^{-j\frac{2n\pi}{5}t}dt+\int_{-2}^{-1}2e^{-j\frac{2n\pi}{5}t}dt$$

$$A=-\frac{1}{j\phi}\left\{te^{-j\phi}\int_{-2}^{-1}\right\}+\frac{1}{\phi^2}e^{-j\phi}\int_{-2}^{-1}+\frac{2}{-j\phi}e^{j}\frac{2n\pi}{5}\int_{-2}^{-1}$$

$$=\frac{5}{j2n\pi}\left(-e^{j\frac{2n\pi}{5}}+2e^{j\frac{4n\pi}{5}}\right)+\frac{25}{4n^2\pi^2}\left(e^{j\frac{2n\pi}{5}}-e^{j\frac{4n\pi}{5}}\right)-\frac{10}{2n\pi j}$$

$$A=\frac{5}{j2n\pi}\left(-e^{j\frac{2n\pi}{5}}+4e^{j\frac{4n\pi}{5}}\right)+\frac{25}{4n^2\pi^2}\left(e^{j\frac{2n\pi}{5}}-e^{j\frac{4n\pi}{5}}\right)$$

$$B=\frac{e^{j\frac{2n\pi}{5}}-e^{-j\frac{2n\pi}{5}}}{j\frac{2n\pi}{5}}=\frac{5}{j2n\pi}\left(e^{j\frac{2n\pi}{5}}-e^{-j\frac{2n\pi}{5}}\right)$$

$$C=\frac{-10}{j2n\pi}\left(e^{-j\frac{4n\pi}{5}}-e^{-j\frac{2n\pi}{5}}\right)+\frac{10}{j2n\pi}e^{-j\frac{4n\pi}{5}}-\frac{5}{j2n\pi}e^{-j\frac{2n\pi}{5}}-\frac{25}{4n^2\pi^2}e^{-j\frac{4n\pi}{5}}+\frac{25}{4n^2\pi^2}e^{j\frac{n2\pi}{5}}$$

$$C_n = \frac{1}{5}\left[\frac{25}{n^2 4\pi^2}\left(e^{\frac{j2n\pi}{5}} - e^{\frac{j4n\pi}{5}}\right) - \frac{25}{4n^2\pi^2}\left(e^{-\frac{j4n\pi}{5}} - e^{-\frac{j2n\pi}{5}}\right)\right]$$

$$C_n = \frac{5}{2n^2\pi^2}\left[\cos\left(\frac{2\pi n}{5}\right) - \cos\left(\frac{4\pi n}{5}\right)\right]$$

Example 10: For the continuous-time periodic signal

$$x(t) = 2 + \cos\left(\frac{2\pi}{3}t\right) + 4\sin\left(\frac{5\pi}{3}t\right)$$

Determine the fundamental frequency w_0 and the Fourier series coefficients C_n such that

$$x(t) = \sum_{n=-\infty}^{\infty} C_n e^{jnw_0 t}$$

SOLUTION:

Given

$$x(t) = 2 + \cos\left(\frac{2\pi}{3}t\right) + 4\sin\left(\frac{5\pi}{3}t\right)$$

The time period of the signal $\cos\left(\frac{2\pi}{3}t\right)$ is

$$T_1 = \frac{2\pi}{w_1} = \frac{2\pi}{2\frac{\pi}{3}} = 3\,\text{sec}$$

The time period of the signal $\sin\left(5\frac{\pi}{2}t\right)$ is

$$T_2 = 2\frac{\pi}{w_2} = \frac{2\pi}{5\frac{\pi}{3}} = \frac{6}{5}\,\text{sec}$$

$\frac{T_1}{T_2} = \frac{3}{\frac{6}{5}} = \frac{5}{2}$ ratio of two integers, rational number, hence periodic.

$$2T_1 = 5T_2$$

The fundamental period of the signal $x(t)$ is

$$T = 2T_1 = 5T_2 = 6\,\text{sec}$$

and the fundamental frequency is

$$w_0 = \frac{2\pi}{T} = \frac{2\pi}{6} = \frac{\pi}{3}$$

$$x(t) = 2 + \cos\left(\frac{2\pi}{3}t\right) + 4\sin\left(\frac{5\pi}{3}t\right)$$

$$= 2 + \cos(2w_0 t) + 4\sin(5w_0 t)$$

$$= 2 + \frac{\left(e^{j2w_0 t} + e^{-j2w_0 t}\right)}{2} + \frac{4\left(e^{j5w_0 t} - e^{-j5w_0 t}\right)}{2j}$$

$$= 2 + 0.5\left(e^{j2w_0 t} + e^{-j2w_0 t}\right) - 2j\left(e^{j5w_0 t} - e^{-j5w_0 t}\right)$$

$$x(t) = 2je^{+j(-5)w_0 t} + 0.5e^{+j(-2)w_0 t} + 2 + 0.5e^{+j2w_0 t} - 2je^{+j5w_0 t}$$

On comparing with $x(t) = \sum_{n=-\infty}^{\infty} C_n e^{jnw_0t}$

$$x(t) = C_{-5}e^{j(-5)w_0t} + C_{-2}e^{j(-2)w_0t} + C_0 + C_2e^{j(2)w_0t} + C_5e^{j(5)w_0t}$$

So we get Fourier series coefficients as

$$C_{-5} = 2j \quad C_{-2} = 0.5 \quad C_0 = 2 \quad C_2 = 0.5 \quad C_5 = -2j$$

Example 11: Obtain the trigonometric F.S. representation for the signal shown

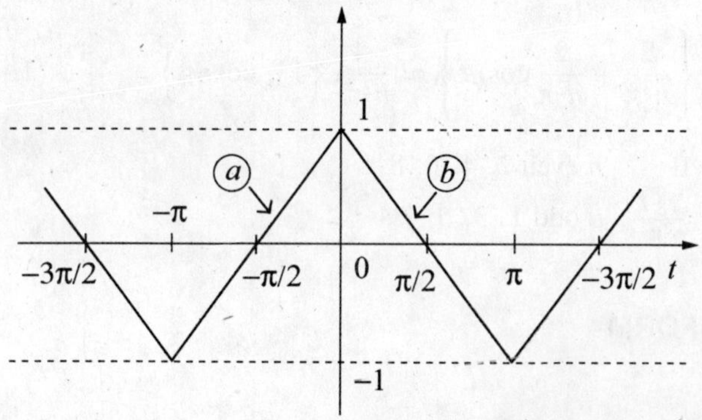

Fig. 2.10.

SOLUTION:

$$T = 2\pi \quad w_0 = 1$$

$$x(t) = \begin{cases} \frac{2t}{\pi} + 1 & -\pi < t < 0 \\ -\frac{2t}{\pi} + 1 & 0 < t < \pi \end{cases}$$

$$a \begin{cases} (-\pi, 1)(0, 1) \\ y - 1 = \frac{-2}{-\pi}(x) \\ y = \frac{2t}{\pi} + 1 \end{cases}$$

$$b \begin{cases} (0, 1)(\pi, -1) \\ y - 1 = \frac{-2}{\pi}(x) \\ y = \frac{-2t}{\pi} + 1 \end{cases}$$

$$a_0 = \frac{1}{T}\left(\int_{-\pi}^{0}\left(\frac{2t}{\pi} + 1\right)dt + \int_{0}^{\pi}\left(\frac{-2t}{\pi} + 1\right)dt\right)$$

$$\frac{1}{T}\left(\frac{t^2}{\pi} + t\int_{-\pi}^{0} + \left(\frac{-t^2}{\pi} + t\right)\int_{0}^{\pi}\right)$$

$$\frac{1}{2\pi}\left\{(-\pi + \pi) + (-\pi + \pi)\right\} = 0$$

$$a_n = \frac{2}{T}\int_{-\pi}^{\pi} x(t)\cos nt dt = \frac{4}{T}\int_{-\pi}^{0}\left(\frac{2t}{\pi}+1\right)\cos nt\, dt$$

$$= \frac{4}{2\pi}\left\{\frac{2t}{n\pi}\sin t + \frac{\sin nt}{n} - \int_{-\pi}^{0}\frac{2}{\pi}\sin nt\, dt\right\}$$

$$= \frac{1}{\pi}\left\{\frac{2t}{\pi}\sin nt + \sin nt + \frac{2}{n^2\pi}\cos nt \int_{-\pi}^{0}\right\}$$

$$= \frac{2}{\pi}\left\{\frac{2}{n^2\pi} + \frac{2}{n^2\pi}\cos nt\right\} = \frac{4}{n^2\pi^2}\left\{1-\cos n\pi\right\} = \frac{4}{n^2\pi^2}(1-(-1)^n)$$

$$a_n = \begin{cases} 0 & n \text{ even } 2,\ 4,\ 6,\ 8,\cdots \\ \frac{8}{n^2\pi^2} & n \text{ odd } 1,\ 3,\ 5,\ 7,\cdots \end{cases}$$

2.5 FOURIER TRANSFORM

2.5.1 Definition

Let $x(t)$ be a signal which is a function of time t. The Fourier transform of $x(t)$ is given as

$$\mathrm{X}\,(jw) = \int_{-\infty}^{\infty} x(t)e^{-jwt}dt \quad (1)$$

Fourier transform or

$$\mathrm{X}\,(if) = \int_{-\infty}^{\infty} x(t)e^{-j2\pi ft}dt \quad (2)$$

Since $w = 2\pi f$

Similarly, $x(t)$ can be recovered from its Fourier transform $X(jw)$ by using Inverse Fourier transform

$$x(t) = \frac{1}{2\pi}\int_{-\infty}^{\infty} X(jw)e^{jwt}dw \quad (3)$$

$$x(t) = \int_{-\infty}^{\infty} X(if)e^{j2\pi ft}dt \quad (4)$$

Fourier transform $X(jw)$ is the complex function of frequency w. Therefore, it can be expressed in the complex exponential form as follows:

$$X(jw) = |X(jw)|e^{j\underline{|X(jw)}}$$

Here $|X(jw)|$ is the amplitude spectrum of $x(t)$ and $\underline{|X(jw)}$ is phase spectrum.

For a real-valued signal

(1) Amplitude spectrum is symmetric about vertical axis c (even function.)

(2) Phase spectrum is anti-symmetrical about vertical axis c (odd function.)

2.5.2 Existence of Fourier transform (Dirichlet's condition)

The following conditions should be satisfied by the signal to obtain its F.T.

(1) The function $x(t)$ should be single valued in any finite time interval T.

(2) The function $x(t)$ should have at the most finite number of discontinuities in any finite time interval T.

(3) The function $x(t)$ should have finite number of maxima and minima in any finite time interval T.

(4) The function $x(t)$ should be absolutely integrable, i.e.

$$\int_{-\infty}^{\infty} |x(t)|dt < \infty$$

- These conditions are sufficient, but not necessary for the signal to be Fourier transformable.
- A physically realizable signal is always Fourier transformable. Thus, physical realizability is the sufficient condition for the existence of F.T.
- All energy signals are Fourier transformable.

$$j\frac{d}{dw}X(jw) = FT(tx(t))$$

$$FT(tx(t)) = j\frac{d}{dw}X(jw)$$

Example 12: Obtain the F.T. of the signal $e^{-at}u(t)$ and plot its magnitude and phase spectrum.

SOLUTION:

$x(t) = e^{-at}u(t)$

$$X(f) = \int_{-\infty}^{\infty} x(t)e^{-j2\pi ft}dt = \int_{0}^{\infty} e^{-(a+j2\pi f)t}dt$$

$$X(f) = \frac{1}{a+j2\pi f}$$

To obtain the magnitude and phase spectrum:

$$|X(f)| = \frac{a-j2\pi f}{a^2+(2\pi f)^2} = \left(\frac{a}{a^2+4\pi^2 f^2}\right)A - j\left(\frac{2\pi f}{a^2+4\pi^2 f^2}\right)B$$

$$|X(f)| = \sqrt{A^2+B^2} = \frac{1}{\sqrt{a^2+4\pi^2 f^2}} = \frac{1}{\sqrt{a^2+w^2}}$$

$$|X(f)| = \tan^{-1}\left[\frac{-2\pi f}{a}\right] = -\tan^{-1}\left(\frac{w}{a}\right)$$

$$\text{for } a = 1,\ |X(f)| = \frac{1}{\sqrt{1+w^2}},\quad \overline{|X(f)} = -\tan^{-1} w$$

w	0	1	2	3	4	5	10	15	25	8
$\|X(w)\|$	1	.707	0.447	0.316	0.242	0.196	0.09	0.066	0.03	0
$\underline{\|X(w)}$	0	45°	−63.4	−71.5	−75.9	−78.6	−84.2	−86.2	−87.7	−90°

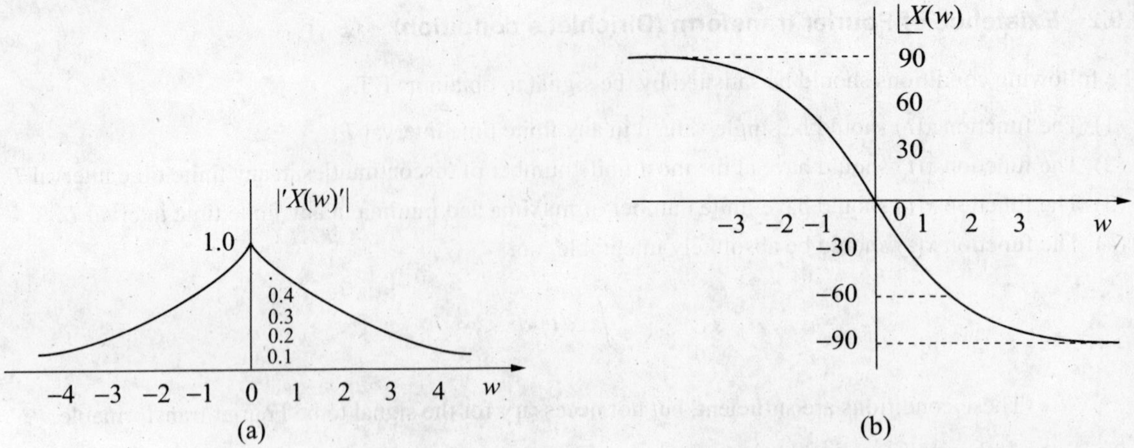

Fig. 2.11. (a) Magnitude plot (b) Phase plot

Example 13: Obtain the F.T. of the following signals:

(i) $x(t) = e^{at}u(-t)$

(ii) $x(t) = e^{-a|t|}$

(iii) $x(t) = e^{-a|t|}\text{sgn}(t)$

SOLUTION:

(i) $$X(w) = \int_{-\infty}^{0} e^{(a-jw)t}dt = \frac{1}{a-jw};\ X(w) = \frac{a+jw}{a^2w^2} = \frac{a}{a^2+w^2} + j\frac{w}{a^2+w^2}$$

$$|X(w)| = \frac{1}{\sqrt{a^2+w^2}};\ \underline{|X(w)} = \tan^{-1}\left(\frac{w}{a}\right)$$

for $a = 1$

w	$-\infty$	−10	−5	−4	−3	−2	−1	0	1	2	3	4	5	10	∞
$\|X(w)\|$	0	.09	.196	242	.316	.447	.707	1	.707	.447	.316	242	.196	.09	0
$\underline{\|X(w)}$	−90°	−84.2	−78.6	−75.9	−71.5	−63.4	−45°	0	+45°	+63.4	+71.5	+75.9	+78.6	+84.2	90°

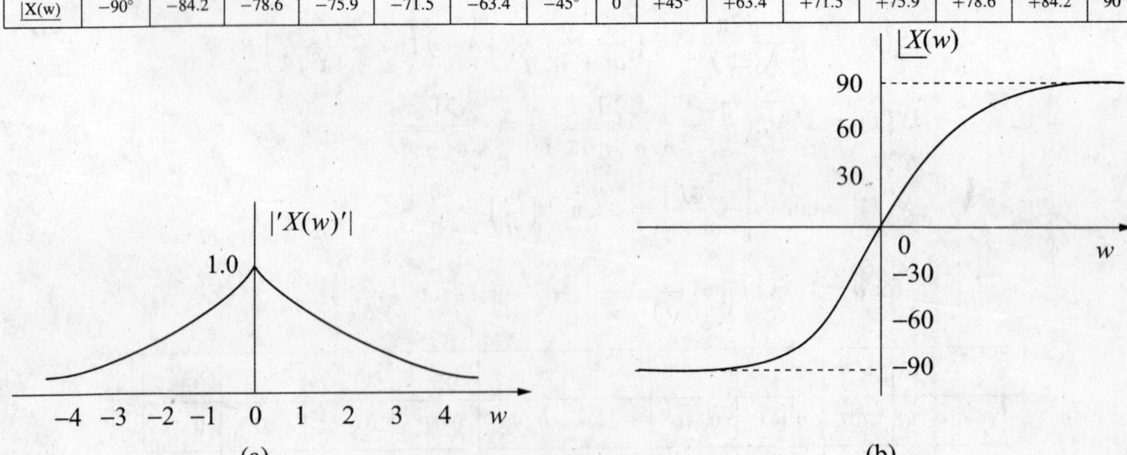

Fig. 2.12. (a) Magnitude plot (b) Phase plot

(ii) $x(t) = e^{-a|t|} = \begin{cases} e^{-at} & t > 0 \\ e^{at} & t > 0 \end{cases}$

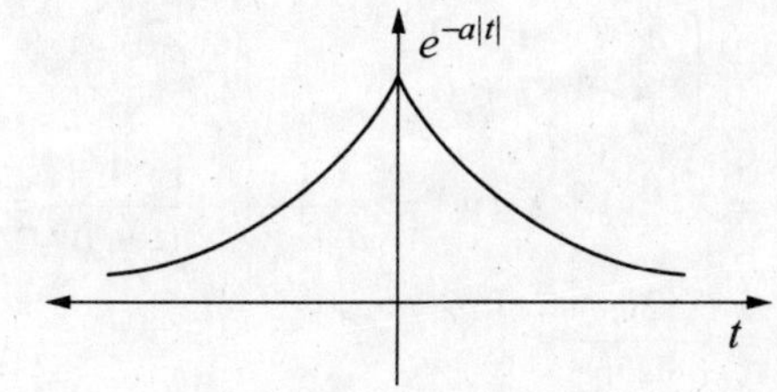

Fig. 2.13. Graphical representation of $e^{-a|t|}$

$$x(w) = \frac{1}{a+jw} + \frac{1}{a-jw} = \frac{2a}{a^2+w^2}$$

$$\text{for } a = 1\ X(w) = \frac{2}{1+w^2}$$

$$|X(w)| = \frac{2}{1+w^2}\ \frac{|X(w)}{} = 0$$

w (in radians)	$-\infty$	-10	-5	-3	-2	-1	0	1	2	3	4	5	10	∞
$\|X(w)\|$	0	0.019	0.0769	0.2	0.4	1	2	1	0.4	0.2	.1176	0.0769	0.019	0

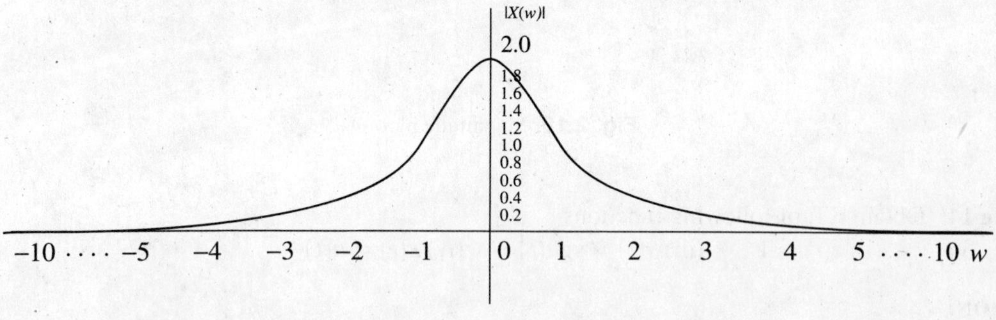

Fig. 2.14. Magnitude plot

(iii) $x(t) = e^{-a|t|}\ \text{sgn}(t)$

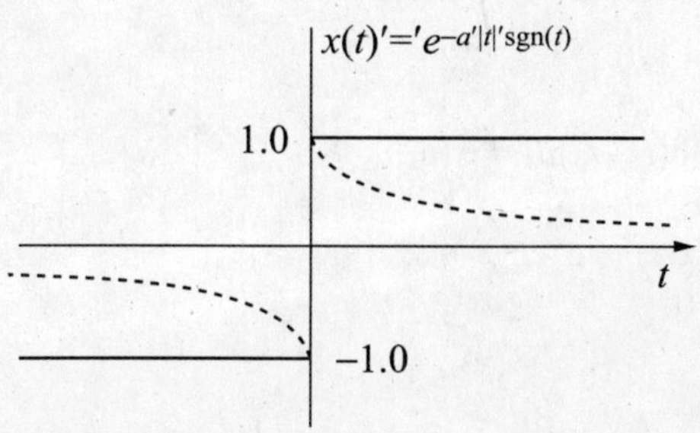

Fig. 2.15. Graphical representation of $e^{-a|t|}\text{sgn}(t)$

$$\text{sgn}(t) = \begin{Bmatrix} 1 & t > 0 \\ -1 & t < 0 \end{Bmatrix}$$

$$x(t) = \begin{Bmatrix} e^{-at} & t > 0 \\ -e^{at} & t < 0 \end{Bmatrix}$$

$$X(w) = X_1(w) + X_2(w) = \frac{1}{a + jw} - \frac{1}{a - jw} = \frac{-2jw}{a^2 + w^2}$$

$$|X(w)| = \frac{2w}{a^2 + w^2}$$

$$\angle X(w) = 90^\circ \quad \text{for } a = 1 \quad |X(w)| = \frac{2w}{1 + w^2}$$

w	$-\infty$	-10	-5	-4	-3	-2	-1	0	1	2	3	4	5	10	∞
$\lvert X(w)\rvert$		.19	.38	.468	0.6	0.8	1	0	1	0.8	0.6	.468	0.38	.19	

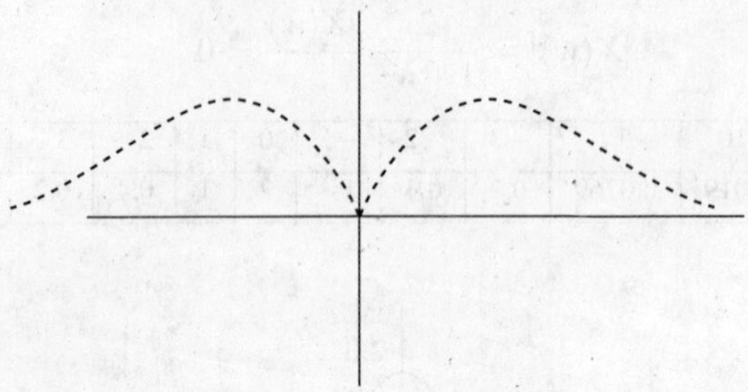

Fig. 2.16. Magnitude plot

Example 14: Obtain F.T. of following functions
(i) $x(t) = \delta(t)$ (ii) $x(t) = 1$ (iii) $x(t) = \text{sgn}(t)$ (iv) $x(t) = u(t)$

SOLUTION:

(i)
$$x(t) = \delta(t)$$

$$X(W) = \int_{-\infty}^{\infty} x(t)e^{-jwt}dt = \int_{-\infty}^{\infty} \delta(t)e^{-jwt}dt \quad (1)$$

By shifting property $\int_{-\infty}^{\infty} x(t)\delta(t - t_0)dt = x(t_0)$

Here in eqn. (1) $t_0 = 0$ and $x(t) = e^{-jwt}$
Hence, $x(t_0) = e^{-jwt_0}|t_0 = 0$
$= 1$

$$X(w) = 1$$

$$\delta(t) \xleftrightarrow{FT} 1$$

(ii)
$$x(t) = 1$$
$$X(w) = \int_{-\infty}^{\infty} e^{-jwt} dt = \infty$$

This means Dirichlet condition is not satisfied. But its F.T. can be calculated with the help of duality property.

$$\delta(t) \xleftrightarrow{FT} 1$$

Duality property states that: $x(t) \xleftrightarrow{FT} X(w)$ then

$$X(t) \xleftrightarrow{FT} 2\pi x(-w)$$

Here $X(t) = 1$, then $x(-w)$ will be

$$x(t) = \delta(t); \quad X(w) = 1$$

then $X(t) = 1; \quad 1 \xleftrightarrow{FT} 2\pi\delta(-w)$

We know that $\delta(w)$ will be an even function of w, since it is impulse function.

Hence, $\delta(-w) = \delta(w)$. Then above equation becomes

$$1 \xleftrightarrow{FT} 2\pi\delta(-w)$$

Thus, if $x(t) = 1$, then $X(w) = 2\pi\delta(w)$

(iii) $x(t) = \text{sgn}(t) \qquad \text{sgn}(t) = \begin{Bmatrix} 1 & t > 0 \\ -1 & t < 0 \end{Bmatrix}$

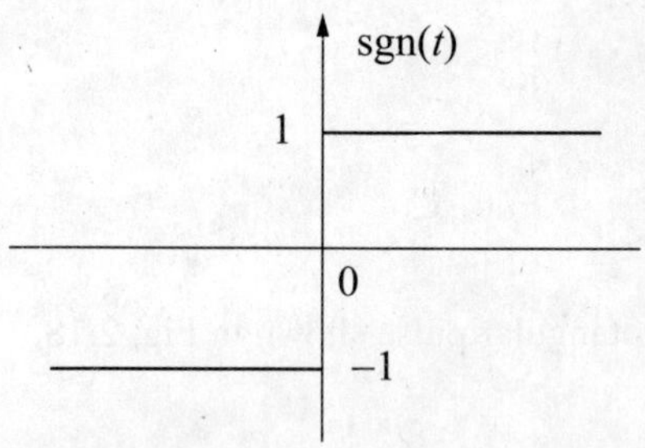

Fig. 2.17. Graphical representation of sgn(t)

$$x(t) = 2u(t) - 1$$

Differentiating both the sides

$$\frac{d}{dt}x(t) = 2\frac{d}{dt}u(t) = 2\delta(t)$$

Taking the F.T. of both sides

$$F\left[\frac{d}{dt}x(t)\right] = 2F[\delta(t)]$$
$$jwX(w) = 2$$
$$X(w) = \frac{2}{jw}$$
$$X(w) = \int_{0}^{\infty} e^{-jwt} dt - \int_{-\infty}^{0} e^{-jwt} dt$$

(iv)

$$x(t) = u(t)$$
$$\text{sgn}(t) = 2u(t) - 1$$
$$2u(t) = 1 + \text{sgn}(t)$$

Taking F.T. of both sides

$$2F[2u(t)] = F(1) + F[\text{sgn}(t)] = 2\pi\delta(w) + \frac{2}{jw}$$

$$2u(t) \stackrel{FT}{\longleftrightarrow} 2\pi\delta(w) + \frac{2}{jw}$$

$$u(t) \stackrel{FT}{\longleftrightarrow} \pi\delta(w) + \frac{1}{jw}$$

Properties of unit impulse:

(1) $\int_{-\infty}^{\infty} x(t)\delta(t) = x(0)$

(2) $x(t)\delta(t - t_0) = x(t_0)\delta(t - t_0)$

(3) $\int_{-\infty}^{\infty} x(t)\delta(t - t_0)dt = x(t_0)$

(4) $\delta(at) = \frac{1}{|a|}\delta(t)$

(5) $\int_{-\infty}^{\infty} x(\tau)\delta(t - x)dt = x(t)$

(6) $\delta(t) = \frac{d}{dt}u(t)$

Example 15: Obtain the F.T. of a rectangular pulse shown in Fig. 2.18.

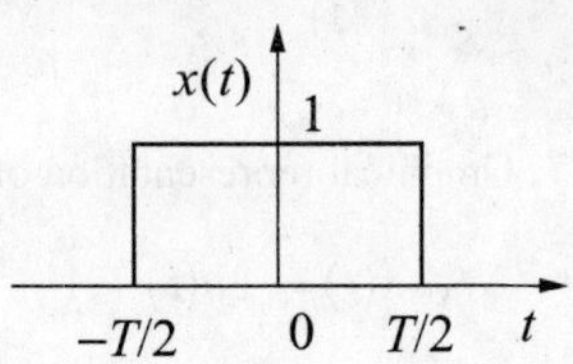

Fig. 2.18. Rectangular pulse

SOLUTION:

$$X(w) = \int_{\frac{-T}{2}}^{\frac{T}{2}} e^{-jwt}dt = \frac{-1}{jw}\left[e^{-jw\frac{T}{2}} - e^{jw\frac{T}{2}}\right] = \frac{2}{w}\sin\left(\frac{wT}{2}\right)$$

$$X(w) = T\frac{\sin\left(\pi\frac{wT}{2\pi}\right)}{\pi\frac{wT}{2\pi}} = \sin c\left(\frac{wT}{2\pi}\right) = T\frac{\sin\left(\pi\frac{wT}{2\pi}\right)}{\pi\frac{wT}{2\pi}}$$

Sampling function or interpolating function or filtering function denoted by $S_a(x)$ or $\sin c(x)$ **as shown in** figure.

$$\sin c(x) = \frac{\sin \pi x}{\pi x}$$

(1) $\sin c(x) = 0$ when $x = \pm n\pi$

(2) $\sin c(x) = 1$ when $x = 0$ (using L'Hospital's rule)

(3) $\sin c(x)$ is the product of an oscillating signal $\sin x$ of period 2π and a decreasing signal $\frac{1}{x}$. **Therefore,** $\sin c(x)$ is making sinusoidal of oscillations of period 2π with **amplified decreasing continuously as** $\frac{1}{x}$.

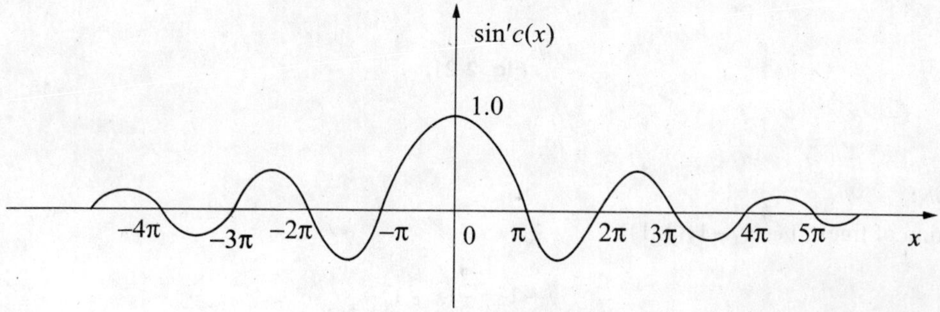

Fig. 2.19. Sine function

$$\sin cx = \frac{\sin \pi x}{\pi x}; \qquad \sin c(0) = \frac{0}{0} = 1 \text{ L'Hospital rule}$$

$$\sin c(1) = \frac{\sin \pi}{\pi} = 0; \quad \sin c(-1) = 0$$

$$\sin c(2) = 0; \qquad \sin c(-2) = 0$$

$\sin c(1/4) = 0.9$ $\qquad$ $\sin c(-1/4) = 0.9$

$\sin c(2/4) = .6366$ $\qquad$ $\sin c(-0.5) = .6366$

$\sin c(3/4) = 0.3$ $\qquad$ $\sin c(-7.5) = .3$

$\sin c(1.5) = -.2122$ $\qquad$ $\sin c(-1.5) = -.2122$

$\sin c(2.5) = .1273$ $\qquad$ $\sin c(2.5) = .1273$

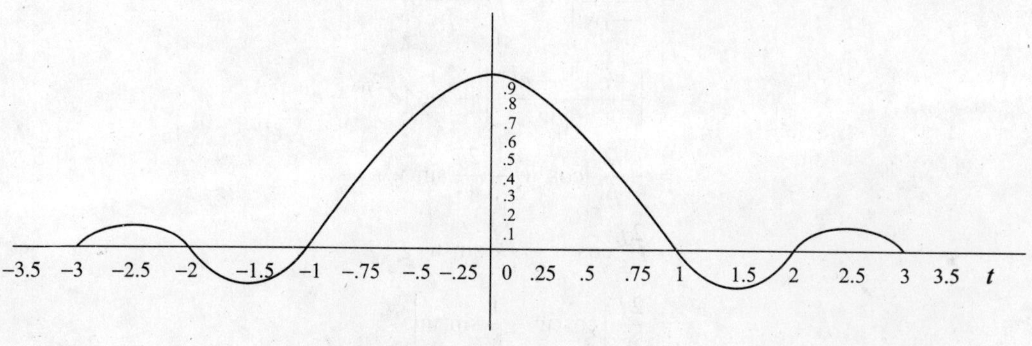

Fig. 2.20. Sine function

Example 16: Find the Fourier transform of the given signal

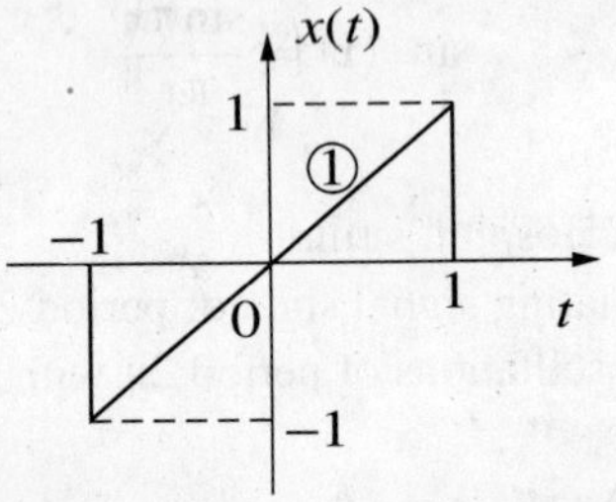

Fig. 2.21.

SOLUTION:

Equation of line (1) $(-1,-1)(1,1)$

$$y+1 = \frac{2}{2}(x+1)$$

$$y = x$$

So equation of line (1) is $x(t) = t$

So

$$x(t) = \begin{cases} t & -1 < t < 1 \\ 0 & \text{elsewhere} \end{cases}$$

$$X(jw) = \int_{-\infty}^{\infty} x(t)e^{-jwt}\,dt$$

$$= \int_{-1}^{1} te^{-jwt}\,dt$$

$$= t\frac{e^{-jwt}}{-jw}\bigg|_{-1}^{1} - \int_{-1}^{1} \frac{e^{-jwt}}{-jw}\,dt$$

$$= \left[\frac{-e^{-jw} - e^{jw}}{-jw}\right] + \frac{1}{w^2}\left[e^{-jw} - e^{jw}\right]$$

$$= -\frac{2}{jw}[\cos w] - \frac{2j}{w^2}\sin w$$

$$= \frac{2j}{w}\cos w - \frac{2j}{w^2}\sin w$$

$$= \frac{2j}{w}\left[\cos w - \frac{1}{w}\sin w\right]$$

Example 17: Obtain F.T. and spectrums of following signals:
(i) $x(t) = \cos w_0 t$ (ii) $x(t) = \sin w_0 t$

SOLUTION:

(i)
$$x(t) = \cos w_0 t = \frac{1}{2} e^{jw_0 t} + \frac{1}{2} e^{-jw_0 t}$$

$$1 \xleftrightarrow{FT} 2\pi\delta(w); \quad \frac{1}{2} \xleftrightarrow{FT} \pi\delta(w)$$

Frequency shifting property states that $e^{j\beta t} x(t) \xleftrightarrow{FT} X(w - \beta)$

$$\frac{1}{2} e^{jw_0 t} \xleftrightarrow{FT} \pi\delta(w - w_0)$$

$$\frac{1}{2} e^{-jw_0 t} \xleftrightarrow{FT} \pi\delta(w + w_0)$$

$$F[x(t)] = FT\left\{\frac{1}{2} e^{jw_0 t} + \frac{1}{2} e^{-jw_0 t}\right\}$$

$$X(w) = \pi[\delta(w - w_0) + \delta(w + w_0)]$$

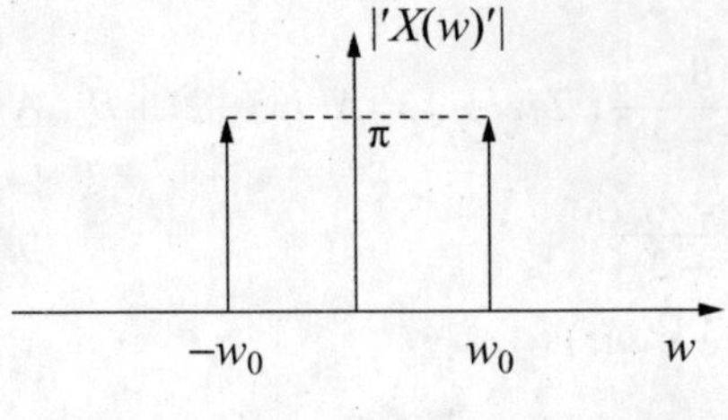

Fig. 2.22. Magnitude plot of $\cos w_0 t$

(ii)
$$x(t) = \sin w_0 t$$

$$X(w) = \frac{\pi}{j}[\delta(w - w_0) - \delta(w + w_0)]$$

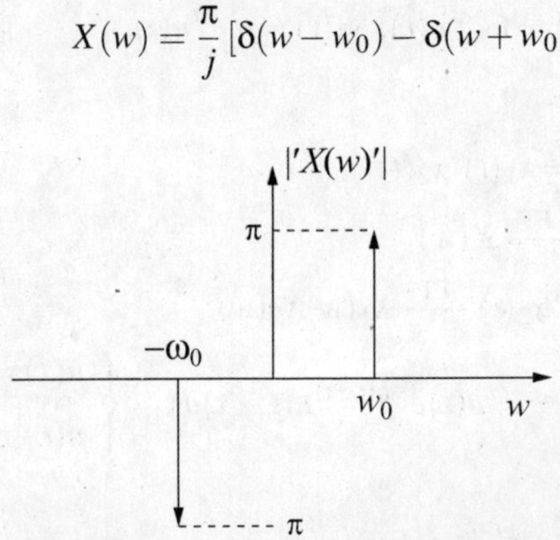

Fig. 2.23. Magnitude plot of $\sin w_0 t$

Example 18: Obtain the F.T. of

$$x(t) = te^{-at}u(t)$$

from property of Fourier transform $\text{FT}[tx(t)] = j\frac{d}{dw}X(w)$

$$\text{FT}\left[e^{-at}\right] = \frac{1}{a+jw}$$

$$\text{FT}(te^{-at}) = j\frac{d}{dw}\left(\frac{1}{a+jw}\right) = j\frac{(a+jw)\frac{d}{dw}(1) - 1\frac{d}{dw}(a+jw)}{(a+jw)^2} = \frac{1}{(a+jw)^2}$$

Inverse Fourier Transform: (IFT)

Example 19: Find the IFT of

(i) $X(w) = \frac{2jw+1}{(jw+2)^2}$ by partial fraction expansions

(ii) $X(w) = \frac{1}{(a+jw)^2}$ by convolution property

(iii) $X(w) = e^{-|w|}$

(iv) $X(w) = e^{-2w}u(w)$

SOLUTION:

(i)

$$X(w) = \frac{A}{jw+2} + \frac{B}{(jw+2)^2};\ 2jw+1 = A(jw+2)+B \quad A=2 \quad 2A+B=1 \quad B=-3$$

$$X(w) = \frac{2}{jw+2} - \frac{3}{(jw+2)^2}$$

$$x(t) = 2e^{-2t}u(t) - 3te^{-2t}u(t)$$

(ii)

$$X(w) = \frac{1}{(a+jw)^2} = \frac{1}{(a+jw)(a+jw)} = X_1(w)X_2(w)$$

$$X_1(w) = \frac{1}{a+jw},\ X_2(w) = \frac{1}{a+jw}$$

$$x_1(t) = e^{-at}u(t),\ x_2(t) = e^{-at}u(t)$$

Using convolution property

$$x(t) = x_1(t)^* x_2(t)$$

$$x(t) \xleftrightarrow{\text{FT}} X(w)$$

$$x_1(t)^* x_2(t) \xleftrightarrow{\text{FT}} X_1(w)X_2(w)$$

$$x(t) = \int_{-\infty}^{\infty} e^{-at}u(t)e^{-a(t-\tau)}u(t-\tau)d\tau \qquad \begin{cases} u(\tau) = 1 \ \tau \le 0 \\ u(t-\tau) = 1 \quad t \le \tau \end{cases}$$

$$= \int_0^t e^{-at}d\tau = te^{-at}u(t)$$

(iii) $X(w) = e^{-|w|}$

We know that FT $[e^{-|t|}] = \frac{2}{1+w^2}$

Using dual property $x(t) \overset{\text{FT}}{\longleftrightarrow} X(w)$

$$X(t) \overset{\text{FT}}{\longleftrightarrow} 2\pi x(-w)$$

$$X(w) = \frac{2}{1+w^2};\ x(t) = e^{-|t|}$$

$$X(t) = \frac{2}{1+t^2};\ x(w) = e^{-|w|}$$

$$\text{FT}\left[\frac{2}{1+t^2}\right] = 2\pi e^{-|w|}$$

$$\frac{1}{\pi(1+t^2)} = \text{FT}^{-1}\left[e^{-|w|}\right]$$

OR

$$x(t) = \frac{1}{2\pi}\int_{-\infty}^{\infty} e^{-|w|}e^{jwt}\,dt$$

$$= \frac{1}{2\pi}\left[\int_{-\infty}^{0} e^{(1+jt)w}dw + \int_{0}^{\infty} e^{-(1-jt)w}dw\right]$$

$$= \frac{1}{2\pi}\left[\frac{1}{1+jt} + \frac{1}{1-jt}\right]$$

$$= \frac{1}{\pi(1+t^2)}$$

(iv)

$$X(w) = e^{-2w}u(w)$$

$$e^{+at}u(-t) \overset{\text{FT}}{\longleftrightarrow} \frac{1}{a-jw}$$

$$x(t)\left(e^{2t}u(t)\right) \rightleftharpoons \left(\frac{1}{2-jw}\right) \leftarrow X(w)$$

$$F[X(t)] = 2\pi x(-w)$$

$$F\left[\frac{1}{2-jt}\right] = 2\pi e^{-2w}u(w)$$

$$e^{-2w}u(w) = \frac{1}{2\pi(2-jt)}$$

OR

$$x(t) = \frac{1}{2\pi}\int_{0}^{\infty} e^{-(2-jt)w}dw$$

$$= \frac{1}{2\pi(2-jt)}$$

Example 20: Find the F.T. of the function

$$x(t-t_0) = e^{-(t-t_0)}u(t-t_0)$$

SOLUTION:

$$\text{If } F[x(t)] = X(w)$$

$$\text{then } \text{FT}[x(t-t_0)] = e^{-jwt_0}X(w)$$

$$F\left[e^{-t}u(t)\right] = \frac{1}{1+jw}$$

$$F\left[e^{-(t-t_0)}u(t-t_0)\right] = \frac{e^{-jwt_0}}{1+jw}$$

Example 21: Find the F.T. of the function

$$x(t) = [u(t+1)-u(t-1)]\cos 2\pi t$$

SOLUTION:

$$\text{FT}(\cos 2\pi t) = \text{FT}\left(\frac{e^{j2\pi t}+e^{-j2\pi t}}{2}\right)$$

$$\text{FT}[1] = 2\pi\delta(w)$$

$$\text{FT}[e^{jw_0t}] = 2\pi\delta(w-w_0)$$

$$F[\cos 2\pi t] = \pi\delta(1w-2\pi)+\pi\delta(w+2\pi) \tag{1}$$

$$F[u(t+1)-u(t-1)] = \int_{-1}^{1} e^{-jwt}dt = -\frac{1}{jw}\left(e^{-jw}-e^{jw}\right) = \frac{2\sin w}{w} \tag{2}$$

$$F[x(t)] = F[\{u(t+1)-u(t-1)\}\cos 2\pi t]$$

$x(t)$ is multiplication of (1) and (2), so by using multiplication property

$$x(t)y(t) \overset{\text{FT}}{\longleftrightarrow} \frac{1}{2\pi}X_1(w)^*Y_1(w) = \frac{1}{2\pi}\int_{-\infty}^{\infty} X(\tau)Y(w-\tau)d\tau$$

$$X(w) = \frac{1}{2\pi}\left[\int_{-\infty}^{\infty}\frac{2\sin\tau}{\tau}\pi\delta(w-2\pi-\tau)+\delta(w+2\pi-\tau)\right]d\tau$$

$$X(w) = \int_{-\infty}^{\infty}\frac{\sin\tau}{\tau}\delta(w-2\pi-\tau)d\tau + \int_{-\infty}^{\infty}\frac{\sin\tau}{\tau}\delta(w+2\pi-\tau)d\tau$$

$$\text{Since} \quad \int_{-\infty}^{\infty} x(t)\delta(t-t_0)dt = x(t_0)$$

$$X(w) = \sin(w-2\pi)/(w-2\pi)+\sin(w+2\pi)/(w+2\pi)$$

Example 22: Determine the Fourier transform of a triangular function as shown in figure.

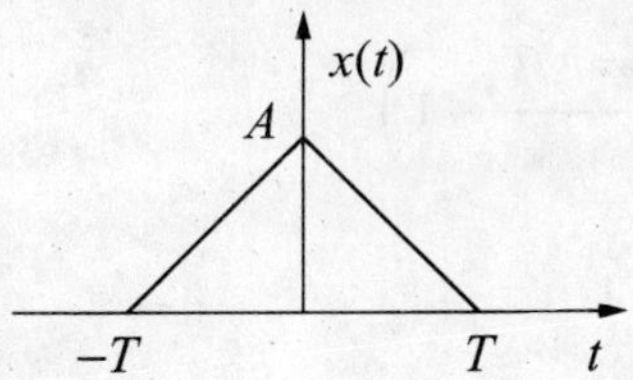

Fig. 2.24. Triangular pulse

SOLUTION:

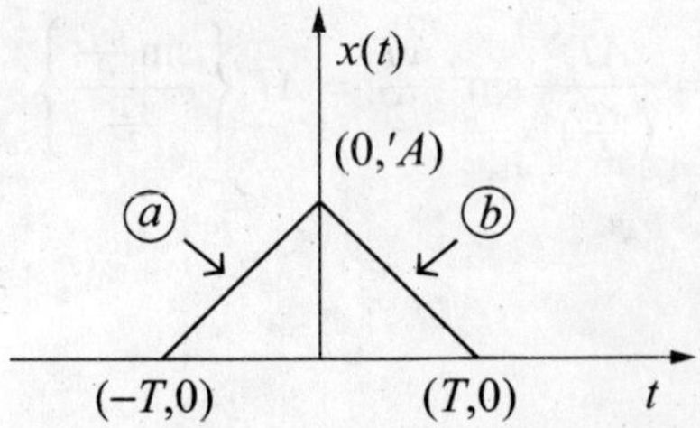

Equation of line (a) is

$$x(t) = A\left(\frac{t}{T}+1\right)$$

Equation of line (b) is

$$x(t) = A\left(1-\frac{t}{T}\right)$$

Mathematically, we can write $x(t)$ as

$$x(t) = A\left(\frac{t}{T}+1\right)[u(t+T)-u(t)] + A\left(1-\frac{t}{T}\right)[u(t)-u(t-T)]$$

$$x(t) = \frac{A}{T}(t+T)[u(t+T)-u(t)] + \frac{A}{T}(T-t)[u(t)-u(t-T)]$$

$$x(t) = \frac{A}{T}\Big\{(t+T)u(t+T)-(t+T)u(t)\Big\} + \frac{A}{T}\Big\{[(T-t)u(t)-(T-t)u(t-T)]\Big\}$$

$$x(t) = \frac{A}{T}\Big\{r(t+T)-tu(t)-Tu(t)\Big\} + \frac{A}{T}\Big\{Tu(t)-tu(t)+r(t-T)\Big\}$$

$$= \frac{A}{T}\Big\{r(t+T)-r(t)-Tu(t)\Big\} + \frac{A}{T}\Big\{Tu(t)-r(t)+r(t-T)\Big\}$$

$$= \frac{A}{T}\left[\Big\{r(t+T)-2r(t)+r(t-T)\Big\}\right]$$

$$X(jw) = \frac{A}{T}\left[\frac{e^{jwT}}{(jw)^2} - \frac{2}{(jw)^2} + \frac{e^{-jwT}}{(jw)^2}\right]$$

$$X(jw) = \frac{A}{T(jw)^2}\left[e^{jwT} + e^{-jwT} - 2\right]$$

$$= \frac{2A}{T(jw)^2}\left(\frac{e^{jwT} + e^{-jwtT}}{2} - 1\right)$$

$$= \frac{2A}{(jw)^2 T}(\cos wT - 1)$$

$$= \frac{2A}{w^2 T}(\cos wT - 1)$$

$$= \frac{2A}{w^2 T}(1 - \cos wT) = \frac{2A}{W^2 T}\left(1 - \left(1 - 2\sin^2 \frac{wT}{2}\right)\right)$$

$$= \frac{4A}{w^2 T}\sin^2 \frac{wT}{2} = \frac{AT}{\left(\frac{wT}{2}\right)^2}\sin^2 \frac{wT}{2} = AT\left\{\frac{\sin \frac{wT}{2}}{\frac{wT}{2}}\right\}^2$$

$$X(jw) = AT\,Sa^2\left(\frac{wT}{2}\right)$$

Since we know that

$$\text{FT}[tu(t)] = \text{FT}[r(t)]$$

$$= \frac{1}{(jw)^2}$$

$$\text{FT}[r(t-T)] = \frac{e^{-jwT}}{(jw)^2}$$

using time shifting property.

Example 23: Show that the F.T. of rect$(t-5)$ is $Sa\left(\frac{w}{2}\right)e^{-j5w}$

SOLUTION:

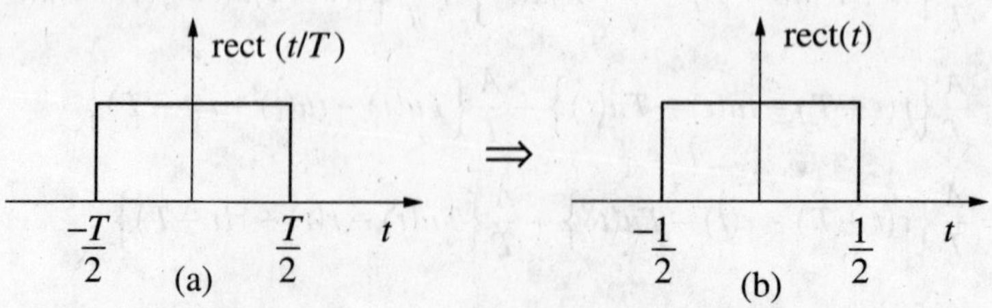

Fig. 2.25. (a) Rectangular pulse (b) Rectangular pulse at $T = 1$

$$\Pi(t) = \text{rect}(t) = \begin{cases} 1 & -\frac{1}{2} < t < \frac{1}{2} \\ 0 & \text{otherwise} \end{cases}$$

$$\text{rect}(t-5) = \begin{cases} 1 & -\frac{1}{2} \le t-5 < \frac{1}{2} \\ 0 & \text{otherwise} \end{cases}$$

$$\text{rect}(t-5) = \begin{cases} 1 & \frac{9}{2} \le t \le \frac{11}{2} \\ 0 & \text{otherwise} \end{cases}$$

$$X(jw) = \int_{-\infty}^{\infty} x(t)e^{-jwt}dt = \int_{-\infty}^{\infty} rect(t-5)e^{-jwt}dt$$

$$= \int_{9/2}^{11/2} e^{-jwt}dt = \left.\frac{e^{-jwt}}{-jw}\right|_{9/2}^{11/2}$$

$$= \frac{e^{-\frac{j11w}{2}} - e^{-\frac{9jw}{2}}}{-jw} = \frac{e^{-9j\frac{w}{2}} - e^{-11j\frac{w}{2}}}{jw}$$

$$= \frac{e^{-5jw}e^{jw/2} - e^{-5jw}e^{-jw/2}}{jw} = \frac{2e^{-5jw}\left(e^{jw/2} - e^{-jw/2}\right)}{w2j}$$

$$= \frac{2e^{-5jw}}{w}\sin\frac{w}{2} = e^{-5jw}\left(\frac{\sin\frac{w}{2}}{\frac{w}{2}}\right)$$

$$X(jw) = e^{-5jw}S_a\left(\frac{w}{2}\right)$$

2.6 PROPERTIES OF CONTINUOUS-TIME FOURIER TRANSFORM

(1) Linearity

If FT $(x_1(t)) = X_1(jw)$
and FT $(x_2(t)) = X_2(jw)$
Then linearity property states that

$$\text{FT}(Ax_1(t) + Bx_2(t)) = AX_1(jw) + BX_2(jw)$$

where A and B are constants.

Proof:

$$\text{Let } r(t) = Ax_1(t) + Bx_2(t)$$

$$\text{FT}(r(t)) = R(jw) = \int_{-\infty}^{\infty} r(t)e^{-jwt}dt$$

$$= \int_{-\infty}^{\infty} (Ax_1(t) + Bx_2(t))\,e^{-jwt}dt$$

$$= \int_{-\infty}^{\infty} (Ax_1(t)e^{-jwt} + Bx_2(t)e^{-jwt})\,dt$$

$$= A\int_{-\infty}^{\infty} x_1(t)e^{-jwt}\,dt + B\int_{-\infty}^{\infty} x_2(t)e^{-jwt}\,dt$$

$$= AX_1(jw) + BX_2(jw)$$

(2) Time scaling

If $\mathrm{FT}(x(t)) = X(jw)$, then time scaling property states that

$$\mathrm{FT}(x(at)) = \frac{1}{|a|}X\left(\frac{jw}{a}\right)$$

Proof:

Let $r(t) = x(at)$

If $a > 0$ $R(jw) = \int_{-\infty}^{\infty} r(t)e^{-jwt}\,dt$

Let $at = 2dt = \frac{a_2}{2}$

$$\mathrm{FT}(x(at)) = \int_{-\infty}^{\infty} x(\tau)e^{-jw\frac{\tau}{a}}\frac{d\tau}{a}$$

$$= \frac{1}{a}\int_{-\infty}^{\infty} x(\tau)e^{-j(\frac{w}{a})\tau}\,d\tau$$

$$\mathrm{FT}(x(at)) = \frac{1}{a}X\left(\frac{jw}{a}\right)$$

(3) Time reversal

FT $(x(t)) = X(jw)$

then FT $(x(-t)) = X(-jw)$

Proof:

Let $r(t) = x(-t)$

$$R(jw) = \int_{-\infty}^{\infty} r(t)e^{-jwt}\,dt = \int_{-\infty}^{\infty} x(-t)e^{-jwt}\,dt$$

Let $-t = \tau$ $\qquad dt = -d\tau$

$$= -\int_{\infty}^{-\infty} t(\tau)e^{jw\tau}\,d\tau$$

$$= \int_{-\infty}^{\infty} x(\tau)e^{-j(-w)\tau}d\tau$$

$$F(x(t)) = X(-jw)$$

(4) Time shifting

If FT $(x(t)) = X(jw)$
then FT $(x(t-t_0)) = e^{-jwt_0}X(jw)$

Proof:

$$\text{Let } r(t) = x(t-t_0)$$

$$R(jw) = \int_{-\infty}^{\infty} r(t)e^{-jwt}dt = \int_{-\infty}^{\infty} x(t-t_0)e^{-jwt}dt$$

$$R(jw) = \text{FT}(x(t-t_0)) = \int_{-\infty}^{\infty} x(t-t_0)e^{-jwt}dt$$

Let $t - t_0 = \tau$ $\qquad dt = d\tau$

$$\text{FT } (x(t-t_0)) = \int_{-\infty}^{\infty} x(\tau)e^{-jw(t_0+\tau)}d\tau$$

$$= \int_{-\infty}^{\infty} x(\tau)e^{-jwt}e^{-jwt_0}d\tau$$

$$= e^{-jwt_0}\int_{-\infty}^{\infty} x(\tau)e^{-jw\tau}d\tau$$

FT $(x(t-t_0)) = e^{-jwt_0}X(jw)$. Similarly, FT $(x(t+t_0)) = e^{jwt_0}X(jw)$

So FT $(x(t \pm t_0)) = e^{\pm jwt_0}X(jw)$

(5) Frequency shifting

$$\text{If FT } (x(t)) = X(jw)$$

$$\text{FT } (e^{jw_0t}x(t)) = X(j(w-w_0))$$

$$\text{Let } r(t) = e^{jw_0t}x(t)$$

$$\text{FT } (r(t)) = \text{FT }\left(e^{jw_0t}x(t)\right) = R(jw) = \int_{-\infty}^{\infty} e^{jw_0t}x(t)e^{-jwt}dt$$

$$\text{FT }\left(e^{jw_0t}x(t)\right) = \int_{-\infty}^{\infty} x(t)e^{-j(w-w_0)t}dt$$

$$\text{Let } w - w_0 = w'$$

$$= \int_{-\infty}^{\infty} x(t)e^{-jw't}dt$$

$$\text{FT}\left(e^{jw_0t}x(t)\right) = X(jw') = X(j(w-w_0))$$

$$\text{Similarly, FT}(e^{-jw_0t}x(t)) = X(j(w+w_0))$$

$$\text{We can write as FT}\left(e^{\pm jw_0t}x(t)\right) = X(j(w \mp w_0))$$

(6) Duality or symmetry property

If FT $(x(t)) = X(jw)$

then FT $(x(t)) = 2\pi x(-jw)$

Proof:

We know that $x(t) = \frac{1}{2\pi}\int_{-\infty}^{\infty} X(jw)e^{jwt}dw$

Replacing t by $-t$, we get

$$x(-t) = \frac{1}{2\pi}\int_{-\infty}^{\infty} X(jw)e^{-jwt}dw$$

$$2\pi\, x(-t) = \frac{2\pi}{2\pi}\int_{-\infty}^{\infty} X(jw)e^{-jwt}dw$$

$$2\pi\, x(-t) = \int_{-\infty}^{\infty} X(jw)e^{-jwt}dw$$

Interchanging t by jw

$$2\pi\, x(-jw) = \int_{-\infty}^{\infty} X(t)e^{-jwt}dt$$

$$2\pi\, x(-jw) = \text{FT}(X(t))$$

(7) Convolution in time domain

If FT $(x_1(t)) = X_1(jw)$ and FT $(x_2(t)) = X_2(jw)$

then FT $(x_1(t)^*x_2(t)) = X_1(jw)X_2(jw)$

i.e., convolution in time domain becomes multiplication in frequency domain.

Proof:

$$r(t) = x_1(t)^* x_2(t) = \int_{-\infty}^{\infty} x_1(\tau)x_2(t-\tau)d\tau$$

$$\text{FT}(r(t)) = R(jw) = \int_{-\infty}^{\infty} r(t)e^{-jwt}dt$$

$$= \int_{-\infty}^{\infty}\left(\int_{-\infty}^{\infty} x_1(\tau)x_2(t-\tau)d\tau\right)e^{-jwt}dt$$

$$= \int_{-\infty}^{\infty}\int_{-\infty}^{\infty} x_1(\tau)x_2(t-\tau)d\tau\, e^{-jwt}dt$$

$$= \int_{-\infty}^{\infty} x_1(\tau)d\tau \int_{-\infty}^{\infty} x_2(t-\tau)\, e^{-jwt}dt$$

Let $t - \tau = \propto$ so $dt = d\propto$

$$\text{FT}\left[x_1(t)^* x_2(t)\right] = \int_{-\infty}^{\infty} x_1(t)d\tau \int_{-\infty}^{\infty} x_2(\propto)\, e^{-jw(\propto+\tau)}d\propto$$

$$= \int_{-\infty}^{\infty} x_1(\tau)d\tau \int_{-\infty}^{\infty} x_2(\propto)\, e^{-jw\propto}e^{-jw\tau}d\propto$$

$$= \int_{-\infty}^{\infty} x_1(\tau)\, e^{-jw\tau}d\tau \int_{-\infty}^{\infty} x_2(\propto)\, e^{-jw\propto}d\propto$$

$$\text{FT}\left[x_1(t)^* x_2(t)\right] = X_1(jw)\, X_2(jw)$$

(8a) Integration in time domain

If FT $(x(t)) = X(jw)$

then FT $\left(\int_{-\infty}^{t} x(\tau)d\tau\right) = \frac{1}{jw} \times (jw)$

Proof: Let $r(t) = \int_{-\infty}^{t} x(\tau)d\tau$

Differentiating w.r.t. t

$$\frac{dr(t)}{dt} = x(t) \Rightarrow \text{FT}(x(t)) = \text{FT}\left(\frac{d}{dt}r(t)\right)$$

From differentiation in time domain

$$X(jw) = jwX(jw)$$

$$R(jw) = \frac{1}{jw}X(jw)$$

$$\text{FT}(r(t)) = \text{FT}\left(\int_{-\infty}^{t} x(\tau)d\tau\right) = \frac{1}{jw}X(jw)$$

(8b) Differentiation in time domain

If FT $(x(t)) = X(jw)$
then $\left(\frac{d}{dt}x(t)\right) = jw \times (jw)$

Proof: We know that $x(t) = \frac{1}{2\pi}\int\limits_{-\infty}^{\infty} X(jw)e^{jwt}\,dw$. Differentiating both sides w.r.t. t

$$\frac{d}{dt}x(t) = \frac{1}{2\pi}\int\limits_{-\infty}^{\infty} X(jw)\left(\frac{d}{dt}e^{jwt}\right)dw$$

$$= \frac{1}{2\pi}\int\limits_{-\infty}^{\infty} jwX(jw)e^{jwt}\,dw$$

$$= j\frac{1}{2\pi}\int\limits_{-\infty}^{\infty} (wX(jw))e^{jwt}\,dw$$

$$\frac{d}{dt}x(t) = j\,\mathrm{FT}^{-1}(wX(jw))$$

yields FT $\left(\frac{d}{dt}x(t)\right) = jwX(jw)$. On generalizing we get FT $\left(\frac{d^n}{dt^n}x(t)\right) = (jw)^n X(jw)$

(9) Differentiation in frequency domain

If FT $(x(t)) = X(jw)$
then FT $(tx(t)) = j\frac{d}{dw}X(jw)$

Proof: We know that $X(jw) = \int_{-\infty}^{\infty} x(t)e^{-jwt}\,dt$
On differentiating both sides w.r.t. w

$$\frac{d}{dw}X(jw) = \int\limits_{-\infty}^{\infty} x(t)\left(\frac{d}{dw}e^{-jwt}\right)dt = -\int\limits_{-\infty}^{\infty} j\,t\,x(t)e^{-jwt}\,dt$$

Multiplying both sides by j

$$j\frac{d}{dw}X(jw) = \int\limits_{-\infty}^{\infty} (tx(t))e^{-jwt}\,dt \quad \text{since } j^2 = -1 \text{ or } -j^2 = 1$$

$$j\frac{d}{dw}X(jw) = \mathrm{FT}\,[t\,x(t)]$$

$$\mathrm{FT}\,[t\,x(t)] = j\frac{d}{dw}X(jw)$$

(10) Convolution in frequency domain (multiplication in time domain (multiplication theorem))

$$\text{If } \mathrm{FT}(x_1(t)) = X_1(jw) \text{ and } \mathrm{FT}\,[x_2(t)] = X_2(jw)$$

$$\mathrm{FT}(x_1(t)x_2(t)) = \frac{1}{2\pi}(X_1(jw)^* X_2(jw))$$

i.e., multiplication in time domain becomes convolution in frequency domain.

Proof:

$$\text{Let } r(t) = x_1(t)x_2(t)$$

$$R(jw) = \int_{-\infty}^{\infty} r(t)e^{-jwt}dt = \int_{-\infty}^{\infty} x_1(t)x_2(t)e^{-jwt}dt$$

$$\text{We know that } x_2(t) = \frac{1}{2\pi}\int_{-\infty}^{\infty} X_2(jw)e^{jwt}dw$$

$$R(jw) = \int_{-\infty}^{\infty} x_1(t)\left(\frac{1}{2\pi}\int_{-\infty}^{\infty} X_2(jw')e^{jw't}dw'\right)e^{-jwt}dt$$

$$= \int_{-\infty}^{\infty} x_1(t)\frac{1}{2\pi}\int_{-\infty}^{\infty} X_2(jw')e^{-j(w-w')t}dw'dt$$

$$\text{Let } \quad w - w' = \lambda$$

$$dw' = -d\lambda$$

$$= \frac{1}{2\pi}\int_{-\infty}^{\infty} x_1(t)dt\int_{-\infty}^{\infty} X_2(j(w-\lambda))e^{-j\lambda t}d\lambda \text{ since } = \int_{+\infty}^{-\infty} -d\lambda = \int_{-\infty}^{\infty} d\lambda$$

$$= \frac{1}{2\pi}\int_{-\infty}^{\infty} x_1(t)e^{-j\lambda t}dt\int_{-\infty}^{\infty} X_2(j(w-\lambda))d\lambda$$

$$= \frac{1}{2\pi}\int_{-\infty}^{\infty} X_2(j(w-\lambda))d\lambda\int_{-\infty}^{\infty} x_1(t)e^{-j\lambda t}dt$$

$$= \frac{1}{2\pi}\int_{-\infty}^{\infty} X_1(j\lambda)X_2(j(w-\lambda))d\lambda$$

$$= \frac{1}{2\pi}[X_1(jw) * X_2(jw)]$$

$$\text{FT}[x_1(t)x_2(t)] = \frac{1}{2\pi}[X_1(jw) * X_2(jw)]$$

Parseval's theorem

If FT $(x(t)) = X(jw)$

Then according to Parseval's theorem

$$E = \int_{-\infty}^{\infty} |x(t)|^2 dt = \frac{1}{2\pi}\int_{-\infty}^{\infty} |X(jw)|^2 dw$$

Proof:

$$E = \int_{-\infty}^{\infty} \left|x(t)^2\right| dt = \int_{-\infty}^{\infty} \left|x(t)x^*(t)dt\right. \qquad (1)$$

$$\text{We know that } x(t) = \frac{1}{2\pi}\int_{-\infty}^{\infty} X(jw)e^{+jwt}dw$$

$$\text{So } x^*(t) = \frac{1}{2\pi}\int_{-\infty}^{\infty} X(jw)e^{-jwt}dw \qquad (2)$$

on putting (1)

$$= \int_{-\infty}^{\infty} x(t)\left[\frac{1}{2\pi}\int_{-\infty}^{\infty} X^*(jw)e^{-jwt}dw\right]dt$$

$$= \frac{1}{2\pi}\int_{-\infty}^{\infty} X^*(jw)\int_{-\infty}^{\infty} x(t)e^{-jwt}dt\,dw$$

$$= \frac{1}{2\pi}\int_{-\infty}^{\infty} X(jw)X^*(jw)dw$$

$$= \int_{-\infty}^{\infty} |x(t)^2|dt = \frac{1}{2\pi}\int_{-\infty}^{\infty} |X(jw)|^2 dw$$

Relation between Laplace Transform and Fourier Transform

Fourier transform $X(jw)$ of a signal $x(t)$ is given as

$$X(jw) = \int_{-\infty}^{\infty} x(t)e^{-jwt}dt \qquad (1)$$

F.T. can be calculated only if $x(t)$ is absolutely integrable

$$= \int_{-\infty}^{\infty} |x(t)|dt < \infty \qquad (2)$$

Laplace transform $X(s)$ of a signal $x(t)$ is given as

$$X(s) = \int_{-\infty}^{\infty} x(t)e^{-st}dt \qquad (3)$$

We know that $s = \sigma + jw$

$$X(s) = \int_{-\infty}^{\infty} x(t)e^{-(\sigma+jw)t}dt$$

$$X(s) = \int_{-\infty}^{\infty} \left[x(t)e^{-\sigma t}\right]e^{-jwt}dt \qquad (4)$$

Comparing (1) and (4), we find that L.T. of $x(t)$ is basically the F.T. of $[x(t)e^{-\sigma t}]$.

If $s = jw$, i.e. $\sigma = 0$, then eqn. (4) becomes $X(s) = \int_{-\infty}^{\infty} x(t)e^{-jwt}dt = X(jw)$

Thus, $X(s) = X(jw)$ when $\sigma = 0$ or $s = jw$

This means L.T. is same as F.T. when $s = jw$. The above equation shows that F.T. is special case of L.T. Thus, L.T. provides broader characterization compared to F.T., $s = jw$ indicates imaginary axis in complex s-plane.

2.7 APPLICATIONS OF FOURIER TRANSFORM OF NETWORK ANALYSIS

Example 24: Determine the voltage $V_{out}(t)$ to a current source excitation $i(t) = e^{-t}u(t)$ for the circuit shown in figure.

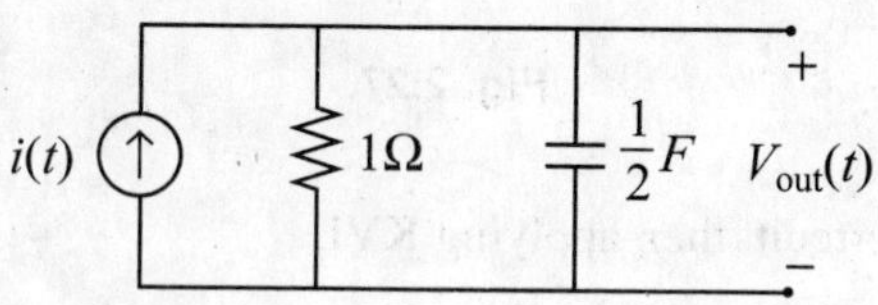

Fig. 2.26.

SOLUTION:

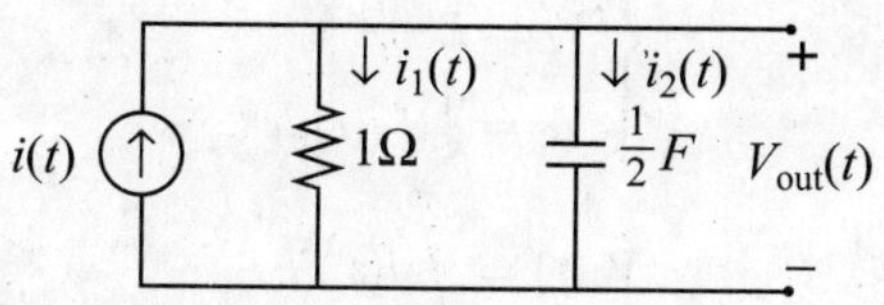

$$i(t) = i_1(t) + i_2(t)$$

$$i(t) = \frac{V_{out}(t)}{1} + \frac{1}{2}\frac{dV_{out}(t)}{dt}$$

$$\left\{ \begin{array}{l} \text{since } i = \frac{V}{R} \\ \text{and } i = c\frac{dv}{dt} \text{ or } v = \frac{1}{c}\int idt \end{array} \right.$$

$$e^{-t}u(t) = V_{out}(t) + \frac{1}{2}\frac{dV_{out}(t)}{dt} \tag{1}$$

On taking the z-transform on both sides

$$\frac{1}{1+jw} = V_{out}(jw)\left\{1 + \frac{jw}{2}\right\} = \frac{(2+jw)}{2}V_{out}(jw)$$

$$V_{out}(jw) = \frac{2}{(1+jw)(2+jw)} = \frac{A}{1+jw} + \frac{B}{2+jw}$$

$$V_{out}(jw) = \frac{2}{1+jw} - \frac{2}{2+jw}$$

$$\left\{ \begin{array}{l} A(2+jw) + B(1+jw) = 2 \\ 2A + B = 2 \\ A + B = 0 \text{ so } A = -B \\ 2A - A = 2; \quad A = 2, B = -2 \end{array} \right.$$

On taking the inverse Fourier transform

$$V_0(t) = 2(e^{-t} - e^{-2t})u(t)$$

Example 25: Determine the voltage $V_0(t)$ for the circuit shown in figure.

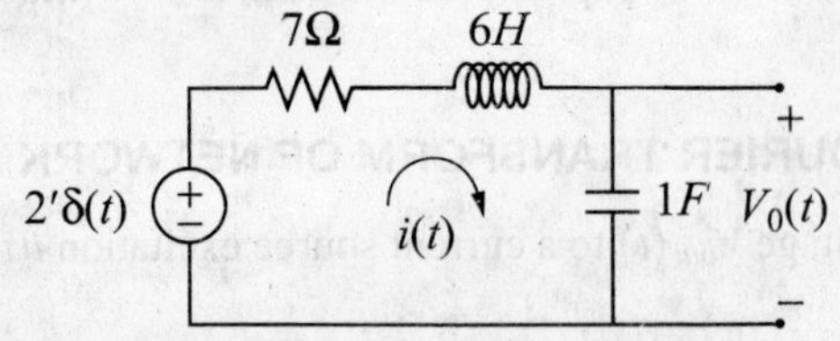

Fig. 2.27.

SOLUTION:

Let $i(t)$ be the current in the circuit, then applying KVL

$$2\delta(t) = 7i(t) + 6\frac{di(t)}{dt} + \int_{-\infty}^{t} i(t)dt \tag{1}$$

$$\begin{cases} \text{since} \\ V_L = L\frac{di(t)}{dt} \\ V_C = \frac{1}{C}\int i(t)dt \end{cases}$$

and

$$V_0(t) = \int_{-\infty}^{t} i(t)dt \tag{2}$$

On taking Fourier transform of (1) and (2)

$$2 = 7I(jw) + 6jwI(jw) + \frac{1}{jw}I(jw)$$

$$2 = \left(7 + 6jw + \frac{1}{jw}\right)I(jw)$$

$$I(jw) = \frac{2}{7 + 6jw + \frac{1}{jw}} = \frac{2jw}{7jw + 6(jw)^2 + 1}$$

since FT $[\delta(t)] = 1$

$$\text{FT}\left[\frac{di(t)}{dt}\right] = jwI(jw)$$

Using differentiation property

$$\text{FT}\left[\int_{-\infty}^{t} i(t)dt\right] = \frac{1}{jw}I(jw)$$

Using integration property

$$I(jw) = \frac{2jw}{6(jw)^2 + 7jw + 1} \tag{3}$$

$$V_0(jw) = \frac{1}{jw}I(jw) \tag{4}$$

$$V_0(jw) = \frac{1}{jw} \times \frac{2jw}{6(jw)^2 + 7jw + 1}$$

$$V_0(jw) = \frac{2}{6(jw)^2 + 7(jw) + 1} = \frac{2}{(6jw+1)(jw+1)}$$

$$V_0(jw) = \frac{1/3}{(jw+1/6)(jw+1)} = \frac{A}{\frac{1}{6} + jw} + \frac{B}{1 + jw}$$

$$V_0(jw) = \frac{2}{5\left(\frac{1}{6} + jw\right)} - \frac{2}{5(1+jw)} \tag{5}$$

Taking inverse Fourier transform, we get

$$V_0(t) = \frac{2}{5}\left(e^{-t/6} - e^{-t}\right)u(t) \tag{6}$$

Example 26: Determine the response of current in the network shown in Fig. 2.28(a) when a voltage having the waveform shown in Fig. 2.28(b) is applied to it by using the Fourier transform.

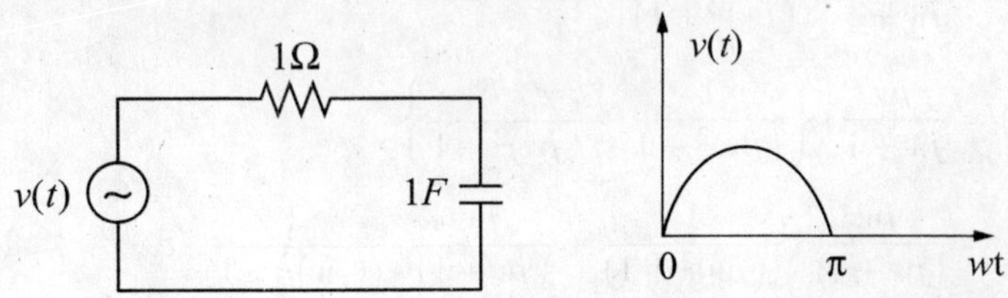

(a) **(b)**

Fig. 2.28.

SOLUTION:

Waveform $V(t)$ is defined as

$$V(t) = \sin t(u(t) - u(t-\pi)) \tag{1}$$

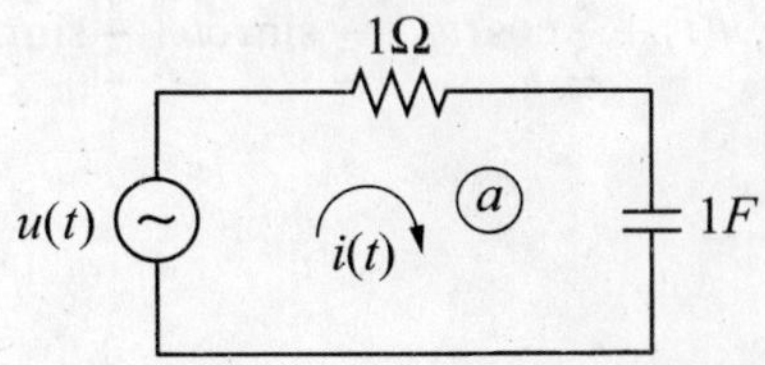

Let $i(t)$ be the current in the loop. Applying KVL in loop

$$V(t) = 1 \cdot i(t) + \frac{1}{1}\int_0^t i(t)dt = i(t) + \int_0^t i(t)dt \tag{2}$$

On taking Fourier transform of

$$V(jw) = \frac{1}{(jw)^2 + 1} + \frac{e^{-j\pi w}}{(jw)^2 + 1}$$

Since
$$\text{FT}\left[\sin t u(t)\right] = \frac{1}{(jw)^2 + 1}$$

$$\text{FT}\left[\sin t u(t-\pi)\right] = \frac{e^{-j\pi w}}{(jw)^2 + 1}$$

Solve using F.T. formula

$$V(jw) = \frac{1+e^{-j\pi w}}{(jw)^2+1} \tag{3}$$

$$V(jw) = I(jw) + \frac{1}{jw} I(jw)$$

$$V(jw) = \left(1+\frac{1}{jw}\right) I(jw) = \frac{jw+1}{jw} I(jw)$$

$$I(jw) = \frac{jw}{jw+1} V(jw) \tag{4}$$

$$I(jw) = \frac{jw}{jw+1} \cdot \frac{(1+e^{j\pi w})}{((jw)^2+1)} \quad \text{From (3)}$$

$$I(jw) = \frac{jw}{jw+1} \cdot \left\{ \frac{1}{(jw)^2+1} + \frac{e^{-j\pi w}}{(jw)^2+1} \right\}$$

$$= \underbrace{\frac{jw}{(jw+1)} \cdot \frac{1}{((jw)^2+1)}}_{I_1(jw)} + \underbrace{\frac{jw}{(jw+1)} \cdot \frac{1}{((jw)^2+1)} \cdot e^{-j\pi w}}_{I_2(jw)}$$

$$I_1(jw) = \frac{A}{jw+1} + \frac{Bjw+c}{((jw)^2+1)}$$

$$= \frac{-1/2}{(jw+1)} + \frac{\frac{1}{2}(jw+1)}{((jw)^2+1)}$$

$$i_1(t) = -\frac{1}{2} e^{-t} u(t) + \frac{1}{2} \cos tut + \frac{1}{2} \sin t \delta t + \frac{1}{2} \sin tu(t)$$

Since IFT $\left\{ \frac{1}{(jw)^2+1} \right\} = \sin tu(t)$

so IFT $\left(\frac{jw}{(jw)^2+1} \right) = \frac{d}{dt} \sin tu(t)$

Using differential in time domain property

$$\text{IFT}\left[\frac{jw}{(jw)^2+1}\right] = \cos tu(t) + \sin t\delta(t)$$

$$I_2(jw) = \frac{jw}{(jw+1)} \cdot \frac{1}{((jw)^2+1)} \cdot e^{-j\pi w}$$

$$I_2(jw) = I_3(jw) \cdot e^{-j\pi w}$$

Since $I_3 = I_1(jw)$

so $i_3(t) = -\frac{1}{2} e^{-t} u(t) + \frac{1}{2} \cos tu(t) + \frac{1}{2} \sin t\delta(t) + \frac{1}{2} \sin tu(t)$

From time shifting property FT $(x(t\pm t_0)) = e^{\pm jwt_0}\times(jw)$

$$i_2(t) = i_3(t-\pi)$$

$$= -\frac{1}{2}e^{-(t-\pi)}u(t-\pi)+\frac{1}{2}\cos(t-\pi)u(t-\pi)+\frac{1}{2}\sin(t-\pi)\delta(t-\pi)+\frac{1}{2}\sin(t-\pi)u(t-\pi)$$

so $$i(t) = \frac{1}{2}-\left[-e^{-t}+\cos t+\sin t\right]u(t)+\frac{1}{2}\sin t\delta(t)+\frac{1}{2}\left[-e^{-(t-\pi)}+\cos(t-\pi)+\sin(t-\pi)\right]u(t-\pi)+\frac{1}{2}\sin(t-\pi)\delta(t-\pi)$$

Example 27: For the *RC* circuit shown in figure.

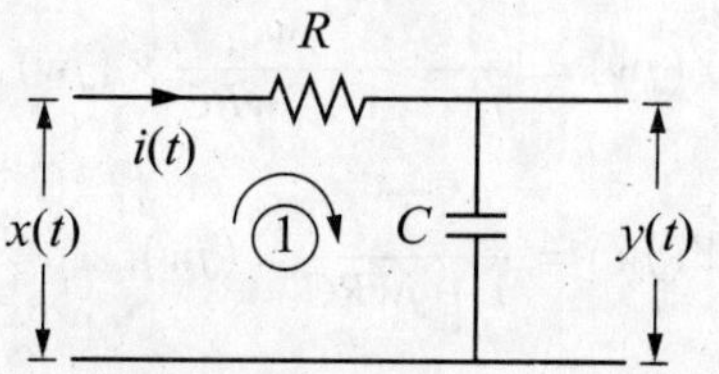

Fig. 2.29.

(a) Determine frequency response of the circuit.

(b) Find impulse response.

(c) Plot the magnitude and phase response for $RC = 1$.

SOLUTION:

Applying KVL in loop (1)

$$x(t) - Ri(t) - \frac{1}{C}\int_{-\infty}^{t} i(t)dt = 0$$

$$x(t) = Ri(t) + \frac{1}{C}\int_{-\infty}^{t} i(t)dt \tag{1}$$

$$\begin{cases} Since \\ V_R = iR \\ V_c = \frac{1}{C}\int i(t)dt \end{cases}$$

and $$y(t) = \frac{1}{C}\int_{-\infty}^{t} i(t)dt \tag{2}$$

Taking Fourier transform of (1) and (2)

$$X(jw) = RI(jw) + \frac{1}{jwc}I(jw)$$

$$X(jw) = \left(R + \frac{1}{jwc}\right)I(jw)$$

$$X(jw) = \frac{(1 + jwRC)}{jwc}I(jw) \tag{3}$$

$$Y(jw) = \frac{1}{jwc}I(jw) \tag{4}$$

On putting the value of $I(jw)$ from (4) in (3)

$$Y(jw) = \frac{1}{jwc} \cdot \frac{jwc}{1 + jwRC} \times (jw)$$

$$Y(jw) = \frac{1}{1 + jwRC} \times (jw).$$

Frequency response

$$H(jw) = \frac{Y(jw)}{X(jw)} = \frac{1}{1 + jwRC} \tag{5}$$

On taking inverse Fourier transform

$$h(t) = \text{IFT}\,[H(jw)] = \text{IFT}\left[\frac{\frac{1}{RC}}{\frac{1}{RC} + jw}\right]$$

$$h(t) = \frac{1}{RC}e^{-t/RC}u(t) \tag{6}$$

When $RC = 1$, eqn. (5) becomes

$$H(jw) = \frac{1}{1 + jw} \tag{7}$$

Magnitude of frequency response

$$|H(jw)| + \frac{|1|}{|1 + jw|}$$

since if $A = BC$

so $|A| = |B||C|$

$$\angle A = \angle B + \angle C$$

and $A = \dfrac{B}{C}$

$$|A| = \frac{|B|}{|C|}$$

$$A = B - C$$

$$\left|H(jw)\right| = \frac{1}{\sqrt{1+w^2}} \tag{8}$$

$$H(jw) = 1 - (1 + jw)$$

$$= \tan^{-1}\frac{0}{1} - \tan^{-1} w = -\tan^{-1} w \tag{9}$$

For different values of w, we find $\left|H(jw)\right|$ and $H(jw)$

S. No	w	$\lvert H(jw)\rvert$	$H(jw)$
1–	$-\infty$	0	90°
2–	−50	0.0199	88.9°
3–	−20	0.0499	87.1°
4–	−10	0.099	84.3°
5–	−5	0.196	78.7°
6–	−2	0.447	63.4°
7–	−1	0.707	45°
8–	0	1	0
9–	1	0.707	−45°
10–	2	0.447	−63.4°
11–	5	0.196	−78.7°
12–	10	0.099	−84.3°
13–	20	0.0499	−87.1°
14–	50	0.0199	−88.9°
15–	∞	0	−90°

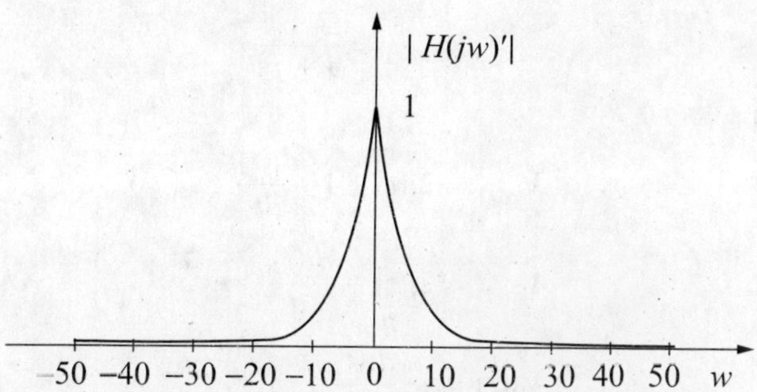

Fig. 2.30. Magnitude plot frequency response of the circuit

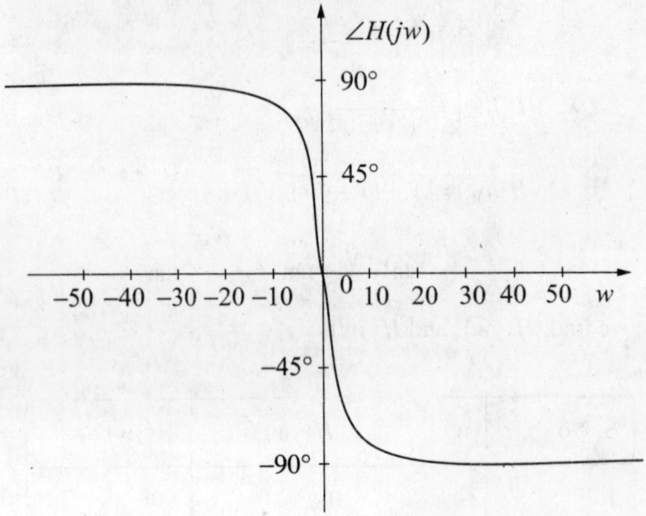

Fig. 2.31. Phase plot

Example 28: For the circuit shown in figure, determine the output voltage $V_0(t)$ to a voltage source excitation $V_{i(t)} = e^{-t}u(t)$ using Fourier transform

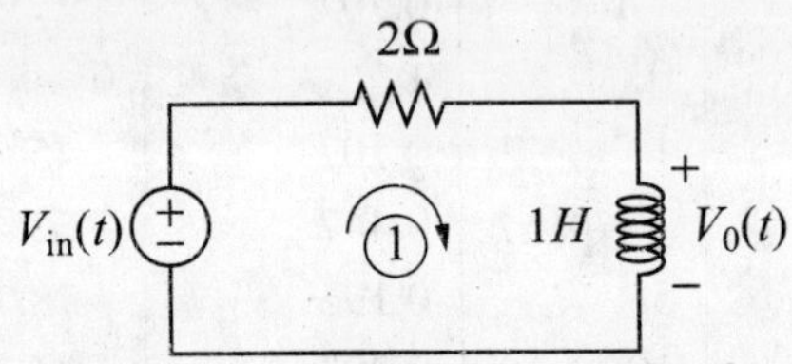

Fig. 2.32.

SOLUTION:

$$\text{Since } V_{in(t)} = e^{-t}u(t) \tag{1}$$

$$V_{in(jw)} = \frac{1}{1+jw} \tag{2}$$

Applying KVL in loop (1)

$$V_{in(t)} = 2i(t) + 1 \cdot \frac{di(t)}{dt}$$

$$V_{in(t)} = 2i(t) + \frac{di(t)}{dt} \tag{3}$$

$$V_0(t) = 1 \cdot \frac{di(t)}{dt}$$

$$V_0(t) = \frac{di(t)}{dt} \tag{4}$$

On taking Fourier transform of (3) and (4)

$$V_{in}(jw) = 2\,I(jw) + jwI(jw)$$

$$V_{in}(jw) = (2 + jw)I(jw)$$

$$\frac{1}{1+jw} = (2 + jw)I(jw)$$

$$I(jw) = \frac{1}{(1+jw)(2+jw)} \tag{5}$$

$$V_0(jw) = jwI(jw) = \frac{jw}{(1+jw)(2+jw)}$$

$$V_0(jw) = \frac{A}{1+jw} + \frac{B}{2+jw}$$

$$V_0(jw) = \frac{-1}{1+jw} + \frac{2}{2+jw}$$

Taking inverse Fourier transform

$$V_0(t) = -e^{-t}u(t) + 2e^{-2t}u(t)$$

$$V_0(t) = (-e^{-t} + 2e^{-2t})u(t)$$

QUESTION BANK

Q1: Find the trigonometric Fourier series for the waveform shown below.

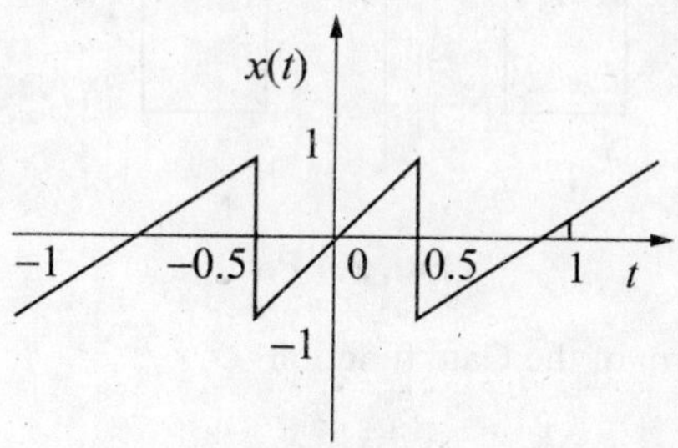

Fig. 2.1 P.

Q2: Determine the Fourier series representation of the following periodic signal.

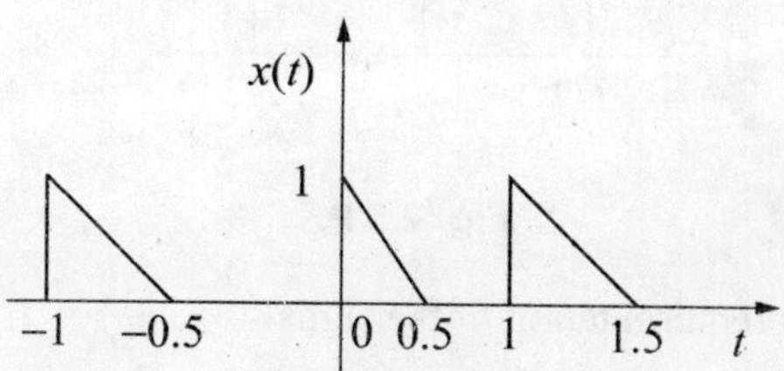

Fig. 2.2 P.

Q3: (i) State and prove the following properties of Fourier series:
(a) Time shifting property (b) Frequency shifting property
(ii) What are Dirichlet's conditions?

Q4: Find the fundamental period T, the fundamental frequency w_0 and the Fourier series coefficients a_n of the following periodic signal;

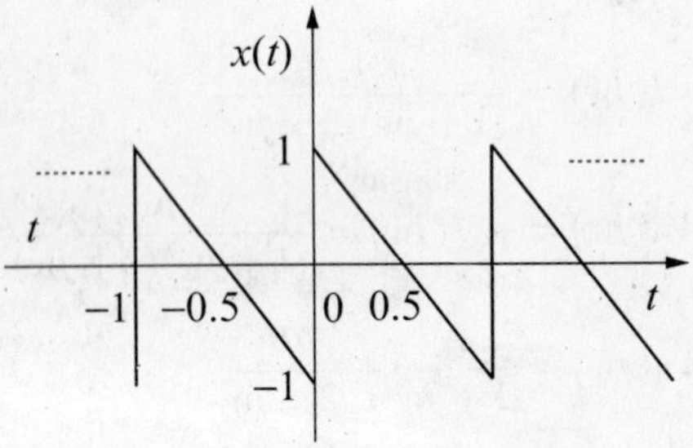

Fig. 2.3 P.

Q5: Obtain the Fourier series component of the periodic square wave signals.

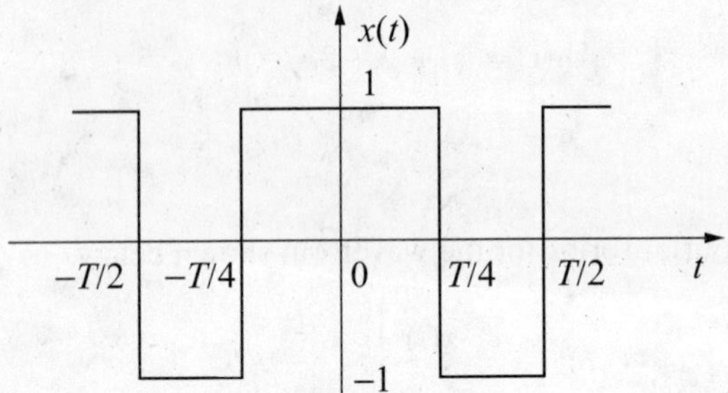

Fig. 2.4 P.

Q6: Determine the Fourier transform of the Gate function

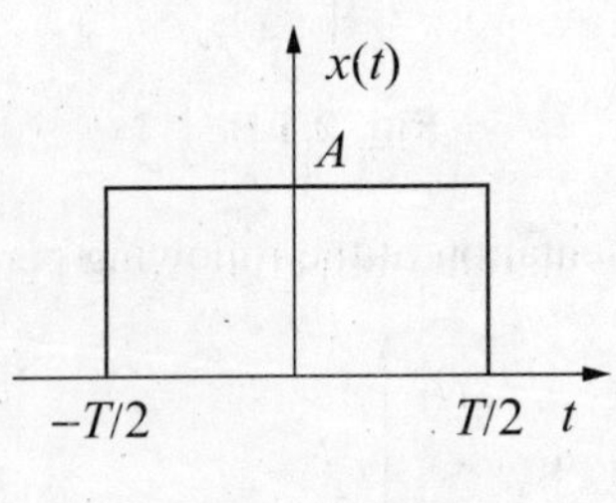

Fig. 2.5 P.

Q7: Determine the Fourier series representation of the signal

$$x(t) = \begin{cases} t - t^2 & \text{for } -\pi \le t \le \pi \\ 0 & \text{elsewhere} \end{cases}$$

Q8: For the continuous-time periodic signal

$$x(t) = 2 + \cos[2\pi t/3] + 4\sin[5\pi t/3]$$

determine the fundamental frequency w_0 and the Fourier series coefficients C_n such that

$$x(t) = \sum_{n=-\infty}^{\infty} C_n e^{jnw_0 t}$$

Q9: Find the Fourier transform of the following signals:

(a) $x(t) = \delta(t)$ (b) $x(t) = 1$ (c) $x(t) = \text{sgn}\ (t)$ (d) $x(t) = u(t)$

(e) $x(t) = \exp(-at)u(t)$ (f) $x(t) = \cos\ [w_0 t] \sin\ [w_0 t]$

Q10: Show that the Fourier transform of rect $(t-5)$ is $Sa(w/2)\exp(j5w)$. Sketch the resulting amplitude and phase spectrum.

Q11: Find the inverse Fourier transform of spectrum shown in figure.

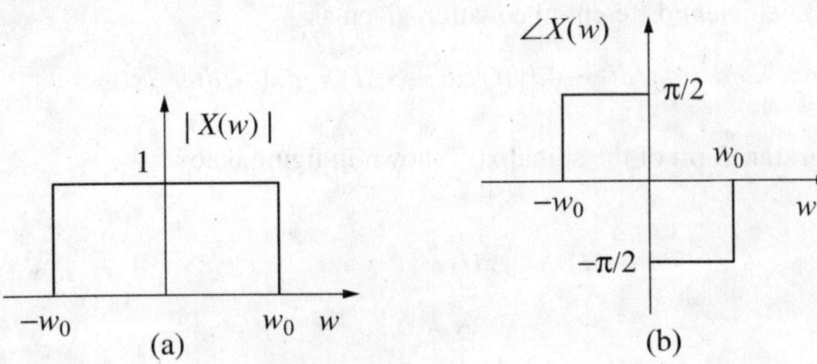

Fig. 2.6 P.

Q12: Find the Fourier transform of the following waveform.

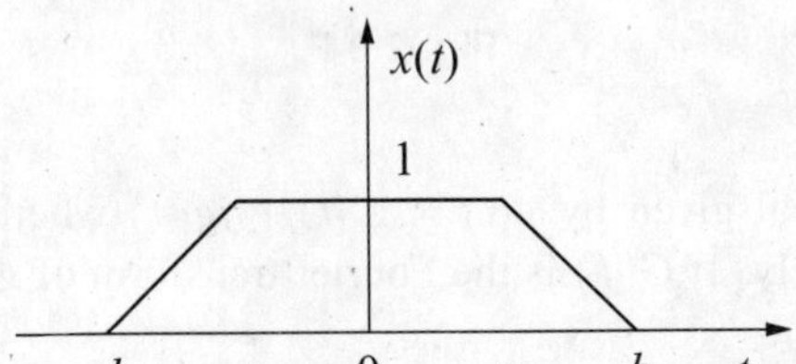

Fig. 2.7 P.

Q13: State and prove duality property of CTFT.

Q14: Determine the Fourier transform of the signal

$x(t) = \{tu(t)^*[u(t) - u(t-1)]\}$, where $u(t)$ is unit step function and * denotes the convolution operation.

Q15: Show that the frequency response of a CTLTIS is $Y(w) = H(w)X(w)$

where $X(w) =$ Fourier transform of the signal $x(t)$

$H(w) =$ Fourier transform of LTIS response $h(t)$

Q16: Find the Fourier transform of the signal $x(t)$ shown in figure below.

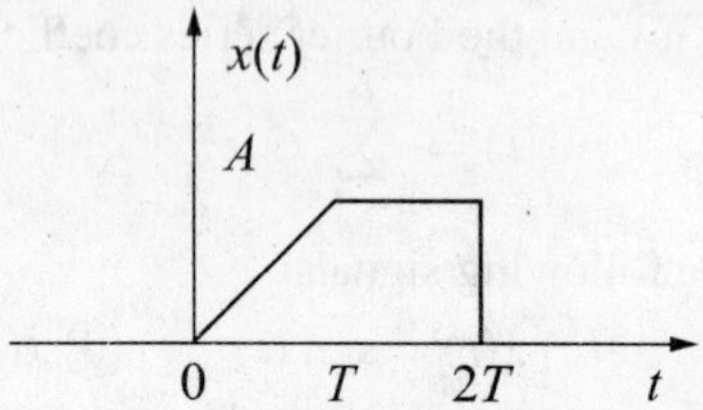

Fig. 2.8 P.

Q17: Determine the frequency response $H(jw)$ and impulse response $h(t)$ for a stable CTLTIS characterized by the linear constant coefficient differential equation given as

$$d^2y(t)/dt^2 + 4dy(t)/dt + 3y(t) = dx(t)/dt + 2x(t)$$

Q18: Find the Fourier transform of the signal $x(t)$ shown in figure below.

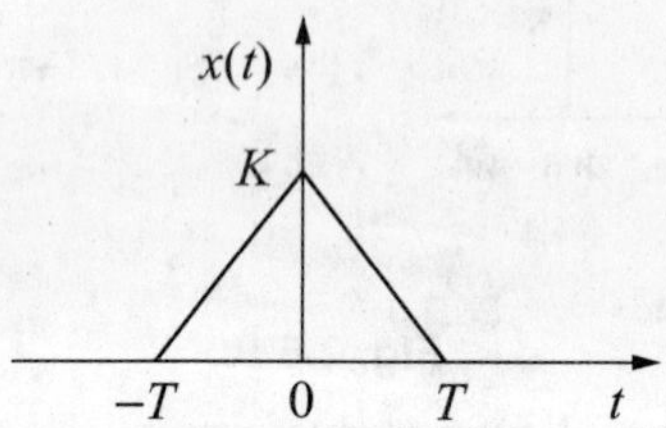

Fig. 2.9 P.

Q19: If $g(t)$ is a complex signal given by $g(t) = g_r(t) + jg_i(t)$ where $g_r(t)$ and $g_i(t)$ are the real and imaginary parts of $g(t)$ respectively. If $G(f)$ is the Fourier transform of $g(t)$, express the Fourier transform of $g_r(t)$ and $g_i(t)$ in terms of $G(f)$.

Q20: Find the coefficients of the complex exponential Fourier series for a half wave rectified sine wave defined by

$$x(t) = \begin{cases} A\sin(w_0 t), \ 0 \le t \le T_0/2 \\ 0, \ T_0/2 \le t \le T_0 \end{cases}$$

with $x(t) = x(t + T_0)$

Q21: (a) Show that the Fourier transform of the convolution of two signals in the time domain can be given by the product of the Fourier transform of the individual signals in the frequency domain.

(b) Determine the Fourier transform of the signal

$$x(t) = \frac{1}{2}\left[\delta(t+1) + \delta(t-1) + \delta\left(t + \frac{1}{2}\right)\delta + \left(t - \frac{1}{2}\right)\right]$$

SOLUTIONS OF QUESTION BANK

Q1: $T = 1, w_0 = \frac{2\pi}{T} = 2\pi$ (a) $(-.5, -1)$
(b) $(0.5, 1)$

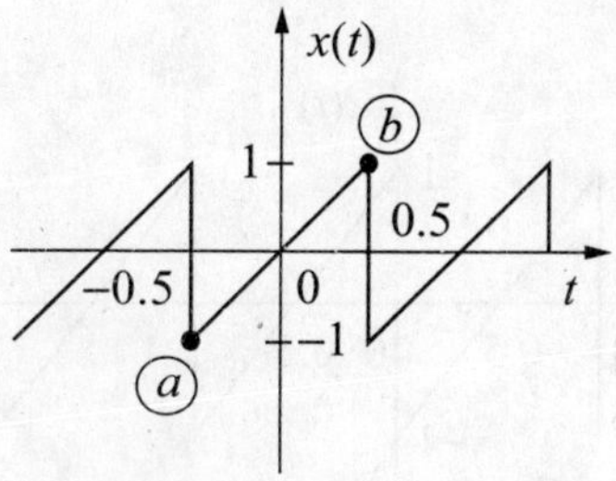

$$y + 1 = \frac{2}{1}(x + 0.5)$$
$$x(t) = 2t(-0.5 < t < 0.5)$$

Since odd function $a_0 = a_n = 0$

$$b_n = \frac{2}{T}\int_{-.5}^{.5} 2t \sin n w_0 t\, dt = \frac{2}{T}\int_{-.5}^{.5} 2t \sin n\pi 2t\, dt$$

$$= 4\int_{-.5}^{.5} t \sin 2n\pi t\, dt = \frac{-2(-1)^n}{n\pi} = \frac{-2}{n\pi} n \rightarrow \text{ even}$$

$$= \frac{2}{n} \quad n \rightarrow \text{ odd}$$

Q2:

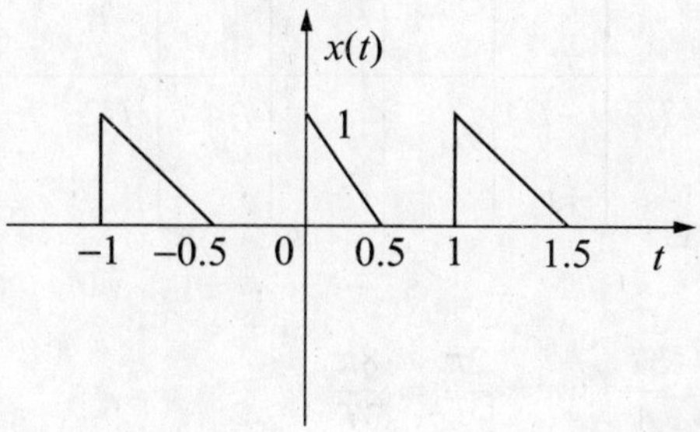

$$T = 1$$
$$w_0 = 2\pi$$
$$x(t) = -2t + 1$$
$$a_0 = \frac{1}{4}$$

$$a_n = \frac{1-(-1)^n}{n^2\pi^2}$$

$$b_n = \frac{1}{n\pi}$$

Q3:

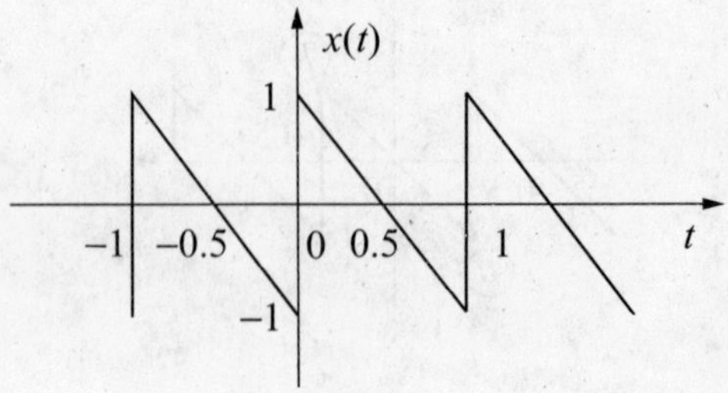

$$T = 1$$

$$w_0 = 2\pi \text{ rad/ sec}$$

$$y - y_1 = \frac{y_2 - y_1}{x_2 - x_1}(x - x_1)$$

$$x(t) = -2t + 1$$

$$a_n = \frac{2}{T}\int_{t_0}^{t_0+T} x(t)\cos n w_0 t \, dt$$

$$a_n = 0$$

Q4:

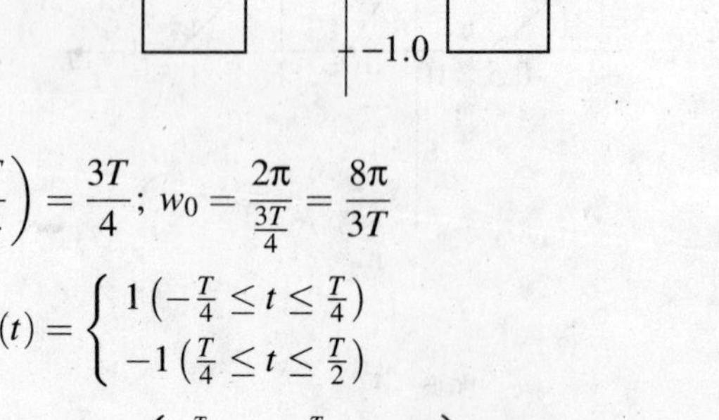

$$\frac{T}{2} - \left(-\frac{T}{4}\right) = \frac{3T}{4};\ w_0 = \frac{2\pi}{\frac{3T}{4}} = \frac{8\pi}{3T}$$

$$x(t) = \begin{cases} 1\left(-\frac{T}{4} \le t \le \frac{T}{4}\right) \\ -1\left(\frac{T}{4} \le t \le \frac{T}{2}\right) \end{cases}$$

$$a_0 = \frac{1}{\frac{3T}{4}}\left\{\int_{-\frac{T}{4}}^{\frac{T}{4}} dt + \int_{\frac{T}{4}}^{\frac{T}{2}} (-1)dt\right\} = \frac{4}{3T}\frac{T}{4} = \frac{1}{3}$$

$$a_n = \frac{8}{3T}\left\{\int_{-\frac{T}{4}}^{\frac{T}{4}} \cos\frac{8n\pi}{3T}dt - \int_{\frac{T}{4}}^{\frac{T}{2}} \cos\frac{8n\pi}{3T}t\,dt\right\}$$

$$a_n = \frac{1}{n\pi}\left[3\sin\frac{2n\pi}{3} - \sin\frac{4n\pi}{3}\right]$$

$b_n = 0$, since even function

$$x(t) = \frac{1}{3} + \frac{1}{\pi}\left[3\sin\frac{2\pi}{3} - \sin\frac{4\pi}{3} + \frac{3}{2}\sin\frac{4\pi}{3} - \frac{1}{2}\sin\frac{8\pi}{3} + \cdots\right]$$

Q5:

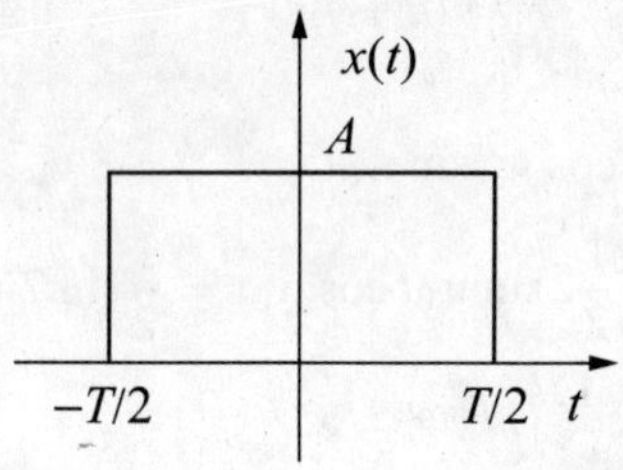

$$x(t) = \begin{cases} A - \frac{T}{2} \le t \le \frac{T}{2} \\ 0 \quad \text{elsewhere} \end{cases}$$

$$X(jw) = A\int_{-\frac{T}{2}}^{\frac{T}{2}} e^{-jwt}dt = \frac{2A}{w}\sin\frac{wT}{2} = \frac{AT}{\frac{wT}{2}}\sin\frac{wT}{2}$$

$$X(if) = AT\,\text{sinc}\,fT$$

Q6:

$$T_0 = 2\pi;$$

$$w_0 = 1;$$

$$a_0 = \frac{1}{2\pi}\int_{-\pi}^{\pi}(t - t^2)\,dt = \frac{-\pi^2}{3}$$

$$a_n = \frac{1}{\pi}\int_{-\pi}^{\pi}(t - t^2)\cos nt\,dt = \frac{-4(-1)^n}{n^2}$$

$$b_n = \frac{1}{\pi}\int_{-\pi}^{\pi}(t - t^2)\sin nt\,dt = \frac{-2(-1)^n}{n}$$

Q7: See Example 10

Q8: See Example 14 for part (a), (b), (c) and (d).

(e) $x(t) = t\exp(-at)u(t)$

Let $x_1(t) = e^{-at}u(t)$

So $X_1(jw) = \frac{1}{a+jw}$

By using the property for differentiation in frequency domain

$$\text{FT}(tx_1(t)) = j\frac{d}{dw}\left(\frac{1}{a+jw}\right) = \frac{j\times -j}{(a+jw)^2}$$

$$\text{FT}\left[te^{-at}u(t)\right] = \frac{1}{(a+jw)^2}$$

(f)

$$x(t) = \cos w_0 \sin w_0 t$$

$$x(t) = \frac{1}{2}2\sin w_0 t\cos w_0 t = \frac{1}{2}\sin 2w_0 t$$

$$x(t) = \frac{1}{4j}\left(e^{j2w_0t} - e^{-j2w_0t}\right)$$

$$X(jw) = \frac{1}{4j}\left(\text{FT}(e^{j2w_0t}) - \text{FT}(e^{-j2w_0t})\right)$$

$$X(jw) = \frac{1}{4j}[2\pi\,\delta(w-2w_0) - 2\pi\delta(w+2w_0)]$$

$$X(jw) = \frac{-j}{2}[\pi\delta(w-2w_0) - \pi\delta(w+2w_0)]$$

Q9: See Example 22

Q10:

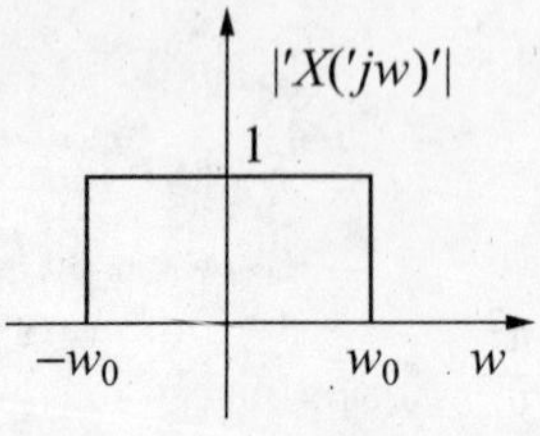

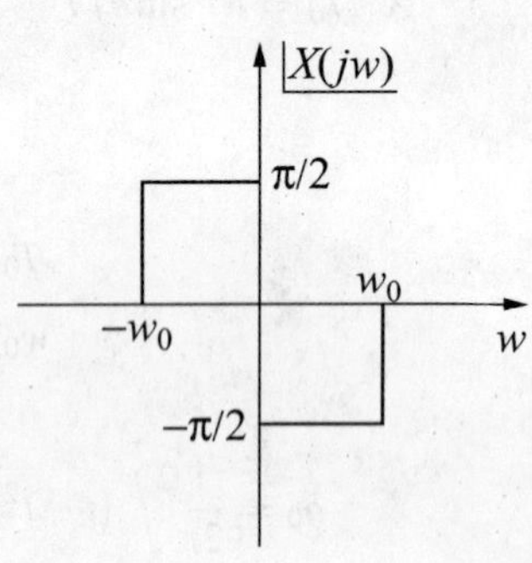

$$|X(jw)| = \begin{cases} 1 & -w_0 < w < w_0 \\ 0 & \text{elsewhere} \end{cases}$$

$$X(jw) = \begin{cases} -\frac{\pi}{2} & 0 < w < w_0 \\ \frac{\pi}{2} & -w_0 < w < 0 \end{cases}$$

$$X_1(jw) = |X_1(jw)|e^{jx_1(jw)} = e^{-j\pi/2} = -j \quad 0 < w < w_0$$

$$X_2(jw) = |X_2(jw)|e^{jX_2(jw)} = e^{j\pi/2} = j \quad -w_0 < w < 0$$

Taking inverse Fourier transform

$$x_1(t) = \frac{1}{2\pi}\int_0^{w_0} -j\,e^{jwt}dw = \frac{1-e^{jw_0t}}{2\pi t}$$

$$x_2(t) = \frac{1}{2\pi}\int_{-w_0}^{0} j\,e^{jwt}dw = \frac{1-e^{-jw_0t}}{2\pi t}$$

$$x(t) = x_1(t)+x_2(t) = \frac{1}{2\pi t}(1-e^{jw_0t}+1-e^{-jw_0t})$$

$$= \frac{1}{2\pi t}(2-2\cos w_0t) = \frac{2\sin^2\frac{w_0t}{2}}{\pi t}$$

Q11:

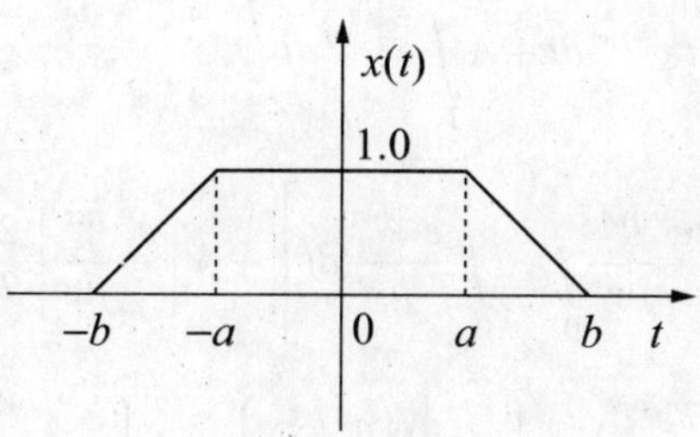

$$x(t) = \begin{cases} \frac{t+b}{b-a} & \text{for} -b<t<-a \\ 1 & \text{for} -a<t<a \\ \frac{t-b}{a-b} & \text{for} \quad a<t<b \end{cases}$$

$$X(jw) = \frac{2}{w^2(b-a)}(\cos wa - \cos wb)$$

Q12:

$$x(t) = tu(t)^*[u(t)-u(t-1)]$$

$$x_1(t) = tu(t) \qquad x_2(t) = u(t)-u(t-1)$$

Differentiating in frequency domain property

$$\text{FT}(tx(t)) = j\frac{d}{dw}X(jw)$$

$$X_1(jw) = \frac{1}{(jw)^2}$$

$$X_2(jw) = \int_0^1 1.e^{-jwt}dt = \frac{1}{jw}(1-e^{-jw})$$

$$X(jw) = X_1(jw)X_2(jw) = \frac{1}{(jw)^3}(1-e^{-jw})$$

Q13: Prove convolution in time domain property.

Q14:

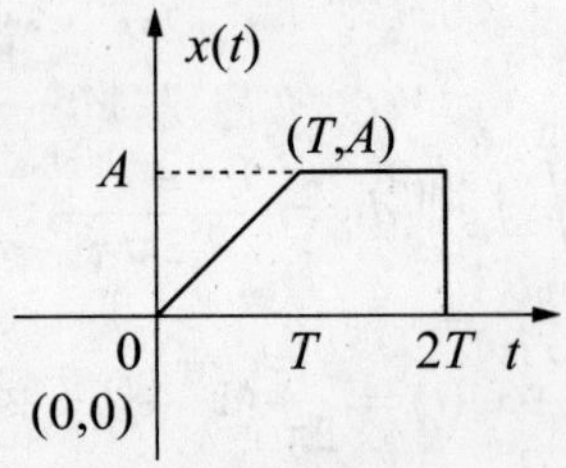

$$x(t) = \begin{cases} \frac{A}{T}t & 0 < t < T \\ A & T < t < 2T \end{cases}$$

$$X(jw) = \frac{A}{T}\int_0^T te^{-jwt}dt + A\int_T^{2T} e^{-jwt}dt$$

$$X(jw) = \frac{A}{T}\left[\frac{te^{-jwt}}{-jw}\Big|_0^T - \int_0^T \frac{e^{-jwt}}{jw}dt\right] + A\left[\frac{e^{-jwt}}{-jw}\Big|_T^{2T}\right]$$

$$= \frac{A}{T}\left\{\frac{Te^{jwt}}{-jw} + \frac{1}{w^2}\left(e^{-jwT}-1\right)\right\} + A\left\{\frac{e^{-j2wT}-e^{-jwT}}{-jw}\right\}$$

$$= A\left\{\frac{e^{-jwt}}{-jw} + \frac{1}{w^2T}\left(e^{-jwT}-1\right)\right\} - \frac{A}{jw}e^{-jwT}\left(e^{-jwT}-1\right)$$

$$= \frac{Ae^{-jwT}}{jw} + \frac{A}{w^2T}e^{-jwT} - \frac{A}{w^2T} - \frac{A}{jw}e^{-j2wT} + \frac{A}{jw}e^{-jwT}$$

$$= \frac{A}{wT}\left(\frac{1}{w}e^{-jwT} - \frac{1}{w} + jTe^{-2jwT}\right)$$

Q15:

$$\frac{d^2y(t)}{dt^2} + 4\frac{dy(t)}{dt} + 3y(t) = \frac{dx(t)}{dt} + 2x(t) \tag{1}$$

Taking Fourier transform on both sides

$$(jw)^2Y(jw) + 4(jw)Y(jw) + 3Y(jw) = (jw)X(jw) + 2X(jw)$$

$$\left((jw)^2 + 4(jw) + 3\right)Y(jw) = \left((jw)+2\right)X(jw) \tag{2}$$

Frequency response $H(jw) = \frac{Y(jw)}{X(jw)} = \frac{2+jw}{(jw)^2+4jw+3}$ (3)

$$H(jw) = \frac{2+jw}{(3+jw)(1+jw)} = \frac{A}{3+jw} + \frac{B}{1+jw}$$

$$A + 3B = 2 \quad (1)$$
$$A + B = 2 \quad (2)$$
$$1 - B + 3B = 2$$
$$1 + 2B = 2$$
$$B = \frac{1}{2}$$
$$A = 1 - \frac{1}{2}$$
$$= \frac{1}{2}$$
$$x(jw) = \frac{\frac{1}{2}}{3 + jw} + \frac{\frac{1}{2}}{1 + jw}$$

Taking inverse Fourier transform, we get

$$\text{Impulse response } h(t) = \frac{1}{2}\left(e^{-3t} + e^{-t}\right)u(t) \quad (4)$$

Q16: See Example 21.

Q17:

$$g(t) = g_r(t) + jg_i(t) \quad (1)$$

Complex conjugate of $g(t)$

$$g^*(t) = g_r(t) - jg_i(t) \quad (2)$$
$$g_r(t) = \frac{1}{2}[g(t) + g^*(t)] \quad (3)$$
$$g_i(t) = \frac{1}{2}[g(t) - g^*(t)] \quad (4)$$

Fourier transform of $g_r(t) = G_r(if)$

$$G_r(jf) = \frac{1}{2}[G(jf) - G^*(-jf)] \quad (5)$$

Fourier transform of $g_i(t) = G_i(jf)$

$$G_i(jf) = \frac{1}{2}[G(jf) - G^*(-jf)] \quad (6)$$

Q18:

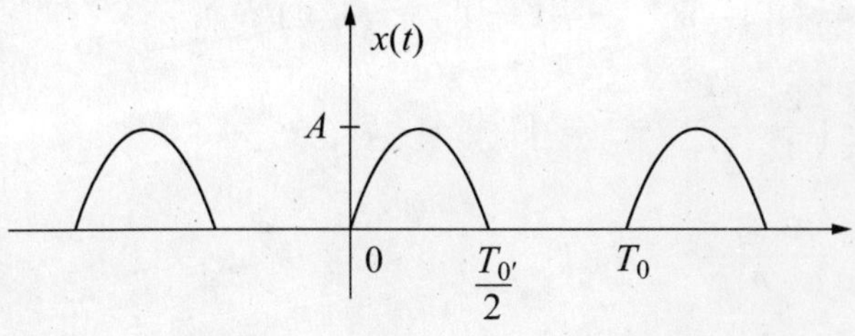

$$x(t) = A\sin w_0 t \text{ for } 0 \le t \le \frac{T_0}{2}$$

$$= 0 \qquad \text{for } \frac{T_0}{2} \le t \le T_0$$

$$C_0 = \frac{1}{T_0}\int_0^{\frac{T_0}{2}} A\sin w_0 t\,dt = \frac{A}{T_0}\left(\frac{-\cos w_0 t}{w_0}\Bigg|_0^{\frac{T_0}{2}}\right)$$

$$= -\frac{A}{T_0\cdot\frac{2\pi}{T_0}}\left[\cos w_0\cdot\frac{T_0}{2} - 1\right] = -\frac{A}{2\pi}[\cos\pi - 1] = \frac{A}{2}$$

$$C_n = \frac{1}{T_0}\int_0^{\frac{T_0}{2}} A\sin w_0 t e^{-jnw_0 t}dt$$

$$= \frac{A}{2jT_0}\int_0^{\frac{T_0}{2}}(e^{jw_0 t} - e^{-jnw_0 t})e^{-jnw_0 t}dt$$

$$= \frac{A}{2jT_0}\int_0^{\frac{T_0}{2}}\left(e^{jw_0 t(1-n)} - e^{-jw_0 t(n+1)}\right)dt$$

$$= \frac{A}{2jT_0}\left(\frac{e^{jw_0 t(1-n)}}{jw_0(1-n)} - \frac{e^{-jw_0 t(n+1)}}{-jw_0(n+1)}\Bigg|_0^{\frac{T_0}{2}}\right)$$

$$= \frac{A}{2jT_0 w_0}\left(\frac{e^{jw_0(1-n)\frac{T_0}{2}}}{1-n} + \frac{e^{-jw_0(n+1)\frac{T_0}{2}}}{(n+1)} - \frac{1}{1-n} - \frac{1}{n+1}\right)$$

$$= -\frac{A}{4\pi}\left[\frac{e^{j\pi(1-n)}}{1-n} + \frac{e^{-j\pi(n+1)}}{n+1} - \frac{1}{1-n} - \frac{1}{n+1}\right]$$

$$= -\frac{A}{4\pi}\left(\frac{e^{j\pi}e^{-jn\pi}}{1-n} + \frac{e^{-jn\pi}\cdot e^{-j\pi}}{n+1} - \frac{1}{1-n} - \frac{1}{n+1}\right)$$

$$= -\frac{A}{4\pi}\left(\frac{-e^{-jn\pi}}{1-n} - \frac{e^{-jn\pi}}{n+1} - \frac{1}{1-n} - \frac{1}{n+1}\right) \qquad \text{Since } e^{j\pi} = -1$$

$$= \frac{A}{4\pi}\left(\frac{2e^{-jn\pi}}{1-n^2} + \frac{2}{1-n^2}\right)$$

$$= \frac{A}{2\pi(1-n^2)}(e^{-jn\pi} + 1)$$

Q19:

$$x(t) = \frac{1}{2}\left(\delta(t+1) + \delta(t-1) + \delta\left(t+\frac{1}{2}\right) + \delta\left(t-\frac{1}{2}\right)\right)$$

Taking Fourier transform on both sides

$$X(jw) = \int_{-\infty}^{\infty} x(t)e^{-jwt}dt \qquad (1)$$

$$X(jw) = \int_{-\infty}^{\infty} \frac{1}{2}\left(\delta(t+1) + \delta(t-1) + \delta\left(t+\frac{1}{2}\right) + \delta\left(t-\frac{1}{2}\right)\right)e^{-jwt}dt$$

$$X(jw) = \frac{1}{2}\left(\int_{-\infty}^{\infty} \delta(t+1)e^{-jwt}dt + \int_{-\infty}^{\infty} \delta(t-1)e^{-jwt}dt + \int_{-\infty}^{\infty} \delta\left(t+\frac{1}{2}\right)e^{-jwt}dt\right.$$

$$\left. + \int_{-\infty}^{\infty} \delta\left(t-\frac{1}{2}\right)e^{-jwt}dt\right)$$

Since $\mathrm{FT}(\delta(t)) = 1$

So $\mathrm{FT}(\delta(t \pm t_0)) = e^{\pm jwt_0}dt$ {using time shifting property}

$$X(jw) = \frac{1}{2}\left(e^{jw} + e^{-jw} + e^{j\frac{w}{2}} + e^{-j\frac{w}{2}}\right)$$

$$X(jw) = \frac{e^{jw} + e^{-jw}}{2} + \frac{e^{j\frac{w}{2}} + e^{-j\frac{w}{2}}}{2}$$

$$X(jw) = \cos w + \cos\frac{w}{2}$$

OBJECTIVE TYPE QUESTIONS

Q1: If the Fourier transform of a function $x(t)$ is $X(jw)$, then $X(jw)$ is defined as

(a) $\int_{-\infty}^{\infty} x(t)e^{jwt}dt$ (b) $\int_{-\infty}^{\infty} \frac{dx(t)}{dt}e^{-jwt}dt$
(c) $\int_{-\infty}^{\infty} x(t)dt$ (d) $\int_{-\infty}^{\infty} x(t)e^{-jwt}dt$

Q2: If $X(jw)$ be the Fourier transform of $x(t)$, then

(a) $x(t) = \frac{1}{2\pi}\int_{-\infty}^{\infty} X(jw)e^{jwt}dw$ (b) $x(t) = \frac{1}{2\pi}\int_{-\infty}^{\infty} X(jw)e^{-jwt}dw$
(c) $x(t) = \frac{1}{2\pi}\int_{0}^{\infty} X(jw)e^{jwt}dw$ (d) $x(t) = \frac{1}{2\pi}\int_{-\infty}^{\infty} X(jw)e^{-jwt}dw$

Q3: Fourier transform of $x(t) = 1$ is

(a) $2\pi\,\delta(w)$ (b) $\pi\,\delta(w)$ (c) $3\pi\,\delta(w)$ (d) $4\pi\,\delta(w)$

Q4: Fourier transform of $x(t - t_0)$ is

(a) $e^{-jwt}{}_0X(jw)$ (b) $e^{jwt}{}_0X(jw)$ (c) $\frac{1}{t_0}X(jw)$ (d) $t_0e^{-jwt}{}_0X(jw)$

Q5: Fourier transform of $\left[\frac{d^n}{dt^n}x(t)\right]$ is
(a) $\frac{1}{(jw)^n}X(jw)$ (b) $(jw)^n X(jw)$ (c) $wX(jw)$ (d) All of these

Q6: If $x(t)$ is a real and odd function, then $X(jw)$ will be
(a) real and even function of w (b) imaginary function of w
(c) imaginary and odd function of w (d) real and odd function of w

Q7: The Fourier transform of $x(t) = \cos\ wt$ is
(a) $X(jw) = \pi[\delta(w-w_0)+\delta(w+w_0)]$ (b) $X(jw) = \frac{1}{\pi}[\delta(w-w_0)$
(c) $X(if) = \frac{1}{2\pi}[\delta(f-f_0)+\delta(f-f_0)]$ (d) $X(if) = 2[(\delta(f-f_0)+\delta(f+f_0)]$

Q8: The Fourier transform of $\frac{dx(t)}{dt}$ is
(a) $jwX(jw)$ (b) $(jw)^2X(jw)$ (c) $j2\pi wX(jw)$ (d) $\frac{1}{jw}X(jw)$

Q9: If $X(jw)$ be the F.T. of $x(t)$, then
(a) $X(0) = \int_{-\infty}^{\infty} x(t)dt$ (b) $X(0) = \int_{0}^{\infty} x(t)dt$ (c) None of these

Q10: Fourier transform of $x(t) = e^{-a(t)}$ is
(a) $\frac{w}{a^2+w^2}$ (b) $\frac{2w}{a^2+w^2}$ (c) $\frac{a}{a^2+w^2}$ (d) $\frac{2a}{a^2+w^2}$

Q11: If FT $[x(t)] = X(jw)$ and $x(t)$ is real, then $\int_{-\infty}^{\infty} x^2(t)$ is
(a) $\int_0^{\infty} |X(jw)|^2 dw$ (b) $\int_{-\infty}^{\infty} |X(jw)| dw$ (c) $\frac{1}{2\pi}\int_{-\infty}^{\infty} |X(jw)|^2 dw$ (d) $\frac{1}{2\pi}\int_0^{\infty} |X(jw)|^2 dw$

Q12: The Fourier transform of [t multiplied exponential function] is
(a) $\frac{1}{a+jw}$ (*b*) $\frac{1}{(a+jw)^2}$ (c) $\frac{1}{(a+jw)^3}$ (d) $\frac{1}{(a+jw)^4}$

Q13: The Fourier transform of a constant 'K' is
(a) $K\delta(w)$ (b) $2\pi\ K\delta(w)$ (c) $\frac{K}{2\pi}\delta(w)$ (d) $\frac{1}{2\pi K}\delta(w)$

Q14: The inverse Fourier transform of $\frac{\sin w}{w}$ is
(a) rectangular pulse (b) triangular wave
(c) ramp function (d) unit step function

Q15: For existence of Fourier transform of $x(t)$ the necessary and sufficiency condition is
(a) $\int_{-\infty}^{\infty} |X(t)| dt < \infty$ (b) $\int_{-\infty}^{\infty} x(t)dt < \infty$ (c) $\int_{-\infty}^{\infty} x(t)dt = 0$ (d) $\int_{-\infty}^{\infty} x(t)dt < \infty$

Q16: The Fourier series expansion of odd function with half wave symmetry will have only odd harmonies
(a) True (b) False

Q17: For expansion in Fourier series, the function $x(t)$ must be/have
(a) single value function of t (b) only finite number of discontinuities
(c) finite number of maxima and minima (d) $\int_{t_1}^{t_2} |f(t)| d < 0$
(e) All of these

Q18: The trigonometric Fourier series of an even function of time does not contain
(a) DC term (b) odd harmonic term
(c) sine term (d) cosine term

Q19: The trigonometric Fourier series of a periodic time function have

(a) sine terms (b) cosine term

(c) both (a) and (b) (d) DC term

Q20: Fourier series is defined as

$$x(t) = a_\circ + \sum_{n=1}^{\infty} (a_n \cos nw_0 t + b_n \sin w_0 t)$$

(a) True (b) False

Answers:

(1) d	(2) a	(3) a	(4) a	(5) b
(6) c	(7) a	(8) a	(9) a	(10) d
(11) c	(12) b	(13) b	(14) a	(15) a
(16) a	(17) e	(18) c	(19) c	(20) a

UNSOLVED PROBLEMS

Q1: Show that the Fourier transform of $x(t) = \delta(t+2) + \delta(t) + \delta(t-2)$ is $(1 + 2\cos 2w)$.

Q2: Show that the inverse Fourier transform of $X(jw) = 2\pi\delta(w) + \pi\delta(w-4\pi) + \pi\delta(w+4\pi)$ is $x(t) = 1 + \cos 4\pi t$.

Q3: Calculate the Fourier transform of $te^{-|t|}$, using the F.T. pair, FT $\left[e^{-|t|}\right] = \frac{2}{1+w^2}$. Also find the Fourier transform of $\frac{4t}{(1+t^2)^2}$ using duality property.

Q4: $X(jw) = \delta(w) + \delta(w-\pi) + \delta(w-5)$; find IFT $x(t)$ and show that $x(t)$ is non-periodic.

Q5: Find the Fourier transform of the triangular pulse as shown in figure.

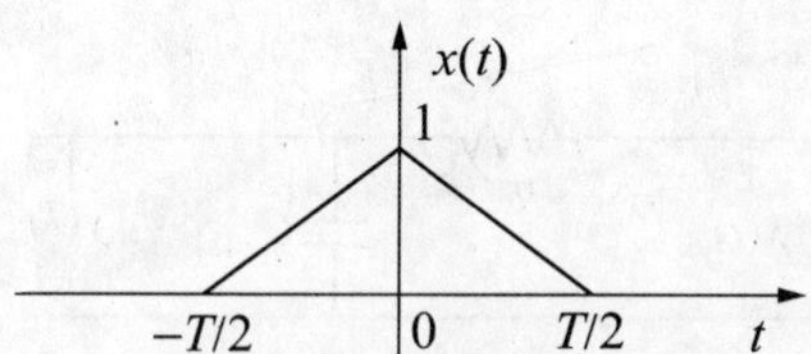

Fig. 2.10 P.

Ans. $X(jw) = \frac{T}{2} \sin c^2\left(\frac{wt}{4}\right)$

Q6: Find the Fourier transform of $x(t) = \text{rect}(t/2)$. Ans. $X(jw) = 2 \sin cw$

Q7: Find the Fourier transform of the signal $x(t) = \cos w_0 t$ by using the frequency shifting property.

Ans: $X(jw) = \pi[\delta(w - w_0) + \delta(w + w_0)]$

Q8: Show that FT $[\sin w_0 t u(t)] = \frac{w_0}{w_0^2 - w^2} + \frac{\pi j}{2}[\delta(w+w_0) - \delta(w-w_0)]$.

Q9: Find inverse Fourier transform of $X(jw) = \frac{jw}{(1+jw)^2}$

Ans. $x(t) = \frac{d}{dt}\left[te^{-t}u(t)\right]$

Q10: Sketch and then find the Fourier transform of following signals

(a) $x_1(t) = \pi\left(t+\frac{3}{2}\right) + \pi\left(t-\frac{3}{2}\right)$ Ans. (a) $X_1(jw) = 2\sin c\frac{w}{2}\cos 3\frac{w}{2}$

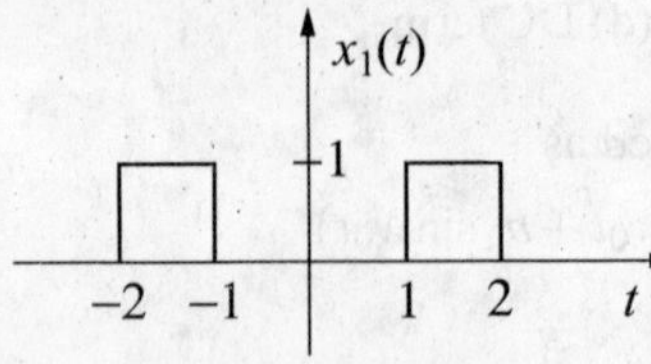

Fig. 2.11 P.

(b) $x_2(t) = \pi\left(\frac{t}{4}\right) + \pi\left(\frac{t}{2}\right)$ Ans. (b) $X_2(jw) = 4\sin c2w + 2\sin cw$

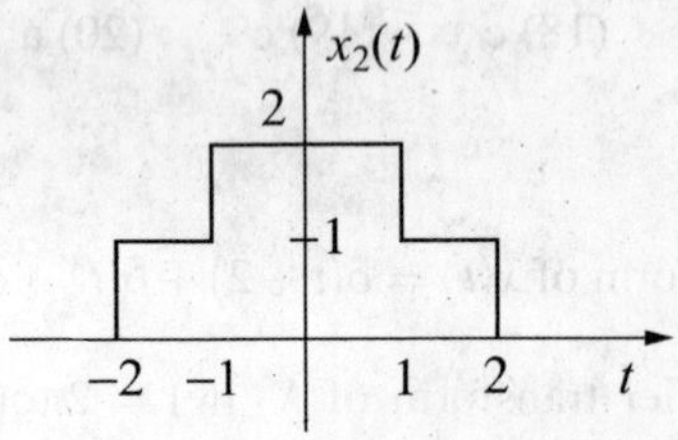

Fig. 2.12 P.

Q11: Find the frequency response $x(jw)$ of the *RC* circuit shown in figure. Plot the magnitude and phase response for $RC = 1$

$$x(jw) = \frac{y(jw)}{x(jw)} = \frac{1}{1 + jwRC}$$

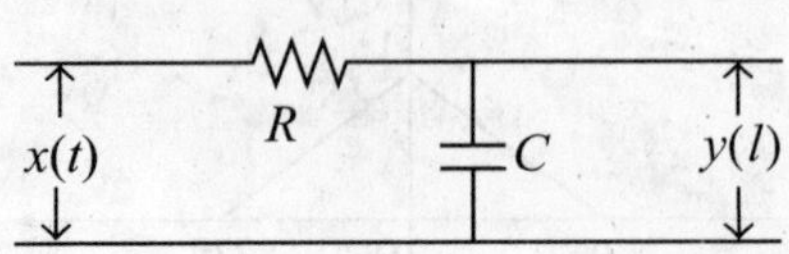

Fig. 2.13 P.

Ans. $|x(jw)| = \dfrac{1}{\sqrt{1+w^2}}$

$x(jw) = -\tan^{-1} w$

Q12: Find the Fourier series of the waveform shown in figure.

$$x(t) = \frac{2A}{jn\pi} \text{ for } n = 1,3,5,7$$

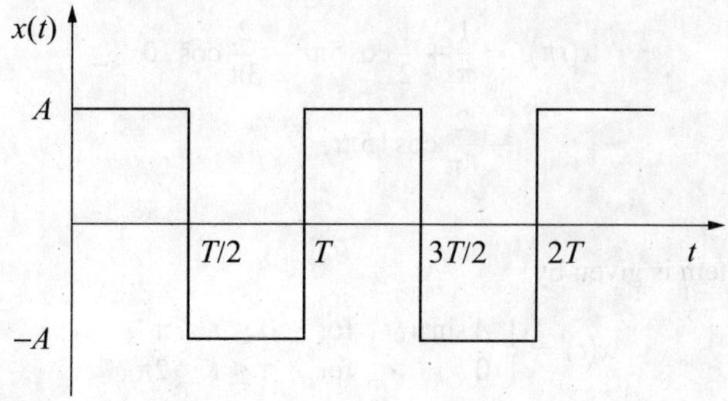

Fig. 2.14 P.

$$\text{Ans. } x(t) = \frac{4A}{\pi}\left(\sin w_0 t + \frac{1}{3}\sin 3w_0 t + \frac{1}{5}\sin 5w_0 t + \ldots\right)$$

Q13: Determine the Fourier series for the wave shown in figure.

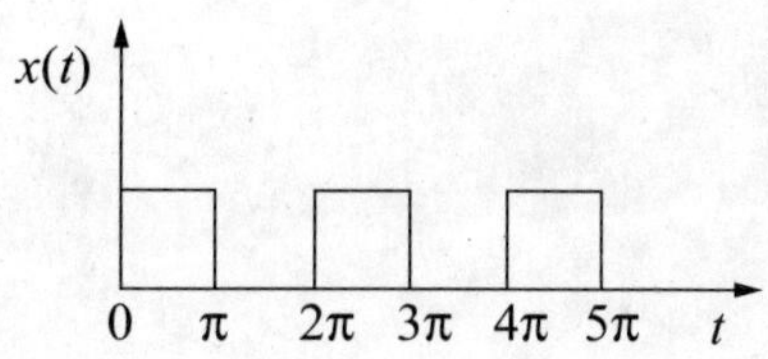

Fig. 2.15 P.

$$x(t) = \frac{1}{2} + \frac{2}{\pi}\left(\sin t + \frac{1}{3}\sin 3t + \frac{1}{5}\sin 5t + \cdots\right)$$

Q14: Find the Fourier series of voltage response shown in figure.

$$T = 0.4,\ w_0\pi = \frac{2\pi}{T} = 5$$

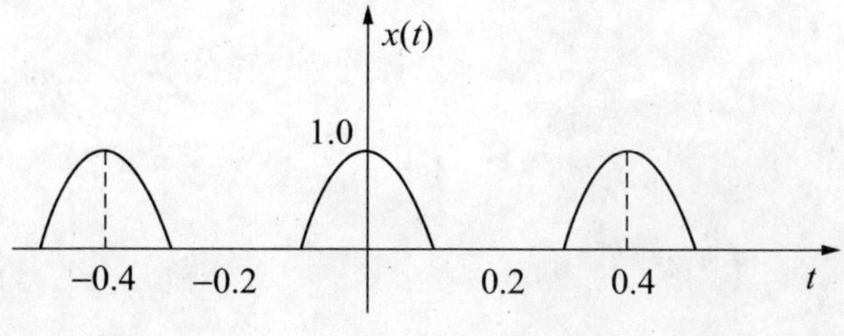

Fig. 2.16 P.

$$x(t\pi)t = \frac{1}{\pi} + \frac{1}{2}\cos 5\pi t - \frac{2}{3\pi}\cos 10$$
$$-\frac{2}{8\pi}\cos 15\pi t$$

Q15: The output of a system is given by

$$x(t) = \begin{cases} A\sin w_0 t & \text{for} \quad 0 \le t \le \pi \\ 0 & \text{for} \quad \pi \le t \le 2\pi \end{cases}$$

Determine trigonometric form of Fourier series of $x(t)$

Ans. $\left[x(t) = \frac{A}{\pi} + \frac{A}{2}\cos(nt - \frac{\pi}{2}) + \sum_{n=2}^{\infty} \frac{2A}{\pi(1-n^2)}\cos nt\right]$

3

Laplace Transform Analysis

3.1 INTRODUCTION

In the previous chapter, we have studied that the Fourier transform of a function $x(t)$ can be represented by continuous sum of exponentials function of the form e^{jwt}.

Since Fourier transform does not exist for many signals. Moreover, the Fourier transform cannot be used to analyze unstable systems.

An alternative approach is to redefine the transform and include an exponential convergence factor σ along with jw. That is, the signal $x(t)$ is represented as a sum of complex exponentials of the form e^{st}, where s is a complex frequency, denoted by $s = \sigma + jw$.

The Fourier transform of a continuous-time signal $x(t)$ can be represented as

$$X(jw) = \int_{-\infty}^{\infty} x(t)e^{-jwt}\,dt \tag{1}$$

where

$$x(t) = \frac{1}{2\pi}\int_{-\infty}^{\infty} X(jw)e^{jwt}\,dw \tag{2}$$

Replacing jw by $\sigma + jw$ in eqn. (1)

$$X(\sigma + jw) = \int_{-\infty}^{\infty} x(t)e^{-(\sigma+jw)t}\,dt$$

$$X(S) = \int_{-\infty}^{\infty} x(t)e^{-st}\,dt \tag{3}$$

3.2 LAPLACE TRANSFORM (L.T.)

Definition of L.T.:

$$X(s) = \int_{-\infty}^{\infty} x(t)e^{-st}\,dt \tag{1}$$

$x(t) \rightarrow$ continuous-time signal

$X(s) \rightarrow$ L.T. of $x(t) \Rightarrow L[x(t)]$

- F.T. represents CT signal in terms of complex sinusoidals, i.e. e^{-jwt}
- L.T. represents CT signal in terms of complex exponentials, i.e. e^{-st}
- L.T. can be applied to the analysis of unstable systems also.

Types of L.T.

Bilateral or two-sided L.T. Unilateral or one-sided L.T.

3.2.1 Bilateral Laplace transform

$s = \sigma + jw$, here the independent variable 's' is complex in nature.
$\sigma \rightarrow$ real part of 's', it is called attenuation constant.
$jw \rightarrow$ imaginary part of 's', it is called complex freq.
From eqn. (1) integration is taken from $-\infty$ to $+\infty$. Hence, it is called bilateral or double-sided L.T.

3.2.2 Unilateral Laplace transform

The ULT is applied to the signals that are causal, that is, for the signals whose magnitude is zero for time $t < 0$.

$$X(s) = \int_0^{\infty} x(t)e^{-st}dt = \int_0^{\infty} [x(t)e^{-\sigma t}]e^{-jwt}dt$$

We know that $|e^{jwt}| = 1$. Integral converges if $\int_0^{\infty} |x(t)e^{-\sigma t}|dt < \infty$

Relationship between F.T. and L.T.:

$$X(jw) = \int_{-\infty}^{\infty} x(t)e^{-jwt}dt \tag{1}$$

F.T. can be calculated only if $x(t)$ is absolutely integrable $\int_{-\infty}^{\infty} |x(t)|dt < \infty$

We know that $s = \sigma + jw$

$$X(s) = \int_{-\infty}^{\infty} x(t)e^{-(\sigma+jw)t} = \int_{-\infty}^{\infty} \{x(t)e^{-\sigma t}\}e^{-jwt}dt \tag{2}$$

Comparing (1) and (2), we find that

L.T. of $x(t)$ is basically the F.T. of $x(t)e^{-\sigma t}$, if $s = jw$, i.e. $\sigma = 0$

Then eqn. (1) becomes

$$X(s) = \int_{-\infty}^{\infty} x(t)e^{-jwt}dt = X(jw)$$

Thus, $X(s) = X(jw)$ when $s = jw$

This means L.T. is same as F.T. when $s = jw$. The above equation shows that F.T. is special case of L.T. Thus, L.T. provides broader characterization compared to F.T. $s = jw$ indicates imaginary axis in complex s-plane.

3.3 CONVERGENCE OF LAPLACE TRANSFORM (ROC) REGION OF CONVERGENCE

From eqn.(2), we know that L.T. is basically the F.T. of $\{x(t)e^{-\sigma t}\}$. Hence, if F.T. of $\{x(t)e^{-\sigma t}\}$ exists, then L.T. of $x(t)$ exists. For F.T. to exist, $x(t)e^{-\sigma t}$ must be absolutely integrable

$$\text{i.e.} \quad \int_{-\infty}^{\infty} |x(t)e^{-\sigma t}|dt < \infty$$

If this condition is satisfied, then F.T. of $x(t)e^{-\sigma t}$ will exist.

- The range of values of 'σ' for which L.T. converges is called region of convergence or ROC.

$$X(jw) = \int_{-\infty}^{\infty} x(t)e^{-jwt}\,dt$$

Replacing jw by $\sigma + jw$

$$X(\sigma + jw) = \int_{-\infty}^{\infty} x(t)e^{-(\sigma+jw)t}\,dt$$

$$x(t) = \frac{1}{2\pi}\int_{-\infty}^{\infty} X(\sigma + jw)e^{(\sigma+jw)t}\,dw$$

$$s = \sigma + jw$$

$$ds = j\,dw$$

$$dw = \frac{1}{j}ds$$

$$x(t) = \frac{1}{2\pi j}\int_{\sigma-j\infty}^{\sigma+j\infty} X(s)e^{st}\,ds$$

3.4 PROPERTIES OF REGION OF CONVERGENCE

Property 1: The ROC does not contain any poles.

Property 2: If $x(t)$ is a finite duration signal, i.e. $x(t) = 0$ except in a finite time interval $t_1 \le t \le t_2$, then ROC is entire s-plane except possibly $s = 0$ or $s = \infty$, e.g. rectangular pulse

Property 3: If $x(t)$ is right-sided signal (causal signal)

i.e. $x(t) = 0$ for $t < 0$ then ROC is $\text{Re}(s) > \alpha$ where α is some constant

e.g. $x(t) = e^{at}u(t)$ ROC: $\text{Re}(s) > a$

$x(t) = e^{-at}u(t)$ ROC: $\text{Re}(s) > -a$

Property 4: If $x(t)$ is left-sided signal (non-causal signal)

i.e. $x(t) = 0$ for $t \ge 0$ then ROC is of the form $\text{Re}(s) < \alpha$

e.g. $x(t) = e^{-at}u(-t)$ ROC: $\text{Re}(s) < a$

$x(t) = e^{-at}u(-t)$ ROC: $\text{Re}(s) < -a$

Property 5: If $x(t)$ is a two-sided signal, then ROC is of the form

$\alpha_1 < \text{Re}(s) < \alpha_2$

e.g. $x(t)e^{-a|t|}$ ROC: $-a < \text{Re}(s) < a$

3.5 PROPERTIES OF LAPLACE TRANSFORM

3.5.1 Linearity

$$\text{Let} \quad x_1(t) \overset{LT}{\longleftrightarrow} X_1(s), \quad \text{ROC: } R_1$$

$$\text{and} \quad x_2(t) \overset{LT}{\longleftrightarrow} X_2(s), \quad \text{ROC: } R_2$$

The linearity property states that:

$$L[ax_1(t) + bx_2(t)] = a\,X_1(s) + b\,X_2(s), \text{ ROC: } R_1 \cap R_2$$

Proof: Let $Z(t) = ax_1(t) + bx_2(t)$

$$Z(s) = \int_{-\infty}^{\infty} z(t)e^{-st}\,dt = \int_{-\infty}^{\infty} \{ax_1(t) + bx_2(t)\}e^{-st}\,dt$$

$$= a\,X_1(s) + b\,X_2(s) \quad \text{ROC: } R_1 \cap R_2$$

Here ROC: $R_1 \cap R_2$ indicates the intersection of R_1 and R_2

3.5.2 Time shifting

$$L[x(t - t_0)] = e^{-st}{}_0X(s) \quad \text{ROC} : R$$

$$\text{Let } Z(t) = x(t - t_0)$$

$$Z(s) = \int_{-\infty}^{\infty} x(t - t_0)e^{-st}\,dt \qquad \text{Let } t - t_0 = \tau$$

$$dt = d_\tau$$

$$= \int_{-\infty}^{\infty} x(\tau)e^{-st_0}e^{-st}\,d_\tau = e^{-st_0}X(s)$$

3.5.3 Shifting in *s*-domain (complex translation)

$$L[e^{s_0t}x(t)] = X(s - s_0), \qquad \text{ROC} : R + R_e(s_0)$$

Let $Z(t) = e^{s_0t}x(t)$

$$Z(s) = \int_{-\infty}^{\infty} e^{s_0t}x(t)e^{-st}\,dt = \int_{-\infty}^{\infty} x(t)e^{-(s-s_0)t}\,dt$$

$$= x(s - s_0) \text{ with ROC} : R + Re(s_0)$$

3.5.4 Time scaling

$$L[x(at)] = \frac{1}{|a|}X(s/a), \text{ ROC} : R/a$$

$$Z(s) = \int_{-\infty}^{\infty} x(at)e^{-st}\,dt \quad \text{Let } at = \tau \quad dt = d\tau/a$$

$$= \frac{1}{a}\int_{-\infty}^{\infty} x(\tau)e^{-st/a}\,d\tau = \frac{1}{a}X(s/a) \quad \text{ROC} : R/a$$

3.5.5 Time reversal

$$x(-t) \stackrel{L}{\longleftrightarrow} X(-s)$$

Let $Z(t) = x(-t)$

$$Z(s) = \int_{-\infty}^{\infty} x(-t)e^{-st}\,dt$$

$$-t = \tau$$

$$= \int_{-\infty}^{\infty} x(\tau)e^{s\tau}\,d\tau = \int_{-\infty}^{\infty} x(\tau)e^{-s(-\tau)}\,d\tau$$

$$= z(s) = X(-s)$$

$$L[x(-t)] = X(-s)$$

3.5.6 Differentiation in time domain

$$\frac{d}{dt}x(t) \stackrel{L}{\longleftrightarrow} sX(s), \quad \text{ROC} : R$$

Proof:

We know that $$x(t) = \frac{1}{2\pi j}\int_{\sigma-j\infty}^{\sigma+j\infty} X(s)e^{st}\,ds$$

On differentiating w.r.t. t $$\frac{d}{dt}x(t) = \frac{1}{2\pi j}\int_{\sigma-j\infty}^{\sigma+j\infty} [sX(s)]e^{st}\,ds$$

$$L\left[\frac{d}{dt}x(t)\right] = sX(s)$$

$$\frac{d^n}{dt^n}x(t) \stackrel{L}{\longleftrightarrow} s^n X(s), \quad \text{ROC containing } R$$

3.5.7 Differentiating in *s*-domain

$$\text{LT }[-tx(t)] \stackrel{L}{\longleftrightarrow} \frac{d}{ds}X(s), \quad \text{ROC} : R$$

$$X(s) = \int_{-\infty}^{\infty} x(t)e^{-st}\,dt$$

$$\frac{d}{ds}X(s) = \int_{-\infty}^{\infty} x(t)\frac{d}{ds}e^{-st}\,dt = \int_{-\infty}^{\infty} [-tx(t)]e^{-st}\,dt$$

$$-tx(t) \stackrel{L}{\longleftrightarrow} \frac{d}{ds}X(s), \quad (-t)^n\, x(t) \stackrel{L}{\longleftrightarrow} \frac{d^n}{ds^n}X(s), \quad \text{ROC} : R$$

3.5.8 Convolution in time domain

$$x_1(t) * x_2(t) \xleftrightarrow{L} X_1(s)X_2(s), \quad \text{ROC} : R$$

$$x_1(t) * x_2(t) = \int_{-\infty}^{\infty} x_1(\tau)\, x_2(t-\tau) d\tau$$

$$L(x_1(t) * x_2(t)) = \int_{-\infty}^{\infty} x_1(t) * x_2(t) e^{-st} dt$$

$$= \int_{-\infty}^{\infty}\int_{-\infty}^{\infty} x_1(\tau)\, x_2(t-\tau) e^{-st} d\tau\, dt$$

$$= \int_{-\infty}^{\infty} x_1(\tau) d\tau \int_{-\infty}^{\infty} x_2(t-\tau) e^{-st} dt$$

Let $t - \tau = \alpha$

$$= \int_{-\infty}^{\infty} x_1(\tau) d\tau \int_{-\infty}^{\infty} x_2(\alpha) e^{-s}(\alpha+\tau)\, d\alpha$$

$$= \int_{-\infty}^{\infty} x_1(\tau) e^{-st} d\tau \int_{-\infty}^{\infty} x_2(\alpha) e^{-s\alpha} d\alpha$$

$$L(x_1(t) * x_2(t)) = X_1(s)X_2(s)$$

3.5.9 Integration in time domain

$$L\left[\int_{-\infty}^{t} x(\tau) d\tau\right] = \frac{X(s)}{s}; \quad \text{ROC} : R \cap [\text{Re}(s) > 0]$$

Proof:

Let $\int_{-\infty}^{t} x(\tau) d\tau = z(t) \quad x(t) = \frac{d}{dt} z(t)$

$$X(s) = sZ(s)$$

$$Z(s) = \frac{X(s)}{s}$$

$$L\left\{\int_{-\infty}^{t} x(\tau) d\tau\right\} = \frac{X(s)}{s}$$

3.5.10 Integration in *s*-domain

$$\frac{x(t)}{t} \overset{L}{\longleftrightarrow} \int_{s}^{\infty} X(s)ds, \text{ ROC:R}$$

Proof:

$$\int_{s}^{\infty} X(s)ds = \int_{s}^{\infty}\left[\int_{-\infty}^{\infty} x(t)e^{-st}dt\right]ds = \int_{-\infty}^{\infty} x(t)\left[\int_{s}^{\infty} e^{-st}ds\right]dt$$

$$= \int_{-\infty}^{\infty} x(t)\frac{e^{-st}}{t}dt = L\left[\frac{x(t)}{t}\right]$$

Initial value theorem:

$$\lim_{t\to 0} x(t) = \lim_{s\to\infty} sX(s)$$

Final value theorem:

$$\lim_{t\to 0} x(t) = \lim_{s\to\infty} sX(s)$$

3.6 LAPLACE TRANSFORM OF A PERIODIC FUNCTION

$$F(s) = \frac{F_1(s)}{1-e^{-sT}}$$

$F_1(s)$–L.T. of the first cycle of the periodic function
T - Period

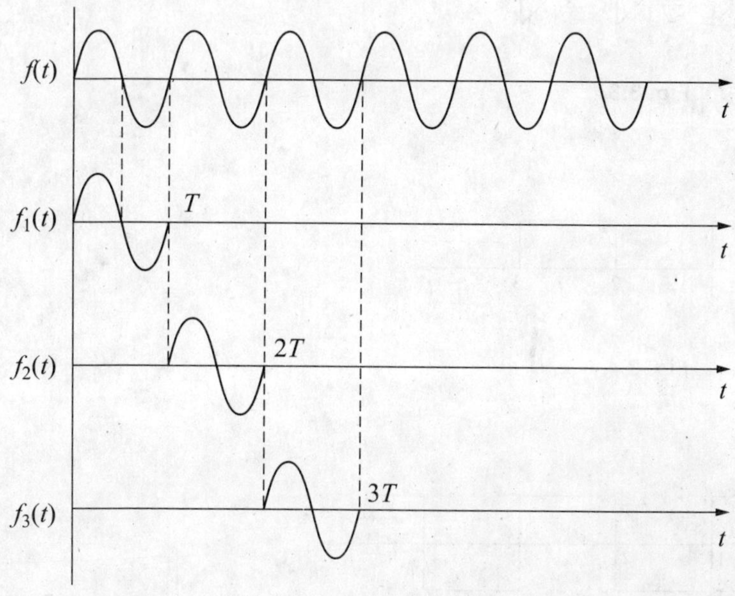

Fig. 3.1. Graphical representation of periodic waveform as time shifted pulses

$$f_2(t) = f_1(t-T)$$
$$f_3(t) = f_1(t-2T)$$
$$f_4(t) = f_1(t-3T)$$
$$f(t) = f_1(t) + f_1(t-T) + f_1(t-2T)$$
$$L[f(t)] = F_1(s) + e^{-sT}F_1(s) + e^{-2sT}F_1(s) + \cdots$$
$$= F_1(s)[1 + e^{-sT} + e^{-2sT} + e^{-3sT} + \cdots]$$
$$F(s) = \frac{F_1(s)}{1 - e^{-sT}}$$

(a) Periodic Waveform

Example 1: Determine the L.T. of following waveform

(1)

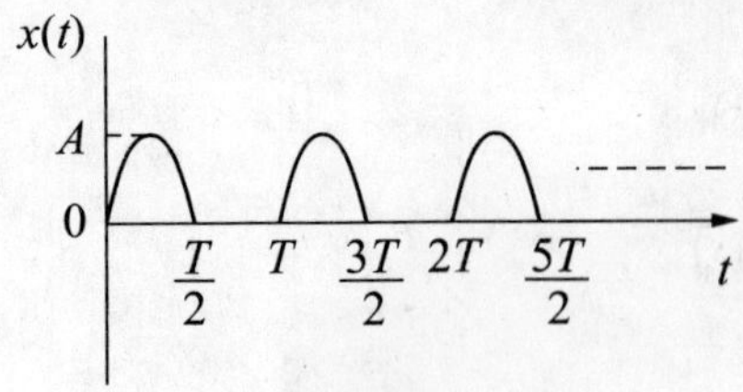

Fig. 3.2.

(2)

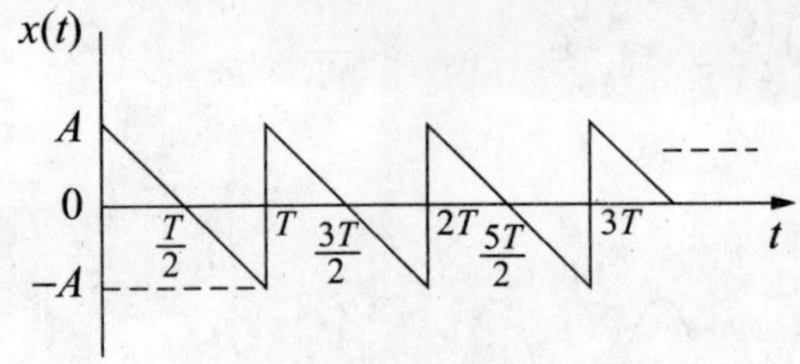

Fig. 3.3.

(3)

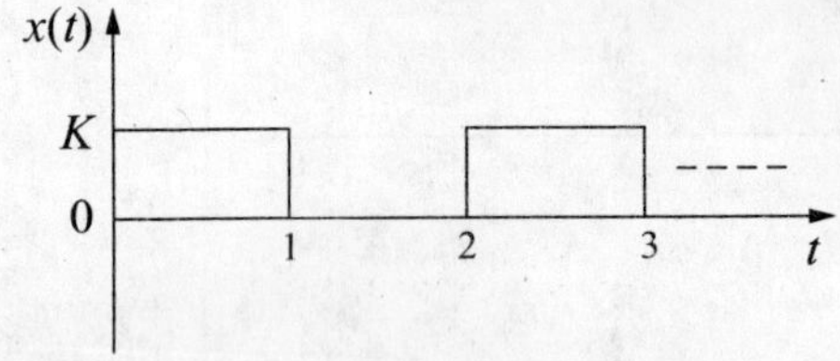

Fig. 3.4.

(4)

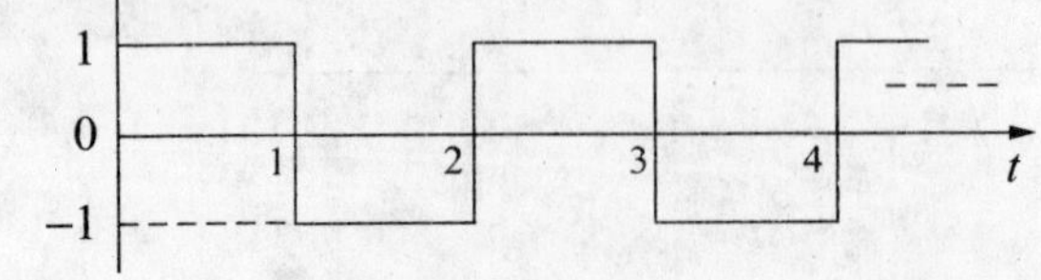

Fig. 3.5.

SOLUTION:

(1)

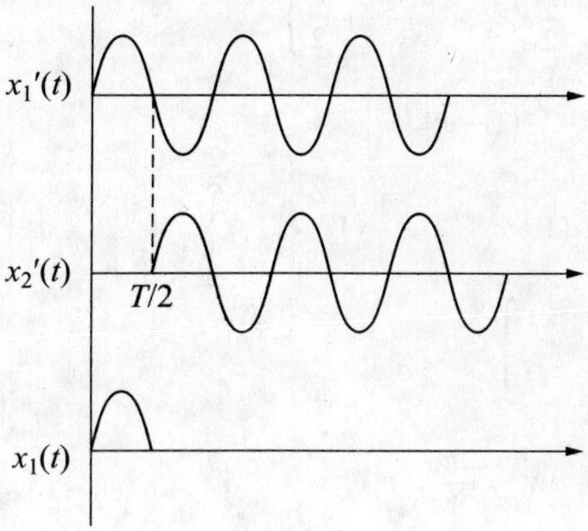

$$x_1(t) = x_1'(t) + x_2'(t)$$

$$= A\sin\frac{2\pi}{T}t + A\sin\frac{2\pi}{T}(t - T/2)$$

$$X_1(s) = \frac{A\frac{2\pi}{T}}{s^2 + (2\pi/T)^2}(1 + e^{-sT/2})$$

$$X(s) = \frac{1}{1 - e^{-sT}}X_1(s)$$

(2)

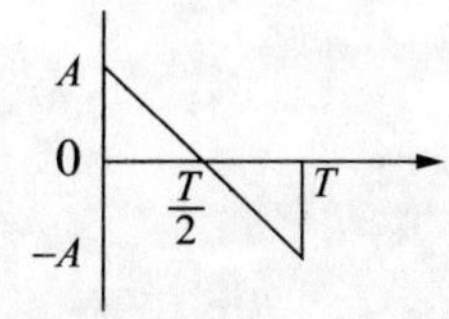

$$(0, A)(T, -A)$$

$$y - A = \frac{-2\pi}{T}(x - 0)$$

Equation of line $x(t) = \frac{-2A}{T}t + A = \frac{-2A}{T}\left(t - \frac{T}{2}\right)$

$$x_1(t) = \frac{-2A}{T}\left(t - \frac{T}{2}\right)[u(t) - u(t - T)]$$

$$= \frac{-2A}{T}\{t\,u(t) - t\,u(t - T)\} + A\{u(t) - u(t - T)\}$$

$$x^1(t) = t\,u(t - T)$$

$$x_1(t) = \frac{-2A}{T}\left(t - \frac{T}{2}\right) \text{ for } 0 < t < T$$

$$X(s) = \int_0^T \frac{-2A}{T}\left(t - \frac{T}{2}\right)e^{-sT}\,dt$$

From property of L.T.

$$L[t\,x(t)] = -\frac{d}{ds}X(s)$$

$$L[tu(t - T)] = -\frac{d}{ds}\frac{e^{-sT}}{s}$$ Since L.T. of $u(t - T) = \frac{e^{-ST}}{s}$

$$= \frac{Tse^{-sT} + e^{-sT}}{s^2}$$ using time shifting property

$$X_1(s) = -\frac{2A}{T}\left\{\frac{1}{s^2} - \frac{Tse^{-sT}}{s^2} - \frac{e^{-sT}}{s^2}\right\} + A\left\{\frac{1}{s} - \frac{e^{-sT}}{s}\right\}$$

$$= \frac{2A}{Ts}\left\{-\frac{1}{s}+Te^{-sT}+\frac{e^{-sT}}{s}\right\}+\frac{2A}{Ts}\left\{\frac{T}{2}-\frac{T}{2}e^{-sT}\right\}$$

$$= \frac{2A}{Ts}\left\{-\frac{1}{s}+Te^{-sT}-\frac{T}{2}e^{-sT}+\frac{e^{-sT}}{s}+\frac{T}{2}\right\}$$

$$= \frac{2A}{Ts}\left\{\frac{T}{2}\left(1+e^{-sT}\right)-\frac{1}{s}\left(1-e^{-sT}\right)\right\}$$

$$X(s) = \frac{1}{1-e^{-sT}}\times\frac{2A}{Ts}\left\{\frac{T}{2}\left(1+e^{-sT}\right)-\frac{1}{s}\left(1-e^{-sT}\right)\right\}$$

$$= \frac{2A}{Ts}\left\{\frac{T}{2}\frac{\left(1+e^{-sT}\right)}{\left(1-e^{-sT}\right)}-\frac{1}{s}\right\}$$

(3)

$$T = 2 \quad w_0 = \frac{2\pi}{2} = \pi$$

$$x_1(t) = K[u(t)-u(t-1)]$$

$$X_1(s) = \frac{K}{s}-\frac{Ke^{-s}}{s}$$

$$X(s) = \frac{X_1(s)}{1-e^{-sT}} = \frac{K(1-e^{-s})}{s(1-e^{-2s})}$$

OR

$$x_1(t) = \begin{cases} K \text{ for } 0<t<1 \\ 0 \text{ for } 0<t<2 \end{cases}$$

$$X_1(s) = \int_0^1 Ke^{-st}dt$$

$$= \frac{K}{s}\left[-e^{-st}\right]_0^1$$

$$= \frac{K}{s}\left[1-e^{-s}\right]$$

(4)

$$T = 2,\ w_0 = \pi$$

$$x_1(t) = u(t)-2u(t-1)+u(t-2)$$

$$X_1(s) = \frac{1}{s}-\frac{2e^{-s}}{s}+\frac{e^{-2s}}{s} = \frac{(1-2e^{-s}+e^{-2s})}{s}$$

$$x_1(t) = \begin{cases} 1 & 0<t<1 \\ -1 & 1<t<2 \end{cases}$$

$$X_1(s) = \int_0^1 e^{-st}dt - \int_1^2 e^{-st}dt = -\frac{1}{s}\left(e^{-s}-1\right)+\frac{1}{s}\left(e^{-2s}-e^{-s}\right)$$

$$= \frac{1}{s}\left\{e^{-2s} - e^{-s} - e^{-s} + 1\right\} = \left(e^{-2s} - 2e^{-s} + 1\right)/s$$

$$X(s) = \frac{(e^{-2s} - 2e^{-s} + 1)}{s(1-e^{-2s})} = \frac{(1-e^{-s})}{s(1+e^{-s})} = \frac{(1-e^{-s})}{s(1+e^{-s})}$$

$$= \frac{1}{s}\tan h\left(\frac{s}{2}\right)$$

3.7 LAPLACE TRANSFORM OF ELEMENTARY SIGNALS

Example 2: Calculate the L.T. of the following signals and plot the ROC
(1) $x(t) = e^{2t}u(t)$ (2) $x(t) = -e^{2t}u(-t)$

SOLUTION:

(1)

$$X(s) = \int_{-\infty}^{\infty} e^{2t}u(t)e^{-st}\,dt = \int_{0}^{\infty} e^{-(s-2)t}\,dt$$

$$= \lim_{t\to\infty}\left[\frac{e^{-(s-2)t}}{-(s-2)}\right] - \lim_{t\to 0}\left[\frac{e^{-(s-2)t}}{-(s-2)}\right] \quad (1)$$

e^x converges if x is negative. Hence, the 1st term in the above equation will converge if $s-2>0$ i.e.

$$\lim_{t\to\infty}\left[\frac{e^{-(s-2)t}}{-(s-2)}\right] = 0 \text{ if } s-2>0$$

$$X(s) = \frac{1}{s-2} \text{ for } s-2>0 \text{ or } s>2$$

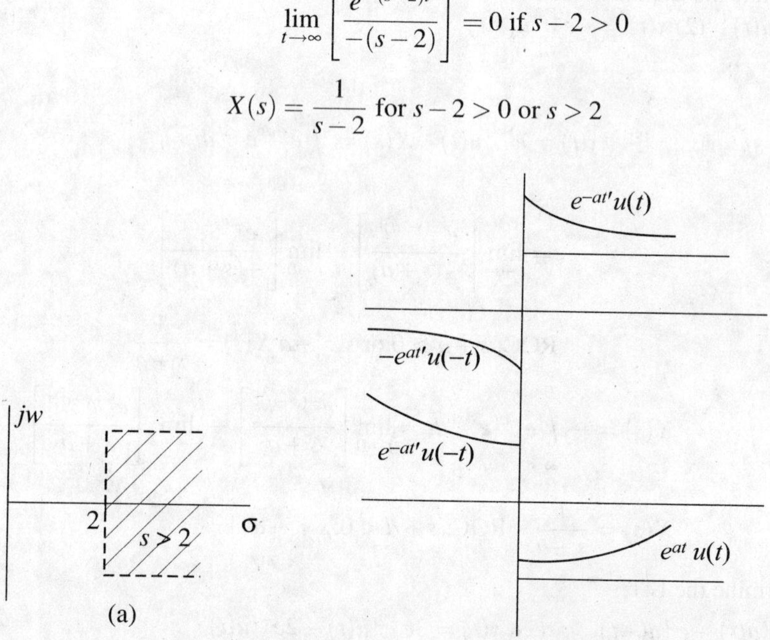

Fig. 3.6. (a) ROC of signal $e^{2t}\,u(t)$ (b) Graphical representation of various signals

(2)
$$X(s) = -\int_{-\infty}^{0} e^{2t} e^{-st} dt = -\int_{-\infty}^{0} e^{-(s-2)} dt$$

$$= \lim_{t \to 0} \left[\frac{e^{-(s-2)t}}{s-2} \right] - \lim_{t \to -\infty} \left[\frac{e^{-(s-2)t}}{s-2} \right]$$

e^x – converges if x is negative

$$s - 2 < 0, \text{ or } s < 2 \qquad X(s) = \frac{1}{s-2}$$

ROC:

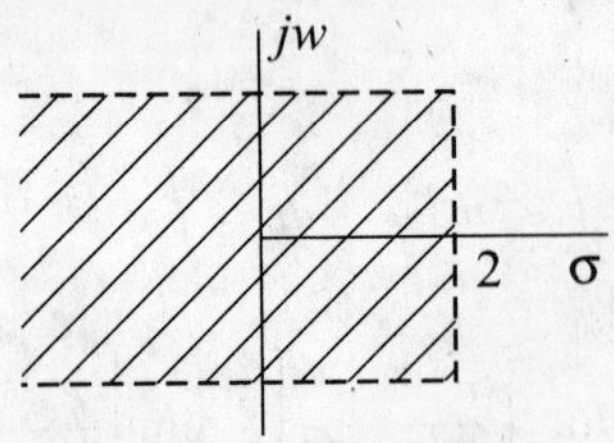

Fig. 3.7. ROC of signal $-e^{2t}u(-t)$

Example 3: Calculate the Laplace transform
(1) $x(t) = e^{-at}u(t)$ (2) $x(t) = -e^{-at}u(-t)$

SOLUTION:

(1)
$$x(t) = e^{-at}u(t) \quad X(s) = \int_{0}^{\infty} e^{-at} e^{-st} dt$$

$$= \lim_{t \to \infty} \left[\frac{e^{-(s+a)t}}{-(s+a)} \right] - \lim_{t \to 0} \left[\frac{e^{-(s+a)}}{-(s+a)} \right]$$

$$\text{ROC: } s + a > 0 \text{ or } s > -a \; X(s) = \frac{1}{s+a}$$

(2)
$$X(s) = -\int_{-\infty}^{0} e^{-at} e^{-st} dt = \lim_{t \to 0} \left[\frac{e^{-(s+a)t}}{s+a} \right] - \lim_{t \to -\infty} \left[\frac{e^{-(s+a)t}}{s+a} \right]$$

$$X(s) = \frac{1}{s+a} \qquad \text{ROC: } s + a < 0; s < -a$$

Example 4: Determine the L.T.
(1) $x(t) = e^{-2t}u(t) - e^{2t}u(-t)$ (2) $x(t) = 3e^{-2t}u(t) - 2e^{-t}u(t)$

SOLUTION:

(1)
$$X(s) = \int_{0}^{\infty} e^{-2t} e^{-st} dt + \int_{-\infty}^{0} e^{2t} e^{-st} dt$$

$$= \frac{1}{s+2}\text{ROC}: s > -2; \frac{1}{s-2}\text{ROC}: s < 2$$

$$= \frac{1}{s+2} + \frac{1}{s-2}, \text{ROC}: s > -2 \text{ and } s < 2$$

$$= \frac{2s}{s^2-4}, \text{ROC} - 2 < s < 2$$

ROC:

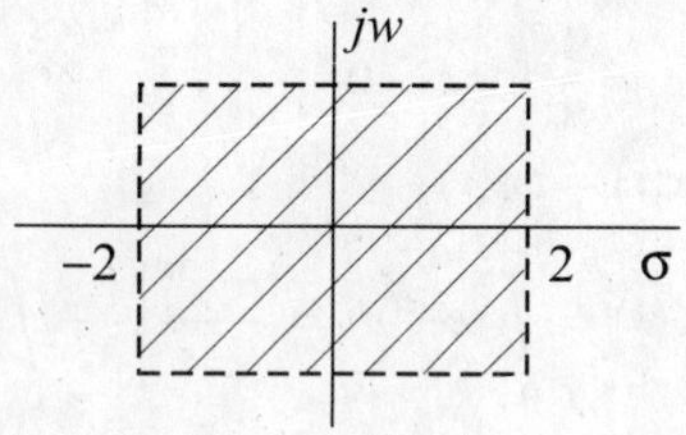

Fig. 3.8. ROC of signal $e^{-2t}u(t) - e^{2t}u(-t)$

(2) $$x(t) = 3\exp(-2t)u(t) - 2\exp(-t)u(t)$$

$$X_1(s) = \frac{3}{s+2}, \text{ROC}: s > -2$$

$$X_2(s) = \frac{2}{s+1}, \text{ROC}: s > -1$$

$$X(s) = \frac{3}{s+2} - \frac{2}{s+1}, \text{ROC}: s > -2 \text{ and } s > -1$$

ROC:

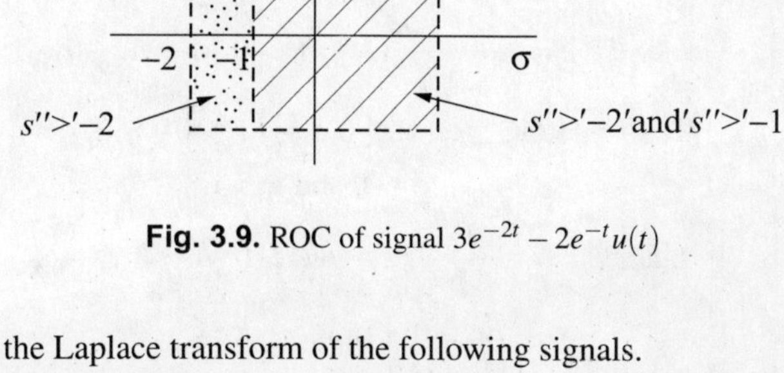

Fig. 3.9. ROC of signal $3e^{-2t} - 2e^{-t}u(t)$

Example 5: Determine the Laplace transform of the following signals.

(1) $x(t) = \delta(t)$, (2) $x(t) = u(t)$, (3) $x(t) = t\,u(t)\{r(t)\}$ (4) $x(t) = A\sin w_0\, tu(t)$,
(5) $x(t) = A\cos w_0 t\; u(t)$

SOLUTION:

(1)
$$X(s) = \int_{-\infty}^{\infty} \delta(t)e^{-st}\,dt = 1 \text{ ROC : entire } s \text{ plane}$$
$$= \int_{t=0} \delta(t)1dt$$

(2)
$$X(s) = \int_0^{\infty} e^{-st}\,dt = \lim_{t\to\infty}\left[\frac{e^{-st}}{-s}\right] - \lim_{t\to 0}\left[\frac{e^{-st}}{-s}\right]$$
$$= \frac{1}{s} \text{ for } s > 0$$

(3)
$$x(t) = t\,u(t)$$
$$X(s) = \int_0^{\infty} te^{-st}\,dt = \left.\frac{te^{-st}}{-s}\right|_0^{\infty} + \int_0^{\infty} \frac{e^{-st}}{s}dt$$
$$= \underset{(1)}{\left.\frac{te^{-st}}{-s}\right|_0^{\infty}} - \underset{(2)}{\left.\frac{1}{s^2}\,e^{-st}\right|_0^{\infty}} \qquad (1)$$

From equation (1), we can see that as we applied the limit, all terms vanish because of limit ∞, so we solve using property of Laplace transform.

$$L[-t\,x(t)] \leftrightarrow \frac{d}{ds}X(s) \text{ \{Property of Laplace transform\}}$$
$$x(t) = u(t) \qquad X(s) = \frac{1}{s}$$
$$L[tu(t)] = \frac{-d}{ds}\frac{1}{s} = -X - \frac{1}{s^2} = \frac{1}{s^2}$$

(4)
$$x(t) = A\sin w_0 t\; u(t)$$
$$\text{Since } \sin w_0 t = \frac{e^{jw_0 t} - e^{-jw_0 t}}{2j}$$
$$X(s) = \frac{A}{2j}\int_0^{\infty} \left[e^{jw_0 t} - e^{-jw_0 t}\right] e^{-st}\,dt$$
$$L[e^{-jw_0 t}] = \frac{1}{s - jw_0},\; s - jw_0 > 0;\; \sigma + jw - jw_0 > 0$$
$$\text{ROC : Re}(s) > 0\; \sigma + jw > 0 + jw_0$$
$$\sigma > 0 \text{ and } w > w_0$$
$$L[e^{-jw_0 t}] = \frac{1}{s + jw_0},\; s + jw_0 > 0$$
$$\sigma + jw > 0 - jw_0$$
$$\sigma > 0$$
$$\text{ROC : Re}(s) > 0$$

$$X(s) = \frac{A}{2j}\left[\frac{1}{s - jw_0} - \frac{1}{s + jw_0}\right] = \frac{Aw_0}{s^2 + w^2{}_0}$$

$$A \sin w_0 t\, u(t) = \frac{Aw_0}{s^2 + w^2{}_0} : \text{ROC Re}(s) > 0$$

(5)

$$x(t) = A \cos w_0 t\, d(t) = \frac{A}{2}\left[e^{jw_0 t} + e^{-jw_0 t}\right] u(t)$$

$$X(s) = X_1(s) + X_2(s)$$

$$X_1(s) = \frac{1}{s - jw_0} \text{Re}(s) > 0$$

$$X_2(s) = \frac{1}{s + jw_0} \text{Re}(s) > 0$$

$$X(s) = \frac{As}{s^2 + w_0^2} : \text{ROC Re}(s) > 0$$

Example 6: L.T. of $x(t) = e^{-at}$ sin wt a damped sine wave

$$x(t) = \frac{e^{-(a-jw)t} - e^{-(a+jw)}t}{2j}$$

$$X(s) = \frac{w}{(s+a)^2 + w^2}, \text{ ROC} : \text{Re}(s) > -a$$

Example 7: $x(t) = e^{-at} \cos wt$

$$X(s) = \frac{s+a}{(s+a)^2 + w^2}$$

Example 8: L.T. $x(t) = t\, e^{-at} u(t)$

$$L\left[e^{-at} u(t)\right] = \frac{1}{s+a} \text{ROC} : R(s) > -a$$

$$\text{LT } [-t\, x(t)] = \frac{d}{ds} X(s)$$

Using differentiation in s-domain property

$$L\left[t\, e^{-at} u(t)\right] = -\frac{d}{ds}\left\{\frac{1}{s+a}\right\} = \frac{1}{(s+a)^2}; \text{ROC} : \text{Re}(s) > -a$$

$$L\left[\frac{t^2}{2} e^{-at} u(t)\right] = \frac{1}{(s+a)^3}$$

$$L\left[\frac{t^{n-1}}{n-1!} e^{-at} u(t)\right] = \frac{1}{(s+a)^n}, \text{ ROC; Re}(s) > -a.$$

Example 9: Find the L.T. of the following waveform

(1)

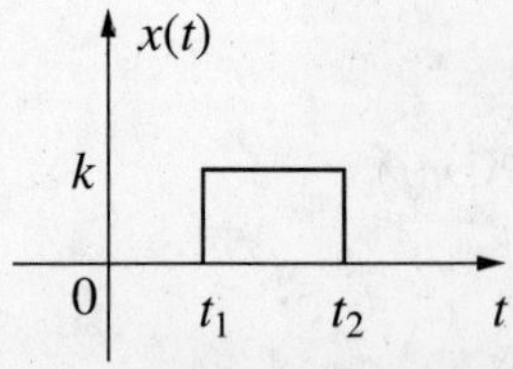

Fig. 3.10

(2)

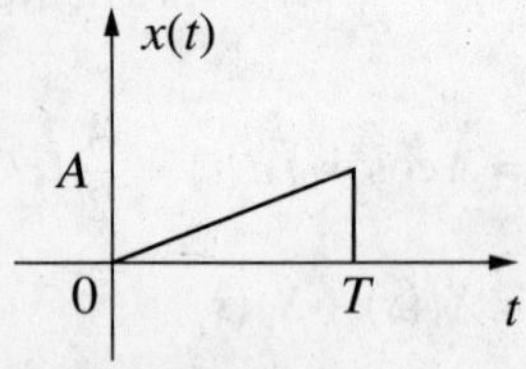

Fig. 3.11

(3)

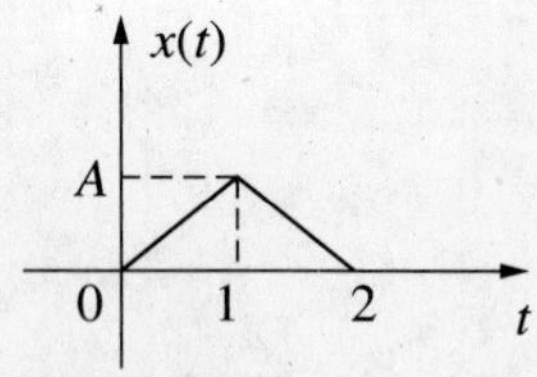

Fig. 3.12

SOLUTION:

(1)

$$x(t) = k[u(t-t_1) - u(t-t_2)]$$

$$X(s) = \left\{K \cdot \frac{1}{s}e^{-st_1} - K \cdot \frac{1}{s}e^{-st_2}\right\}$$

$$= \frac{K}{s}[e^{-st_1} - e^{-st_2}]$$

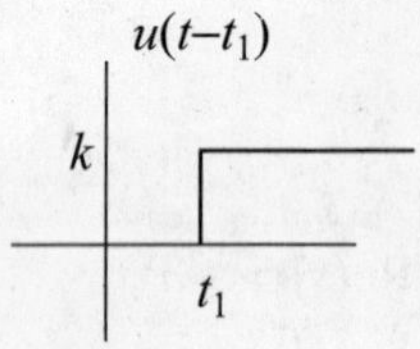

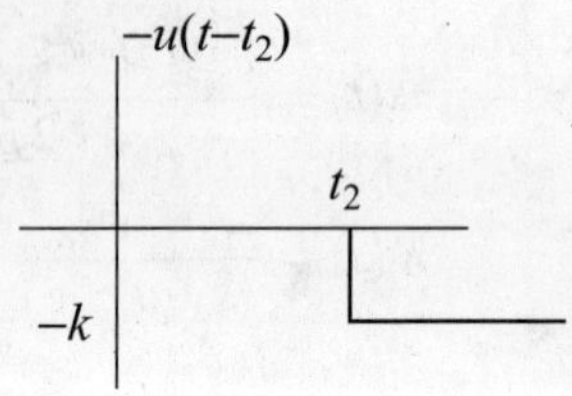

(2) $(0,0)(T,A)$; $y = \frac{A}{T}x \quad x^1(t) = \frac{At}{T}$

$$x(t) = \frac{At}{T}[u(t) - u(t-T)]$$

$$X_1(s) = L[t\,u(t)] = \frac{1}{s^2}$$

$$X_2(s) = L[t\,u(t-T)]$$

$$\uparrow$$

$$x_3(t)$$

$$L[x_3(t)] = L[u(t-T)] = \frac{e^{-sT}}{s}$$

$$[tx(t)] = -\frac{d}{ds}X(s) = -\frac{d}{ds}\frac{e^{-sT}}{s}$$

$$= \frac{Tse^{-sT} + e^{-sT}}{s^2}$$

$$X(s) = \frac{A}{T}\left\{\frac{1}{s^2} - sTe^{-sT} - e^{-sT}\right\}$$

$$= \frac{A}{S^2T}\left\{1 - e^{-sT} - sTe^{-sT}\right\}$$

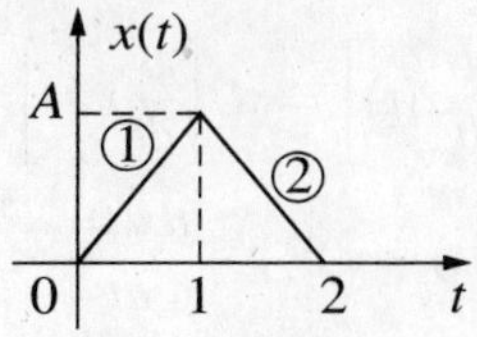

(3) $(0,0)(1,A);\ (1,A)(2,0)$

$$\text{Equation of line (1) } y = Ax \qquad y - A = \frac{-A}{1}(x-1)$$

$$y - A = -Ax + A$$

$$\text{Equation of line (2)} \qquad y = -A(t-2)$$

$$\begin{aligned} x(t) &= A\,t[u(t)-u(t-1)] + A(t-2)[u(t-1)-u(t-2)] \\ &= A\,r(t) - 2A\,r(t-1) + A\,r(t-2) \\ X(s) &= \frac{A}{s^2} - \frac{2A}{s^2}e^{-s} + \frac{A}{s^2}e^{-2s} \\ &= \frac{A}{s^2}[1 - 2e^{-s} + e^{-2s}] \end{aligned}$$

Unilateral L.T. $X(s) = \int_0^{\infty} x(t)e^{-st}\,dt$

3.8 UNILATERAL LAPLACE TRANSFORM (ULT)

Unilateral L.T. can be used for causal signals, that is, the value is zero for $t < 0$

$|h(t)| = 0 \quad t < 0$

- The lower limit is taken as 0 to indicate that initial conditions at $t = 0$ are also considered.
- Bilateral and unilateral L.T. are same for causal signals.
- Since lower limit of integration is '0' the ROC will be always RHS of s-plane, i.e. the L.T. will be always convergent for $\text{Re}(s) > 0$. Hence, ROC need not be specified with ULT.
- Most of the properties of ULT are same as BLT except differentiation property.

3.8.1 Differentiation in time domain

$$\frac{dx(t)}{dt} \xleftrightarrow{L} sX(s) - x(0)$$

$$L\left[\frac{d}{dt}x(t)\right] = \int_0^{\infty} \underbrace{\frac{d}{dt}x(t)}_{v}\underbrace{e^{-st}\,dt}_{u} \qquad \left[\int u\,dv = vu - \int v\,du\right.$$

$$= e^{-st}x(t)\Big|_0^{\infty} - \int_0^{\infty} x(t)(-se^{-st})dt$$

$$L\left[\frac{d}{dt}x(t)\right] = -x(0) + s\int_0^{\infty} x(t)e^{-st}\,dt = sX(s) - x(0)$$

$$L\left[\frac{d^2}{dt^2}x(t)\right] = \frac{d}{dt}\left[\frac{d}{dt}x(t)\right] = sL\left[\frac{d}{dt}x(t)\right] - \frac{d}{dt}x(0)$$

$$= s\{sX(s) - x(0)\} - \frac{dx(0)}{dt}$$

$$= s^2X(s) - sx(0) - \frac{dx(0)}{dt}$$

$$L\left[\frac{d^n}{dt^n}x(t)\right] = s^nX(s) - s^{n-1}x(0) - s^{n-2}\frac{dx(0)}{dt}\cdots\frac{d^{n-1}x(0)}{dt^{n-1}}$$

3.8.2 Initial value theorem (IVT)

The IVT allows us to calculate initial value of $x(t)$, i.e. $x(0)$ directly from the transform $X(s)$ without the need for finding the inverse transform of $X(s)$

$$L\left[\frac{d}{dt}x(t)\right] = sX(s) - x(0-) = \int_0^\infty \frac{d}{dt}x(t)e^{-st}\,dt$$

$$\lim_{s\to\infty}\int_0^\infty \frac{d}{dt}x(t)e^{-st}\,dt = \lim_{s\to\infty}\{sX(s) - x(0-)\}$$

$$\lim_{s\to\infty} e^{-st} = 0; \qquad \lim_{s\to\infty} sX(s) = x(0-)$$

$x(0-)$ indicates the value of $x(t)$ just before $t = 0$ and $x(0+)$ indicates the value of $x(t)$ just after $t = 0$. If the function $x(t)$ is continuous at $t = 0$, then its value just before and after $t = 0$ will be the same, i.e. $x(0+) = x(0-)$

$$x(0+) = \lim_{s\to\infty} sX(s)$$

$$\lim_{t\to 0} x(t) = \lim_{s\to\infty} sX(s)$$

3.8.3 Final value theorem

$$L\left[\frac{d}{dt}x(t)\right] = sX(s) - x(0-)$$

$$\lim_{s\to 0}\int_0^\infty \frac{d}{dt}x(t)e^{-st}\,dt = \lim_{s\to 0}\{sX(s) - x(0-)\}$$

$$\lim_{s\to 0} e^{-st} = 1$$

$$\int_0^\infty \frac{dx(t)}{dt}\,dt = \lim_{s\to 0} sX(s) - x(0) \tag{1}$$

$$[x(t)]_0^\infty = \lim_{t\to\infty} x(t) - x(0-) \tag{2}$$

On comparing

$$\lim_{t\to\infty} x(t) = \lim_{s\to 0} sX(s)$$

3.9 INVERSE LAPLACE TRANSFORM (ILT)

(1) Using partial fraction expansion
(2) Using convolution

3.9.1 LIT using partial fraction expansion

Find the partial fraction expansion and hence LIT (Laplace Inverse Transform) of the following function.

$$X(s) = \frac{s^2+2s-2}{s(s+2)(s-3)}, \text{ ROC : Re}(s) > 3$$

$$X(s) = \frac{A}{s} + \frac{B}{s+2} + \frac{C}{s-3}$$

$$s^2+2s-2 = A(s+2)(s-3) + B\,s(s-3) + C\,s(s+2)$$

$$A+B+C = 1$$

$$-A-3B+2C = 2$$

$$-6A = -2$$

$$A = \frac{1}{3}$$

$$B = -\frac{1}{5};\ C = \frac{13}{15}$$

$$X(s) = \frac{1}{3s} - \frac{1}{5(s+2)} + \frac{13}{15(s-3)}$$

$$x(t) = \frac{1}{3}u(t) - \frac{1}{5}e^{-2t}u(t) + \frac{13}{15}e^{+3t}u(t)$$

$$X_1(s) = \frac{1}{s}, \qquad \text{ROC : Re}(s) > 0$$

$$X_2(s) = \frac{1}{s+2}, \text{ ROC : Re}(s) > -2$$

$$X_3(s) = \frac{1}{s-3}, \text{ ROC : Re}(s) > 3$$

Example 10: Obtain bilateral ILT of the function

$$X(s) = \frac{3s+7}{s^2-2s-3}$$

For ROC of

(1) Re $(s) > 3$
(2) Re $(s) < -1$
(3) $-1 <$ Re $(s) < 3$

SOLUTION:

$$X(s) = \frac{3s+7}{(s-3)(s+1)} = \frac{4}{s-3} - \frac{1}{s+1}$$

(1) ROC: Re $(s) > 3$

$$x(t) = 4e^{3t}u(t) - e^{-t}u(t)$$

(2) for ROC: Re $(s) < -1$

$$x(t) = -e^{at}u(-t);\ X(s) = \frac{1}{s-a} \text{ for } s < a$$

$$x(t) = -e^{at}u(-t);\ X(s) = \frac{1}{s+a} \text{ for } s < -a$$

$$x(t) = 4e^{3t}u(-t) + e^{-t}u(-t)$$

$$= (e^{-t} - 4e^{3t})u(-t)$$

(3) $-1 < \text{Re}(s) < 3$

$$\text{Re}(s) > -1\,\{x(t) = e^{-at}u(t); X(s) = \frac{1}{s+a} \text{ for } s > -a$$

$$X_2(s) = \frac{1}{s+1} \quad x_2(t) = e^{-t}u(t)$$

$$\text{Re}(s) < 3\,\{x(t) = -e^{at}u(-t); X(s) = \frac{1}{s-a} s < a$$

$$X_1(s) = \frac{1}{s-3} \quad x_1(t) = -e^{3t}u(-t)$$

$$x(t) = -4e^{3t}u(-t) - e^{-t}u(t)$$

Example 11: Find the ILT of $X(s) = \dfrac{-3}{(s+2)(s-1)}$

If the ROC is (1)$-2 < \text{Re}(s) < 1$ (2) $\text{Re}(s) > 1$ (3) $\text{Re}(s) < -2$

SOLUTION:

$$X(s) = \frac{1}{s+2} - \frac{1}{s-1}$$

(1)
$$\text{Re}(s) > -2\left\{X(s) = \frac{1}{s+a} \text{ for Re}(s) > -a; x(t) = e^{-at}u(t)\right.$$

$$\text{Re}(s) < 1\left\{X(s) = \frac{1}{s-a} \text{ for Re}(s) < a; x(t) = -e^{at}u(-t)\right.$$

$$x(t) = e^{-2t}u(t) + e^{t}u(-t)$$

(2)
$$\text{Re}(s) > 1\,\{X(s) = \frac{1}{s-a} \text{ for Re}(s) > a; x(t) = e^{at}u(t)$$

$$x(t) = e^{-2t}u(t) - e^{t}u(t) = (e^{-2t} - e^{t})u(t)$$

(3)
$$\text{Re}(s) < -2\,\{X(s) = \frac{1}{s+a} \text{ for Re}(s) < -a; x(t) = -e^{-at}u(-t)$$

$$x(t) = -e^{-2t}u(-t) + e^{t}u(-t) = (e^{t} - e^{-2t})u(-t)$$

Example 12: Find the ILT of $X(s) = \dfrac{s-2}{s(s+1)^3}$ for ROC $\text{Re}(s) < -1$

SOLUTION:

$$X(s) = \frac{s-2}{s(s+1)^3} = \frac{A}{s} + \frac{B}{s+1} + \frac{C}{(s+1)^2} + \frac{D}{(s+1)^3}$$
$$X(s) = \frac{-2}{s} + \frac{2}{s+1} + \frac{2}{(s+1)^2} + \frac{3}{(s+1)^3}$$

$$\frac{t^{n-1}}{\underline{|n-1}} e^{-at} u(t) \stackrel{L}{\longleftrightarrow} \frac{1}{(s+a)^n}, \text{ ROC} : \text{Re}(s) > -a$$
$$\frac{-t^{n-1}}{\underline{|n-1}} e^{-at} u(-t) \stackrel{L}{\longleftrightarrow} \frac{1}{(s+a)^n}, \text{ ROC} : \text{Re}(s) < -a$$
$$x(t) = \left\{2 - \frac{3t^2}{2} e^{-t} - 2t\,e^{-t} - 2e^{-t}\right\} u(-t) \text{ for Re}(s) < -1$$

3.9.2 ILT using convolution integral

Example 13: Using convolution integral find $L^{-1} \dfrac{1}{s^2(s+2)}$

SOLUTION:

$$X(s) = \frac{1}{s^2(s+2)} = X_1(s)X_2(s)$$
$$X_1(s) = \frac{1}{s^2} \qquad X_2(s) = \frac{1}{s+2}$$
$$x_1(t) = tu(t) \quad x_2(t) = e^{-2t}u(t)$$

Since $u(\tau) = \begin{cases} 1 & \tau \geq 0 \\ 0 & \text{otherwise} \end{cases}$

$$u(t-\tau) = \begin{cases} 1 \text{ for } & t-\tau \geq 0 \\ \text{or} & t \geq \tau \\ & \tau < t \end{cases}$$

$$x(t) = x_1(t) * x_2(t) = \int_{-\infty}^{\infty} x_1(\tau)x_2(t-\tau)d\tau$$

So limit of integration for $u(\tau)u(t-\tau) = \begin{cases} 1 & \text{for } 0 \leq \tau < t \\ 0 & \text{elsewhere} \end{cases}$

$$= \int_{-\infty}^{\infty} \tau\, u(\tau)e^{-2(t-\tau)}u(t-\tau)d\tau = \int_0^t \tau\, e^{-2(t-\tau)}d\tau = e^{-2t}\int_0^t \tau\, e^{+2\tau}d\tau$$
$$= e^{-2t}\left[\frac{\tau\, e^{2\tau}}{2}\bigg|_0^t - \int_0^t \frac{e^{2\tau}}{2}d\tau\right]$$
$$= e^{-2t}\left\{\frac{t\, e^{2t}}{2} - \frac{1}{4}(e^{2t}-1)\right\} = \frac{t}{2} - \frac{1}{4} + \frac{1}{4}e^{-2t}$$
$$x(t) = \frac{t}{2}u(t) - \frac{1}{4}u(t) + \frac{1}{4}e^{-2t}u(t)$$

Example 14: Find the ILT of $X(s) = \frac{1}{s^2(s^2-a^2)}$ using convolution integral.

SOLUTION:

$$X(s) = \frac{1}{s^2(s^2-a^2)} \quad X_1(s) = \frac{1}{s^2} \quad X_2(s) = \frac{1}{s^2-a^2}$$

$$X_2(s) = \frac{1}{(s-a)(s+a)} = \frac{1}{2a}\left[\frac{1}{s-a} - \frac{1}{s+a}\right]$$

$$x_1(t) = t\,u(t) \quad x_2(t) = \frac{1}{2a}\left[e^{at} - e^{-at}\right]$$

$$x(t) = x_1(t) * x_2(t) = \int_{-\infty}^{\infty} x_1(\tau)x_2(t-\tau)d\tau$$

$$= \frac{1}{2a}\int \tau\left(e^{a(t-\tau)} - e^{-a(t-\tau)}\right)d\tau$$

$$= \frac{1}{2a}\int \tau\, e^{a(t-\tau)}d\tau - \frac{1}{2a}\int_0^t \tau\, e^{-a(t-\tau)}d\tau$$

$$= \frac{1}{2a}e^{at}\int_0^t \tau\, e^{-a\tau}d\tau - \frac{1}{2a}e^{-at}\int_0^t \tau\, e^{a\tau}d\tau$$

$$= \frac{1}{2a}\left[\frac{-2t}{a} + \frac{1}{a^2}\left(e^{at} - e^{-at}\right)\right]$$

Inversion of unilateral Laplace transform

Defined as $x(t) = \frac{1}{2\pi j}\int_{\sigma-j\infty}^{\sigma+j\infty} X(s)e^{st}ds$

The direct method of finding Inverse L.T. using the above equation requires an understanding of complex variables.

There are other methods to find ILT that use one-to-one relationship between a signal and its unilateral L.T. One such method is partial fraction expansion.

To find partial fraction expansion

$$\text{Let} \quad X(s) = \frac{b_2s^2 + b_1s + b_0}{(s-p_1)(s-p_2)(s-p_3)}$$

Higher power of numerator < highest power of denominator

Case 1

Distinct poles, i.e. $p_1 \neq p_2 \neq p_3$

We can write as $X(s) = \frac{C_1}{s-p_1} + \frac{C_2}{s-p_2} + \frac{C_3}{s-p_3}$

$$\text{where} \quad C_1 = (s-p_1)X(s)|_{s=p_1}$$

$$C_2 = (s-p_2)X(s)|_{s=p_2}$$

$$C_3 = (s-p_3)X(s)|_{s=p_3}$$

On generalizing

$$C_i = (s-p_i)X(s)\big|_{s=p_i}$$

Example 15: Find ILT of $X(s) = \dfrac{4}{(s-1)(s-2)}$

SOLUTION:

$X(s) = \frac{4}{(s-1)(s-2)}$ then we can write $X(s) = \frac{C_1}{s-1} + \frac{C_2}{s-2}$

$$\text{where}\quad C_1 = (s-1)X(s)\big|_{s=1} = (s-1)\frac{4}{(s-1)(s-2)}\bigg|_{s=1} = -4$$

$$C_2 = (s-2)X(s)\big|_{s=2} = (s-2)\frac{4}{(s-1)(s-2)}\bigg|_{s=2} = 4$$

$$X(s) = \frac{-4}{s-1} + \frac{4}{s-2}$$

Case 2

Multiple poles

$$X(s) = \frac{b_2s^2 + b_1s + b_0}{(s-p)^3}$$

$$p_1 = p_2 = p_3 = p$$

$$= \frac{C_1}{(s-p)} + \frac{C_2}{(s-p)^2} + \frac{C_3}{(s-p)^3}$$

$$C_3 = (s-p)^3X(s)\big|_{s=p}$$

$$C_2 = \frac{1}{2-1!}\frac{d}{ds}\left[(s-p)^3X(s)\right]\big|_{s=p}$$

$$C_1 = \frac{1}{3-1!}\frac{d^2}{ds^2}\left[(s-p)^3X(s)\right]_{s=p}$$

Example 16: Find ILT of $X(s) = \dfrac{2(s+1)}{(s+2)^3}$.

SOLUTION:

$$X(s) = \frac{2(s+1)}{(s+2)^3}$$

$$= \frac{C_1}{s+2} + \frac{C_2}{(s+2)^2} + \frac{C_3}{(s+2)^3}$$

$$C_3 = (s+2)^3X(s)|_{s=-2} = (s+2)^3\frac{(2s+1)}{(s+2)^3}|_{s=-2} = -3$$

$$C_2 = \frac{1}{\underline{|2-1}}\frac{d}{ds}\left[(s+2)^3X(s)\right]|_{s=-2} = \frac{d}{ds}\left[(s+2)^3\frac{(2s+1)}{(s+2)^3}\right]|_{s=-2}$$

$$= \left[2\frac{d}{ds}(s) + \frac{d}{ds}(1)\right]|_{s=-2} = 2$$

$$C_1 = \frac{1}{\underline{|3-1}}\frac{d^2}{ds^2}\left[(s+2)^3\frac{2s+1}{(s+2)^3}\right]|_{s=-2} = \frac{1}{2}\frac{d^2}{ds^2}[2s+1]|_{s=-2}$$

$$= \frac{1}{2}\left[2\frac{d^2}{ds^2}(s) + \frac{d^2}{ds^2}(1)\right]\Big|_{s=-2} = 0$$

$$X(s) = \frac{2}{(s+2)^2} - \frac{3}{(s+2)^3}$$

Case 3

Complex roots: If $X(s)$ has complex roots (poles), then partial fraction of $X(s)$ can be expressed as

$$X(s) = \frac{C}{s-p} + \frac{C^*}{s-p^*}$$

where C* is complex conjugate of C
p^* is complex conjugate of p
Find the C as in Case 1.

Example 17: Find the ILT of $X(s) = \dfrac{s}{s^2+2s+2}$

SOLUTION:

$$X(s) = \frac{s}{s^2+2s+2}$$

Roots of $s^2+2s+2=0$

$$s = \frac{-2 \pm \sqrt{4-8}}{2} = \frac{-2 \pm 2j}{2} = -1 \pm j$$

$$X(s) = \frac{C}{s-(-1-j)} + \frac{C^*}{s-(-1+j)} = \frac{C}{s+1+j} + \frac{C^*}{s+1-j}$$

$$C = (s+1+j)X(s)\Big|_{s=-1-j} = (s+1+j)\frac{s}{(s+1+j)(s+1-j)}\Big|_{s=-1-j}$$

$$= \frac{-1-j}{-2j} = \frac{1+j}{2j} = -0.5j + 0.5 = 0.5 - 0.5j$$

$$C^* = 0.5 + 0.5j$$

$$X(s) = \frac{0.5-0.5j}{s+1+j} + \frac{0.5+0.5j}{s+1-j}$$

Example 18: Find the Laplace transform of the following:
(a) $x(t) = u(t-2)$ (b) $x(t) = t^2e^{-2t}u(t)$

SOLUTION:

(a)
$$X(s) = \int_0^\infty x(t)e^{-st}\,dt = \int_0^\infty u(t-2)e^{-st}\,dt$$

$$= \int_2^\infty e^{-st}\,dt = \frac{e^{-st}}{-s}\Big|_{-2}^{\infty} \qquad u(t-2) = \begin{cases} 1 & \text{for } t \geq 2 \\ 0 & \text{otherwise} \end{cases}$$

$$= \underbrace{\frac{e^{-st}}{-s}\Big|_{t=\infty}}_{(1)} + \underbrace{\frac{e^{-st}}{s}\Big|_{t=2}}_{(2)}$$

For $s > 0\,(1)$ term is zero and for $s < 0\,(1)$ term is ∞ so ROC is $\text{Re}(s) > 0$ and $X(s) = \dfrac{e^{-2s}}{s}$

or by using time shifting property, L.T. $(x(t-t_0)) = e^{-st0}X(s)$ and we know that $\text{LT}[u(t)] = \frac{1}{s}$

$$\therefore \quad L[u(t-2)] = \frac{e^{-2s}}{s}$$

(b)

$$x(t) = t^2e^{-2t}u(t)$$
$$= t^2x(t)$$
$$x(t) = e^{-2t}u(t)$$
$$X(s) = \frac{1}{s+2}$$

Note: Always solve the signal of $\{t^n x^1(t)\}$ form using property, b/c by simply integrating, we get a term of t and on putting the value of t as ∞ whole integration vanishes.

Using the property differentiation in s-domain

$$L[(-t)^n x(t)] = \frac{d^n X(s)}{ds^n}$$
$$L\left[(-t)^2 x(t)\right] = \frac{d^2 X(s)}{ds^2}$$
$$X(s) = \frac{d^2}{ds^2}\left[\frac{1}{s+2}\right]$$
$$= \frac{d}{ds}\left[\frac{-1}{(s+2)^2}\right]$$
$$X(s) = \left[\frac{2}{(s+2)^3}\right]$$

Example 19: For the following transform pair

$$L[x(t)] = X(s) = \frac{2s}{s^2-2}$$

determine the L.T. of $x(2t)$.

SOLUTION:

We know that the time scaling property states that

$$x(at) = \frac{1}{|a|}X\left(\frac{s}{a}\right)$$

$$\text{Given} \quad X(s) = \frac{2s}{s^2-2} \Rightarrow X\left(\frac{s}{2}\right) = \frac{2\cdot\frac{s}{2}}{\left(\frac{s}{2}\right)^2-2} = \frac{4s}{s^2-8}$$

$$\text{LT}\left[x(2t)\right] = \frac{1}{2}\cdot\frac{4s}{s^2-8} = \frac{2s}{s^2-8}$$

Example 20: A signal has L.T. $X(s) = \dfrac{s+2}{s^2+4s+5}$

Find L.T. of the following signals

(a) $y(t) = t\,x(t)$ (b) $y(t) = e^{-2t}x(t)$ (c) $y(t) = x(t)*x(t)$

SOLUTION:

Given $X(s) = \dfrac{s+2}{s^2+4s+5}$

(a) $L[y(t)] = L[t\,x(t)]$

By using differentiation property in s-domain

$$Y(s) = \frac{-d}{ds}X(s) = \frac{-d}{ds}\left[\frac{s+2}{s^2+4s+5}\right]$$

$$= -\left[\frac{(s^2+4s+5)-(2s^2+8s+8)}{(s^2+4s+5)^2}\right] = \frac{s_2+4s+3}{(s^2+4s+5)^2}$$

(b) $y(t) = e^{-2t}x(t)$

$$\text{LT}[y(t)] = \text{LT}\left[e^{-2t}x(t)\right]$$

By using frequency shifting property

$$\text{LT}\left[e^{-at}x(t)\right] = X(s+a)$$

$$Y(s) = X(s-2) = \frac{(s+2)+2}{(s+2)^2+4(s+2)+5} = \frac{s+4}{s^2+8s+17}$$

(c) $L(y(t)) = L[x(t)*x(t)]$

By using convolution in time domain property

$$x_1(t)*x_2(t) = X_1(s)\,X_2(s)$$

$$Y(s) = X(s)X(s) = X^2(s) = \frac{(s+2)^2}{(s^2+4s+5)^2}$$

3.10 SYNTHESIZE OF THE WAVEFORM

Following steps are involved in synthesizing a waveform.

Step 1:

Separate each and every part of waveform in the form of Gate function

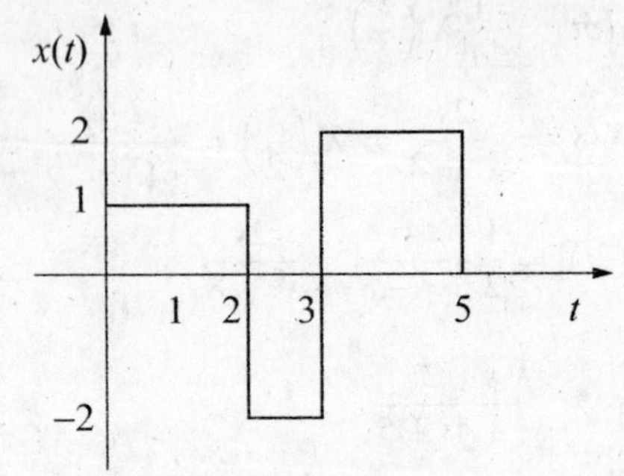

Fig. 3.13.

Example:

For given function, we form three Gate functions.

First from $0 \rightarrow 2$

Second from $2 \rightarrow 3$

Third from $3 \rightarrow 5$

Step 2: Define each and every Gate function

First Gate function is defined as $x_1(t) = u(t) - u(t-2)$

Second Gate function is defined as $x_2(t) = u(t-2) - u(t-3)$

Third Gate function is defined as $x_3(t) = u(t-3) - u(t-5)$

Step 3: Multiply each and every Gate function by

(1) Amplitude of waveform in case of square wave

(2) Equation of line in case of triangular wave.

$$x(t) = 10\,x_1(t) - 2\,x_2(t) + 2\,x_3(t)$$
$$= [u(t) - u(t-1)] - 2[u(t-2) - u(t-3)] + 2[u(t-3) - u(t-5)]$$
$$= u(t) - 3u(t-2) + 4u(t-3) - 2u(t-5)$$

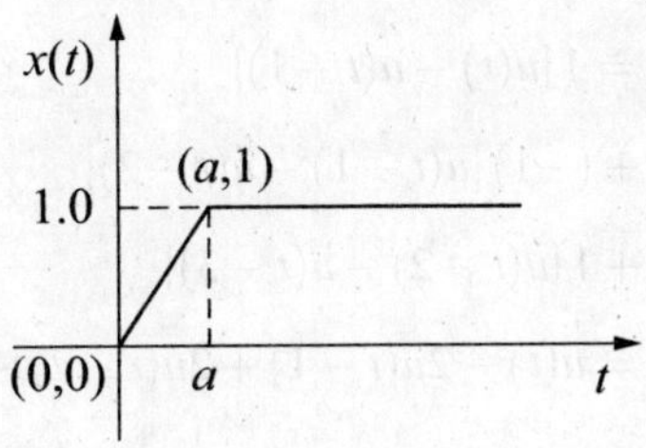

Fig. 3.14.

(1) First Gate function from 0 to a

(2) Second Gate function from a to ∞

$$x_1(t) = u(t) - u(t-a)$$
$$x_2(t) = u(t-a)$$

Here first Gate function is multiplied by equation of line with points $(0,0)$ and $(a,1)$ and second Gate function by amplitude.

Equation of line

$$y - y_1 = \frac{y_2 - y_1}{x_2 - x_1}(x - x_1)$$

$$x(t) = \frac{1}{a}t$$

$$x(t) = \text{equation of line } [x_1(t)] + \text{ amplitude } [x_2(t)]$$
$$= \frac{t}{a}[u(t) - u(t-a)] + 1.0\,[u(t-a)]$$

$$= \frac{t}{a}u(t) - \frac{t}{a}u(t-a) + u(t-a)$$

$$= \frac{1}{a}t\,u(t) - \frac{1}{a}u(t-a)(t-a)$$

$$= \frac{1}{a}r(t) - \frac{1}{a}r(t-a)$$

Since $tu(t) = r(t)$
$(t-a)u(t-a) = r(t-a)$

Synthesise the following waveforms

(a)

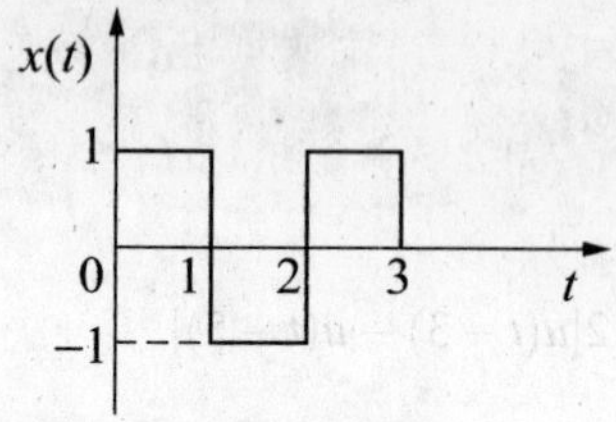

Fig. 3.15.

$$x(t) = 1\,[u(t) - u(t-1)]$$
$$+ (-1)\,[u(t-1) - u(t-2)]$$
$$+ 1\,[u(t-2) - u(t-3)]$$
$$= u(t) - 2u(t-1) + 2u(t-2) - u(t-3)$$

(b)

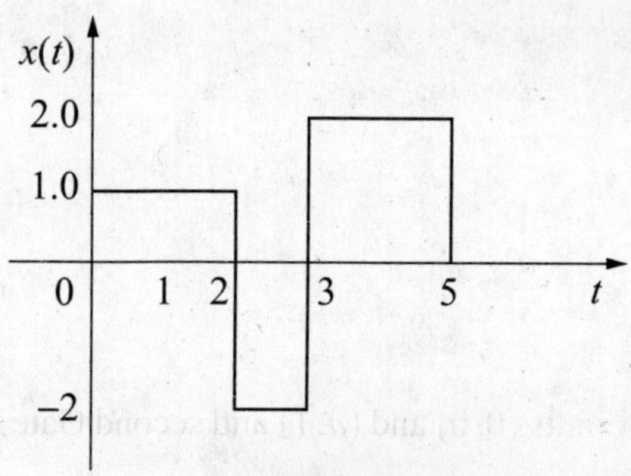

Fig. 3.16.

$$x(t) = [u(t) - u(t-2)]$$
$$+ (-2)\,[u(t-2) - u(t-3)]$$
$$+ 2\,[u(t-3) - u(t-5)]$$
$$= u(t) - 3u(t-2) + 4u(t-3) - 2u(t-5)$$

(c)

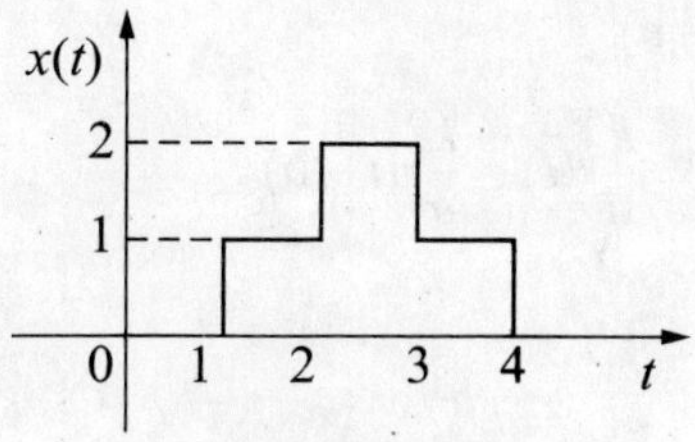

Fig. 3.17.

$$\begin{aligned} x(t) &= 1\,[u(t-1)-u(t-2)] \\ &\quad +2\,[u(t-2)-u(t-3)] \\ &\quad +1\,[u(t-3)-u(t-4)] \\ &= u(t-1)+u(t-2)-u(t-3)-u(t-4) \end{aligned}$$

(d)

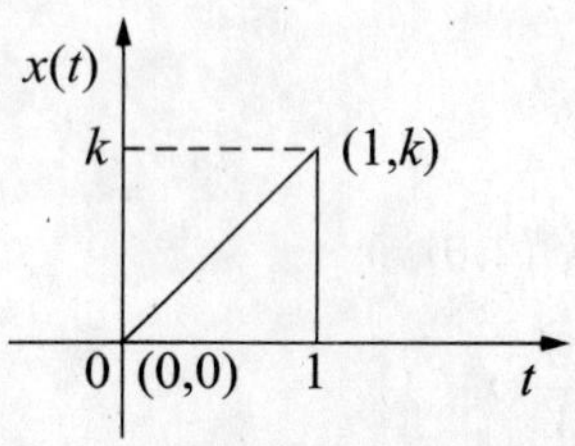

Fig. 3.18.

$$y-0=\frac{k-0}{1-0}(x-0) \qquad y=kx \quad x(t)=kt$$

$$x(t)=(1)\,[u(t)-u(t-1)]=kt\,[u(t)-u(t-1)]$$

(e)

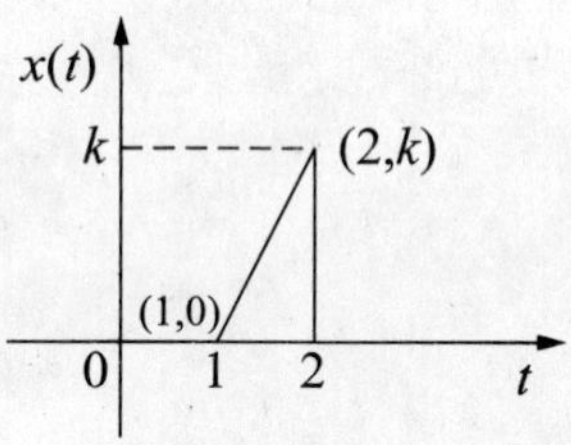

Fig. 3.19.

$$y-0=\frac{k}{1}(x-1) \quad x(t)=k(t-1)$$

$$x(t)=k(t-1)\,[u(t-1)-u(t-2)]$$

(f)

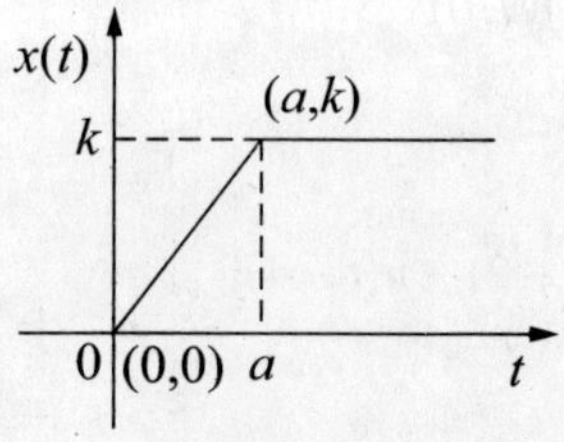

Fig. 3.20.

$$x(t) = \frac{k}{a}t\,[u(t) - u(t-a)] + ku(t-a)$$

$$= \frac{k}{a}t\,u(t) - \frac{k}{a}(t-a)u(t-a) = \frac{k}{a}r(t) - \frac{k}{a}r(t-a)$$

$$\text{Since } y - 0 = \frac{k-0}{a-0}(x-0)$$

$$x(t) = \frac{k}{a}t$$

(g)

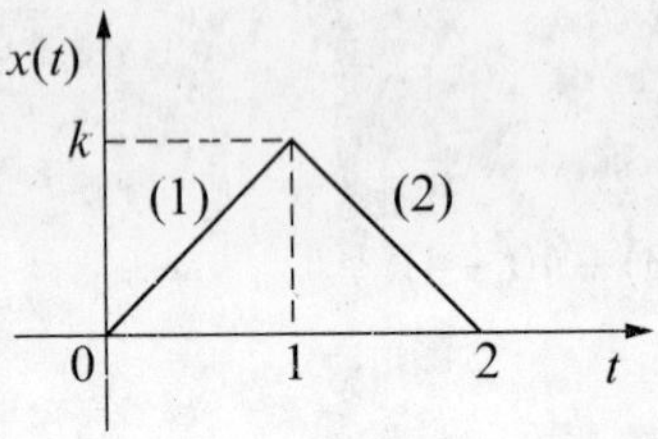

Fig. 3.21.

(1) $(0,0)(1,k)$ (2) $(1,k)(2,0)$

$$y = \frac{k}{1}x \qquad y - k = \frac{-k}{1}(x-1)$$

$$x(t) = kt \qquad x(t) = -kt + 2k$$

$$= k(2-t)$$

$$x(t) = kt\,[u(t) - u(t-1)] + k(2-t)\,[u(t-1) - u(t-2)]$$

(h)

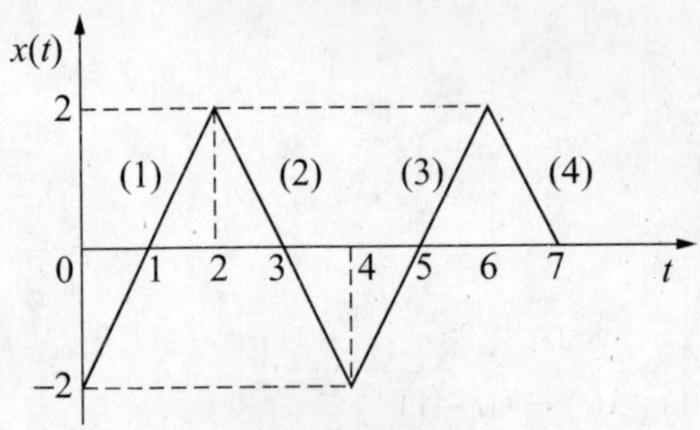

Fig. 3.22.

(1) $(0,-2)(2,2)$ (2) $(2,2)(4,-2)$ (3) $(4,-2)(6,2)$ (4) $(6,2)(7,0)$

$x(t) = 2(t-1)$ $x(t) = (-2t+6)$ $x(t) = (2t-10)$

$$x(t) = 2(t-1)$$

$$x(t) = (2t-2)\,[u(t) - u(t-2)] + (-2t+6)\,[u(t-2) - u(t-4)]$$

$$+ (2t-10)\,[u(t-4) - u(t-6)] + \ldots$$

(i)

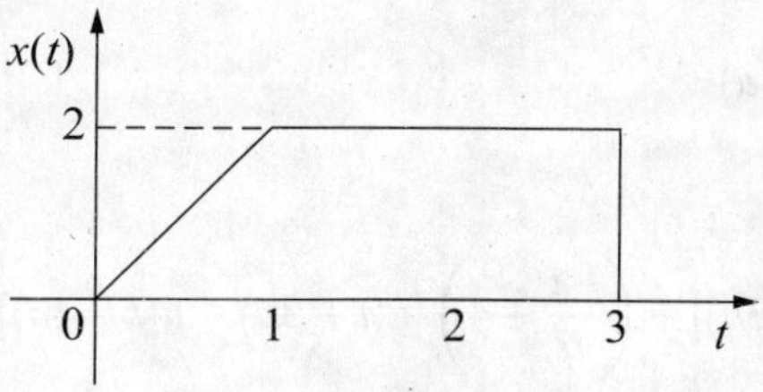

Fig. 3.23.

$$x(t) = 2t\,[u(t) - u(t-1)] + 2\,[u(t-1) - u(t-3)]$$

$$(0,0)(1,2)$$

$$= \frac{2}{1}x$$

$$(t) = 2t$$

(j)

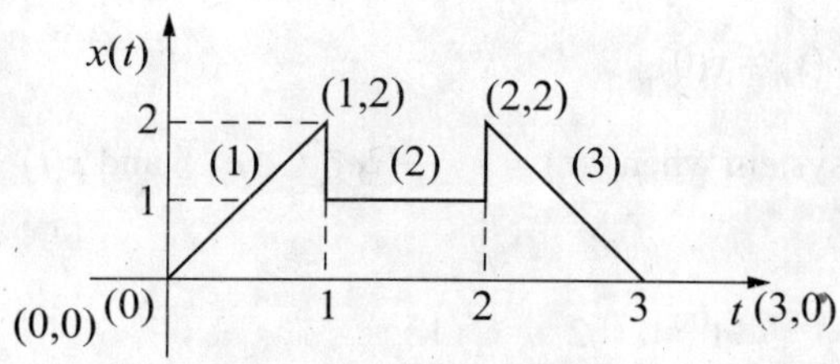

Fig. 3.24.

$$(0,0)(1,2);(2,2)(3,0)$$

$$y = 2t \qquad\qquad y - 2 = \frac{-2}{1}(x-2)$$

$$x(t) - 2 = -2t + 4$$

$$x(t) = (6-2t)$$

$$x(t) = 2t\,[u(t) - u(t-1)] + 1\,[u(t-1) - u(t-2)] + (6-2t)\,[u(t-2) - u(t-3)]$$

$$= 2t\,u(t) + (2t+1)u(t-1) + u(t-2)\{-1 + 6 - 2t\} - (6-2t)u(t-3)$$

$$= 2t\,u(t) + (2t+1)u(t-1) + (5-2t)u(t-2) + (2t-6)u(t-3)$$

(k)

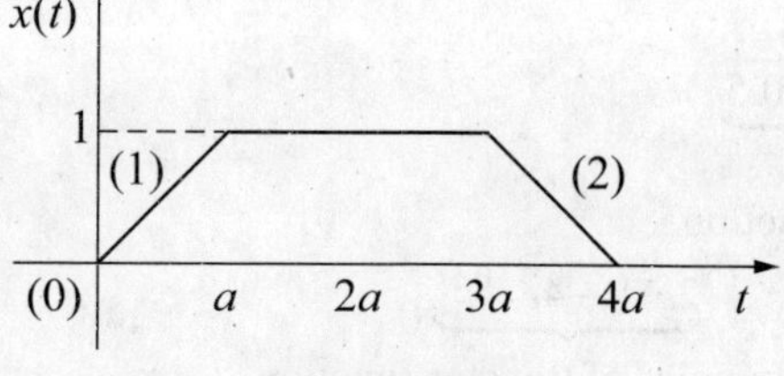

Fig. 3.25.

$$(1)\quad (0,0)(a,1) \qquad\qquad (2)\quad (3a,1)(4a,0)$$

$$y=\frac{1}{a}x \qquad\qquad y-1=\frac{-1}{a}(x-3a)$$

$$x(t)=\frac{t}{a} \qquad\qquad x(t)=\frac{-t}{a}+4$$

$$x(t)=\frac{t}{a}[u(t)-u(t-a)]+[u(t-a)-u(t-3a)]+\left(\frac{-t}{a}+4\right)[u(t-3a)-u(t-4a)]$$

$$=\frac{1}{a}tu(t)+u(t-a)\left\{1-\frac{t}{a}\right\}+u(t-3a)\left\{-1-\frac{t}{a}+4\right\}+\frac{-1}{a}(4a-t)u(t-4a)$$

$$x(t)=\frac{1}{a}r(t)+r(t-a)-\frac{1}{a}r(t-3a)+\frac{1}{a}r(t-4a)$$

3.11 ANALYSIS AND CHARACTERIZATION OF LTI SYSTEM

$$L\left\{\frac{d^2y(t)}{dt^2}\right\}=s^2Y(s)-sy(0)-\frac{dy(0)}{dt}$$

$$L\left\{\frac{dy(t)}{dt}\right\}=sY(s)-y(0)$$

Example 21: Obtain the transform function of the system when $y(t)=e^{-t}-2e^{-2t}+e^{-3t}$ and $x(t)=e^{-0.5t}$

SOLUTION:

$$X(s)=\frac{1}{s}+0.5 \quad Y(s)=\frac{1}{s+1}-\frac{2}{s+2}+\frac{1}{s+3}$$

$$=\frac{2}{(s+1)(s+2)(s+3)}$$

$$H(s)=\frac{Y(s)}{X(s)}=\frac{2(s+.5)}{(s+1)(s+2)(s+3)}$$

Example 22: Find the forced response of the system with transfer function
$H(s)=\dfrac{1}{(s+0.5)(s+1)}$ for a unit step input.

SOLUTION:

$$x(t)=u(t);\ X(s)=\frac{1}{s}$$

$$Y(s)=H(s)X(s)=\frac{1}{s(s+1)(s+0.5)}=\frac{A}{s}+\frac{B}{s+1}+\frac{C}{s+0.5}$$

$$Y(s)=\underbrace{\frac{2}{s}}_{\text{Due to pole of input}}+\underbrace{\frac{2}{s+1}-\frac{4}{s+0.5}}_{\text{Due to poles of transfer function}}$$

$$y(t)=\underbrace{2}_{\substack{\text{Due to I/P pole}\\ \text{steady state response}}}+\underbrace{2e^{-t}-4e^{-0.5t}}_{\substack{\text{Due to poles of transfer function}\\ \text{transient response}}}$$

Example 23: Solve the differential equation

(1) $\frac{dy(t)}{dt} + 5y(t) = x(t)$ with initial conditions $y(0) = -2$ and input $x(t) = 3e^{-2t}u(t)$

(2) $\frac{d^2y(t)}{dt^2} + 11\frac{dy(t)}{dt} + 24y(t) = 5\frac{dx(t)}{dt} + 3x(t)$

SOLUTION:

(1)
$$sY(s) - y(0) + 5Y(s) = X(s)$$
$$X(s) = \frac{3}{s+2}$$
$$sY(s) + 2 + 5Y(s) = X(s)$$
$$Y(s)(s+5) = \frac{3}{s+2} - 2$$
$$Y(s) = \frac{3}{(s+2)(s+5)} - \frac{2}{(s+5)} = \frac{1}{s+2} - \frac{1}{s+5} - \frac{2}{s+5}$$
$$= \frac{1}{s+2} - \frac{3}{s+5}$$
$$y(t) = e^{-2t}u(t) - 3e^{-5t}u(t) = (e^{-2t} - 3e^{-5t})u(t)$$

(2)
$$(s^2 + 11s + 24)Y(s) = (5s+3)X(s)$$
$$H(s) = \frac{Y(s)}{X(s)} = \frac{5s+3}{s^2 + 11s + 24}$$

Example 24: By using L.T., solve the following differential equation

$$\frac{d^2y(t)}{dt^2} + 3\frac{dy(t)}{dt} + 2y(t) = \frac{d}{dt}x(t) \quad \text{if } y(0) = 2\, \frac{dy(0)}{dt} = 1$$
$$\text{and } x(t) = e^{-t}u(t)$$

SOLUTION:

$$\left(s^2Y(s) - sy(0) - \frac{dy(0)}{dt}\right) + 3(sY(s) - y(0)) + 2Y(s) = sX(s) - x(0)$$
$$Y(s)(s^2 + 3s + 2) - 2s - 1 - 3 \times 2 = sX(s)$$
$$Y(s)(s^2 + 3s + 2) = 2s + 7 + sX(s);\ X(s) = \frac{1}{s+1}$$
$$Y(s) = \frac{2s+7}{s^2+3s+2} + \frac{1}{(s+1)(s^2+3s+2)} = \frac{2s^2+10s+7}{(s+1)(s^2+3s+2)}$$
$$= \frac{2s^2+10s+7}{(s+1)^2(s+2)} = \frac{7}{s+1} - \frac{1}{(s+1)^2} - \frac{5}{s+2}$$
$$y(t) = 7e^{-t}u(t) - te^{-t}u(t) - 5e^{-2t}u(t)$$

Example 25: Find the impulse response and the step response of (i) $H(s) = \dfrac{s+2}{s^2+5s+4}$ (ii) $H(s) = \dfrac{10}{(s+3)^2+1}$.

SOLUTION:

(i)
$$H(s) = \frac{s+2}{s^2+5s+4} = \frac{s+2}{(s+4)(s+1)} = \frac{A}{(s+4)} + \frac{B}{s+1}$$
$$H(s) = \frac{1}{3(s+1)} + \frac{2}{3(s+4)}$$
$$h(t) = \frac{1}{3}e^{-t}u(t) + \frac{2}{3}e^{-4t}u(t)$$

OR $[Y(s) = H(s)X(s)$ for impulse response

$Y(s) = H(s)$ $x(t) = \delta(t)$

$X(s) = 1$

$$y(t) = h(t) = \frac{1}{3}e^{-t}u(t) + \frac{2}{3}e^{-4t}u(t)]$$

(b) Step response $s(t) = u(t)*h(t)$
for step response

$$s(t) = \int_{-\infty}^{\infty} h(\tau)u(t-\tau)d\tau$$
$$s(t) = \int_{-\infty}^{\infty} \left(\frac{1}{3}e^{-\tau} + \frac{2}{3}e^{-4\tau}\right) u(\tau)u(t-\tau)d\tau$$
$$s(t) = \int_{0}^{t} \left(\frac{1}{3}e^{-\tau} + \frac{2}{3}e^{-4\tau}\right) d\tau$$
$$= -\frac{1}{3}(e^{-t}-1) - \frac{1}{6}(e^{-4t}-1)$$
$$s(t) = \frac{1}{2}u(t) - \frac{1}{3}e^{-t}u(t) - \frac{1}{6}e^{-4t}u(t)$$

OR $x(t) = u(t)$ $X(s) = 1/s$

$$H(s) = \frac{Y(s)}{X(s)};\ Y(s) = H(s)X(s) = \frac{H(s)}{s} = \frac{s+2}{s(s+1)(s+4)}$$
$$Y(s) = \frac{A}{s} + \frac{B}{s+1} + \frac{C}{s+4} = \frac{1}{2s} - \frac{1}{3(s+1)} - \frac{1}{6(s+4)}$$
$$y(t) = \frac{1}{2}u(t) - \frac{1}{3}e^{-t}u(t) - \frac{1}{6}e^{-4t}u(t)$$

(ii) $H(s) = \dfrac{10}{s^2+6s+10}$

(1)

$$H(s) = \frac{10}{(s+3)^2+1}$$
$$h(t) = 10e^{-3t}\sin t u(t)$$

(2) 2nd method

$$H(s) = \frac{Y(s)}{X(s)};\ Y(s) = H(s)X(s)$$

$$x(t) = \delta(t) \quad Y(s) = H(s)$$

$$X(s) = 1 \quad y(t) = h(t)$$

$$y(t) = 10e^{-3t} \sin t\, u(t)$$

(1)

$$s(t) = h(t) * u(t)$$

$$= \int_0^t 10e^{-3\tau} \sin\tau d\tau$$

$$s(t) = \left[1 - e^{-3t}(\cos t + 3\sin t)\right] u(t)$$

OR

(2)

$$x(t) = u(t),\ X(s) = 1/s$$

$$Y(s) = \frac{10}{s((s+3)^2+1)}$$

$$= \frac{A}{s} + \frac{Bs+C}{s^2+6s+10}$$

$$= \frac{1}{s} - \frac{(s+6)}{(s+3)^2+1}$$

$$= \frac{1}{s} - \frac{(s+3)}{(s+3)^2+1} + \frac{3}{(s+3)^2+1}$$

$$= \left[1 - e^{-3t}(\cos t + 3\sin t)\right] u(t)$$

Example 26: (a) Determine the initial value $x(0^+)$ for the following L.T.

(i) $X(s) = \dfrac{3}{s^2+5s-1}$ (ii) $X(s) = \dfrac{2s+3}{s(s^2+5s+6)}$

(b) Find the initial and final value for the following transform.

(i) $\dfrac{s+5}{s^2+3s+2}$ (ii) $\dfrac{s^2+5s+7}{s^2+3s+2}$

SOLUTION:

(a) (i) $\lim\limits_{t\to 0} x(t) = \lim\limits_{s\to\infty} sX(s) = \lim\limits_{s\to\infty} \dfrac{3s}{s^2+5s-1} = \lim\limits_{s\to\infty} \dfrac{\frac{3}{s}}{1+\frac{5}{s}-\frac{1}{s^2}} = 0$

(ii) $\lim\limits_{t\to 0} x(t) = \lim\limits_{s\to\infty} sX(s) = \dfrac{2s^2+3s}{s(s^2+5s+6)} = \dfrac{\frac{2}{s}+\frac{3}{s^2}}{1+\frac{5}{s}+\frac{6}{s^2}} = 0$

(b) (i) $x(0) = \lim_{s\to\infty} \dfrac{s(s+5)}{s^2+3s+2} = \lim_{s\to\infty} \dfrac{1+\frac{5}{s}}{1+\frac{3}{s}+\frac{2}{s^2}} = 1$

$$x(\infty) = \lim_{s\to 0} \frac{s(s+5)}{s^2+3s+2} = 0$$

(ii) $x(0) = \lim_{s\to\infty} \dfrac{s(s^2+5s+7)}{s^2+3s+2} = \dfrac{1+\frac{5}{s}+\frac{7}{s^2}}{\frac{1}{s}+\frac{3}{s^2}+\frac{2}{s^3}} = \infty$

The initial value of time function does not exist b/c

$$\lim_{s\to\infty} \frac{s\,(s^2+5s+7)}{s^2+3s+2} \text{ does not exist}$$

$$x(\infty) = \lim_{s\to 0} \frac{s\,(s^2+5s+7)}{s^2+3s+2} = 0$$

Property	$x(t)$	$X(s)$
	$x_1(t)$	$X_1(s)$
	$x_2(t)$	$X_2(s)$
Linearity	$a\,x_1(t)+b\,x_2(t)$	$a\,X_1(s)+b\,X_2(s)$
Time shifting	$x(t-t_0)$	$e^{-st0}X(s)$
Time reversal	$x(-t)$	$X(-s)$
Frequency shifting	$e^{st}x(t)$	$X(s-s_0)$
Time differentiation	$\dfrac{dx(t)}{dt}$	$sX(s)-x(0)$
	$\dfrac{d^2x(t)}{dt^2}$	$s^2X(s)-sx(0)-\dfrac{dx(0)}{dt}$
	$\dfrac{d^nx(t)}{dt^n}$	$s^nX(s)-s^{n-1}x(0)-\dfrac{d^{n-1}}{dt^{n-1}}x(0)$
Time integration	$\int_0^t x(\tau)d\tau$	$\dfrac{X(s)}{s}$
	$\int_{-\infty}^t x(\tau)d\tau$	$\dfrac{X(s)}{s}+\dfrac{1}{s}\int_{-\infty}^0 x(\tau)d\tau$
Frequency differentiation	$-tx(t)$	$\dfrac{dX(s)}{ds}$
Frequency integration	$\dfrac{x(t)}{t}$	$\int_s^\infty X(s)ds$
Scaling	$x(at)$	$\dfrac{1}{\lvert a\rvert}X\left(\dfrac{s}{a}\right)$
Time convolution	$x_1(t)*x_2(t)$	$X_1(s)X_2(s)$
Frequency convolution	$x_1(t)x_2(t)$	$X_1(s)*X_2(s)$
Initial value theorem	$x(0^-)$	$\lim_{s\to\infty} sX(s)$
Final value theorem	$x(\infty)$	$\lim_{s\to 0} sX(s)$

3.11.1 Solution of differential equation

To explain clearly let us take a second order differential equation

$$a_2\frac{d^2y(t)}{dt^2}+a_1\frac{dy(t)}{dt}+a_0\,y(t)=b_2\frac{d^2x(t)}{dt^2}+b_1\frac{dx(t)}{dt}+b_0x(t) \tag{1}$$

Now taking the Laplace transform on both sides, we get

$$a_2\left\{s^2Y(s)-sy(0)-\frac{dy(0)}{dt}\right\}+a_1\{sY(s)-y(0)\}+a_0Y(s)$$

$$=b_2\left\{s^2X(s)-sx(0)-\frac{dx(0)}{dt}\right\}+b_1\{sX(s)-x(0)\}+b_0X(s)$$

$$Y(s)\left\{a_2s^2+a_1s+a_0\right\}-a_2sy(0)-a_2\frac{dy(0)}{dt}-a_1y(0)$$

$$=X(s)\left\{b_2s^2+b_1s+b_0\right\}-b_2sx(0)-b_2\frac{dx(0)}{dt}-b_1x(0) \tag{2}$$

For given initial conditions, we can find a relation between $X(s)$ and $Y(s)$.

3.11.2 Transfer function of the system

We know that the output of the linear time invariant continuous-time system is given as

$$y(t)=h(t)*x(t) \tag{1}$$

Taking the Laplace transform of the above equation

$$Y(s)=H(s)X(s) \tag{2}$$

$$\text{Transfer function}\qquad H(s)=\frac{Y(s)}{X(s)} \tag{3}$$

$H(s)$ is called the system transfer function.

3.11.3 Frequency response of the system

We know that frequency response of the system is given by Fourier transform function $H(jw)$.

Frequency response $\quad H(jw)=H(s)|_{s=jw}$

3.11.4 Step response of the system

Step response is defined as $\quad s(t)=h(t)*u(t)$

where $h(t)$ is impulse response and $u(t)$ is unit step function.

$$s(t)=\int_{-\infty}^{\infty}h(\tau)u(t-\tau)d\tau$$

So $s(t)=\int_{-\infty}^{t}h(\tau)d\tau$

For causal system $s(t)=\int_0^t h(\tau)d\tau$

$$[u(t-\tau)=1 \text{ for } t-\tau\geq 0$$
$$=0 \text{ otherwise}$$
$$u(t-\tau)=1 \text{ for } t\geq\tau$$
$$\text{or } 1 \text{ for } \tau\leq t]$$

3.12 APPLICATIONS OF LAPLACE TRANSFORM TO NETWORK ANALYSIS, CT SYSTEM ANALYSIS USING LAPLACE TRANSFORM

Analysis involves the study of basic analog elements like resistance, inductances and capacitances connected in some manner along with some voltage or current sources. The ratio of the transform of voltage to the transform of current, is called the 'transform impedance' or the 'Laplace impedance'.

S.No		Resistance	Inductance	Capacitance
1.	Time-domain equations	$v(t) = Ri(t)$	$v(t) = L\frac{di(t)}{dt}$	$i(t) = \left(c\frac{dv(t)}{dt}\right)$ Or $v(t) = \frac{1}{c}\int_{-\infty}^{t} i(\tau)d\tau$
2.	Transform equations	$V(s) = RI(s)$	$V(s) = sLI(s)$	$V(s) = \frac{1}{sc}I(s)$
3.	Transform equation with initial condition	–	$V(s) = L[sI(s) - i(0)]$	$I(s) = c[sV(s) - V(0)]$ Or $V(s) = \frac{1}{sc}I(s) + \frac{v(0)}{s}$
4.	Transform impedance	$z(s) = R$	$z(s) = sL$	$Z(s) = \frac{1}{sc}$
5.	Transform admittance	$Y(s) = \frac{1}{R}$	$Y(s) = \frac{1}{sL}$	$Y(s) = sc$
6.	Diagrammatic representation	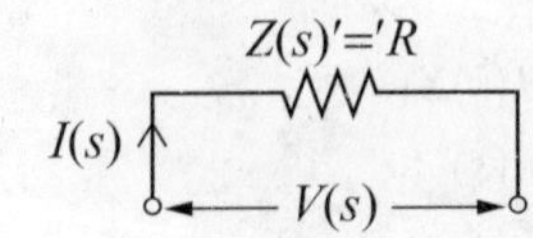	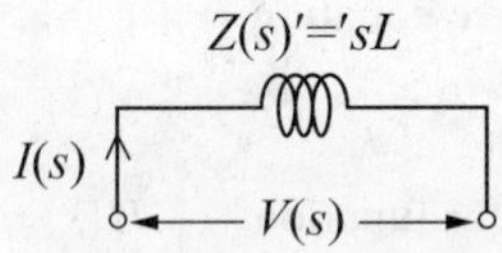	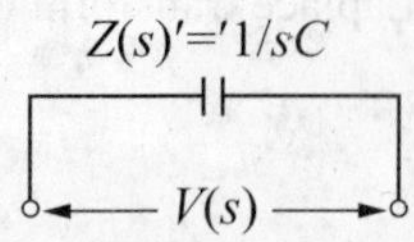
7.	Diagrammatic representation with initial conditions	-	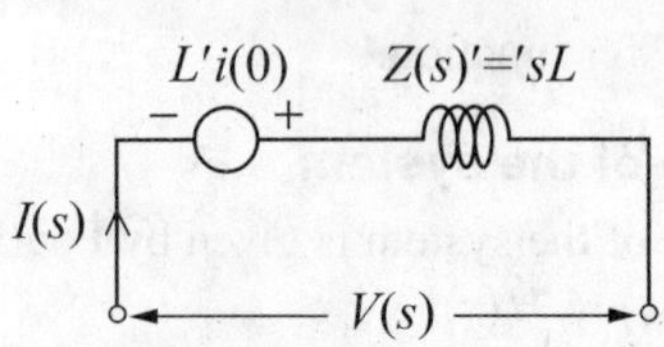	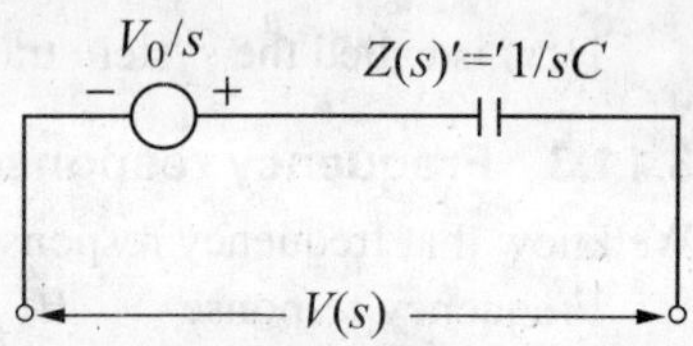

Example 27: Consider the *R-L* circuit with $R = 8\Omega$ and $L = 2H$ excited by a 48 volt dc source. Assume the initial current through the inductor is 3A, using the L.T. determine the current $i(t); t > 0$. Also draw the *s*-domain representation of the circuit.

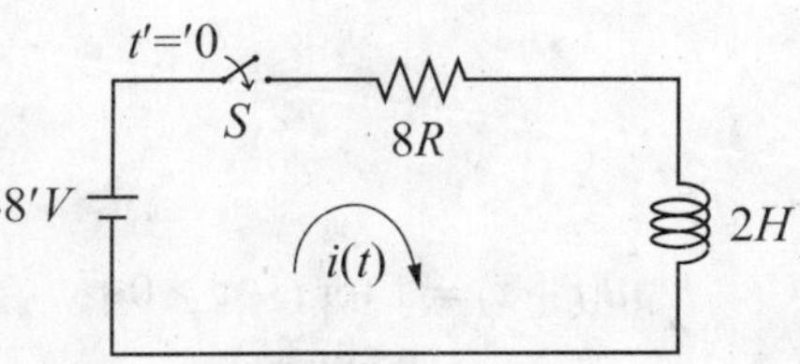

Fig. 3.26.

SOLUTION:

s-domain representation of the circuit

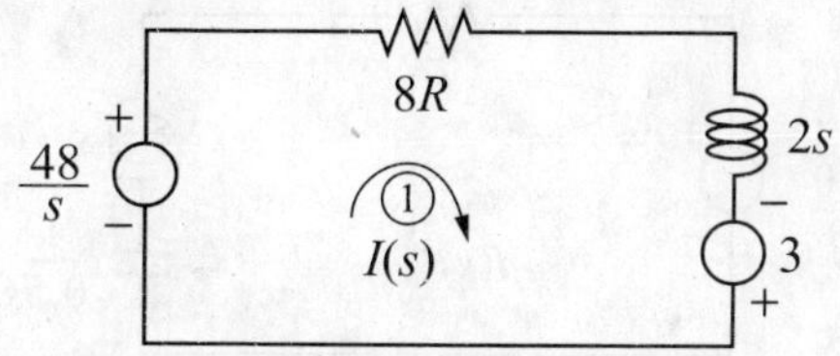

Applying KVL in loop (1)

$$\frac{48}{s} - 8\,I(s) - 2\,[sI(s) - i(0)] = 0$$

$$8\,I(s) + 2\,[sI(s) - 3] = \frac{48}{s}$$

$$4\,I(s) + sI(s) - 3 = \frac{24}{s}$$

$$I(s)(s+4) = \frac{24}{s} + 3 = \frac{24+3s}{s}$$

$$I(s) = \frac{24+3s}{s(s+4)} = \frac{A}{s} + \frac{B}{s+4}$$

On solving using partial fraction $A = 6; B = -3$

$$I(s) = \frac{6}{s} - \frac{3}{s+4}$$

$$\mathrm{LT}^{-1}[I(s)] = \mathrm{LT}^{-1}\left[\frac{6}{s} - \frac{3}{s+4}\right]$$

$$i(t) = 6 - 3e^{-4t}$$

Example 28: In the series R-L-C circuit shown in Fig. 3.27, there is no initial charge on the capacitor. If the switch S is closed at $t = 0$, determine the resulting current.

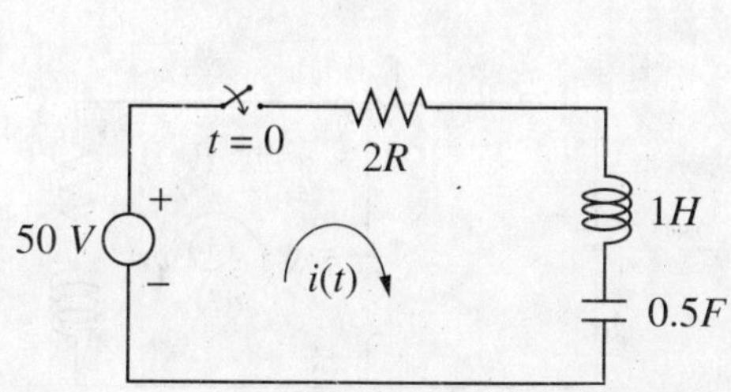

Fig. 3.27.

SOLUTION:

First we draw the s-domain representation of the circuit and then solve for $i(t)$.

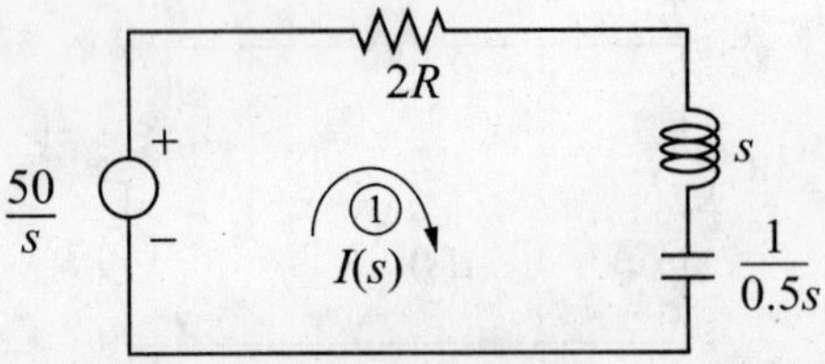

Applying KVL in loop (1)

$$\frac{50}{s} - 2\,I(s) - [sI(s) - i(0)] - \frac{1}{0.5}\left[\frac{1}{s}I(s) + \frac{v(0)}{s}\right] = 0$$

Since initial conditions are zero, therefore, $i(0) = v(0) = 0$

$$2I(s) + sI(s) + \frac{2}{s}I(s) = \frac{50}{s}$$

$$I(s)\frac{(s^2+2s+2)}{s} = \frac{50}{s}$$

$$I(s) = \frac{50}{s^2+2s+2} = \frac{50}{s^2+2s+1+1} = \frac{50}{(s+1)^2+1}$$

On taking ILT

$$i(t) = 50e^{-t}\sin t$$

$$\text{Since } \text{LT}[\sin t] = \frac{1}{s^2+1}$$

$$\text{LT}[e^{-at}\sin t] = \frac{1}{(s+a)^2+1}$$

Example 29: The switch in Fig. 3.28 has been in position 1 for a long time, it is moved to position 2 at $t = 0$. Obtain the current $i(t)$; for $t > 0$. Use L.T.

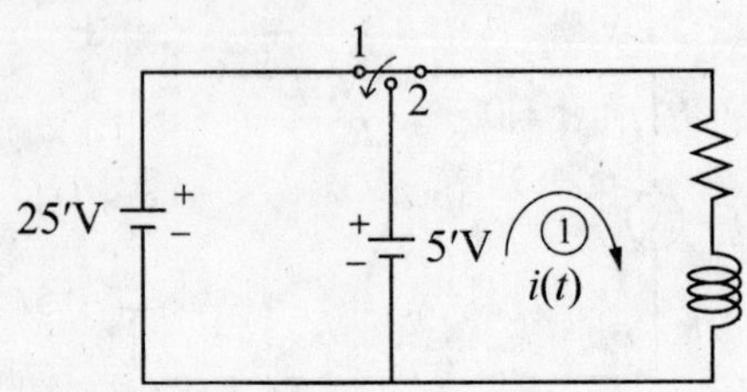

Fig. 3.28.

SOLUTION:

S-domain equivalent of Fig. 3.28.

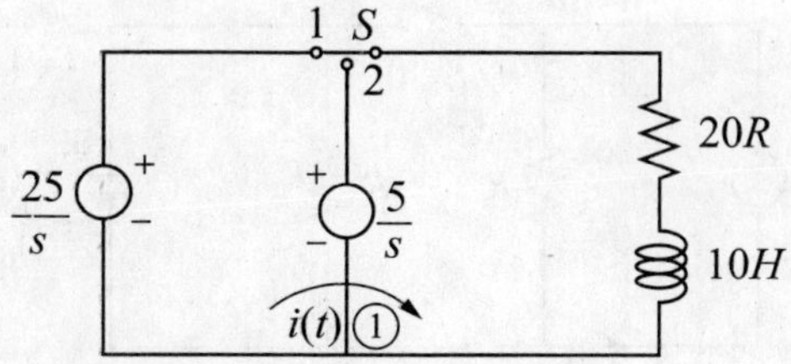

Applying KVL in loop (1)

$$\frac{25}{s} - (20+10s)I'(s) = 0 \qquad \text{Since } i'(0) = 0$$

$$I'(s) = \frac{2.5}{s(s+2)} = \frac{A}{s} + \frac{B}{s+2}$$

$$I'(s) = \frac{1.25}{s} - \frac{1.25}{s+2}$$

On taking ILT

$$i^1(t) = 1.25(1-e^{-2t}) \tag{1}$$

Since switch has been in position 1 for a long time

$$i(t)|_{t\to\infty} = 1.25A$$

With switch on position 2

$$\frac{5}{s} - 20I(s) - 10[sI(s) - i(0)] = 0$$

$$I(s)(20+10s) - 10i(0) = \frac{5}{s}$$

Since $i'(\infty) = i(0) = 1.25$

$$(20+10s)I(s) = \frac{10}{s} + 10\times 1.25$$

$$s(s+2)I(s) = 1 + 1.25s$$

$$I(s) = \frac{(1+1.25s)}{s(s+2)} = \frac{A}{s} + \frac{B}{s+2}$$

$$I(s) = \frac{0.5}{s} + \frac{0.75}{s+2}$$

On taking ILT

$$i(t) = 0.5 + 0.75e^{-2t}$$

Example 30: Switch S is closed at $t = 0$ in Fig. 3.29. Find $v_c(t)$ and $i_c(t)$. All initial conditions are zero.

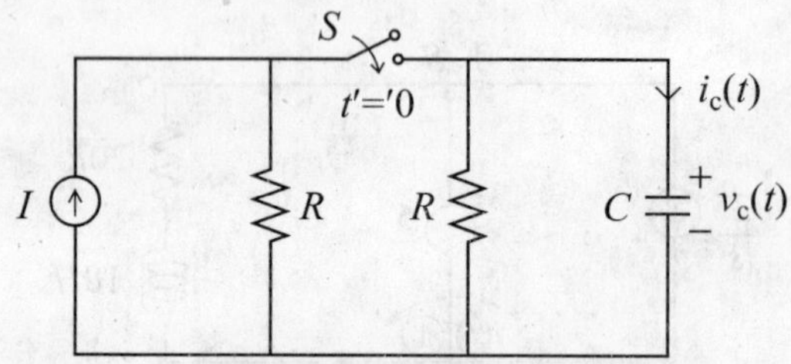

Fig. 3.29.

SOLUTION:

S-domain equivalent of Fig. 3.29.

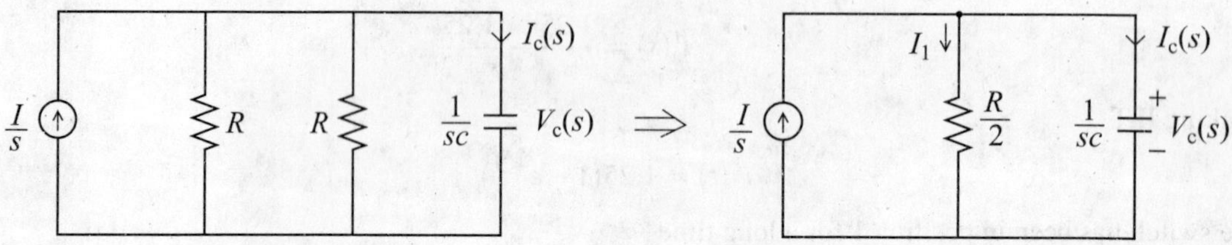

$$KCL \text{ at(a)} \frac{I}{s} = \frac{V_c(s)}{\frac{R}{2}} + I_c(s) \quad \text{where } I_c(s) = sc\,V_c(s)$$

$$\frac{I}{s} = \frac{2V_c(s)}{R} + scV_c(s) = \frac{(2+scR)}{R}V_c(s)$$

$$V_c(s) = \frac{IR}{s(2+scR)} = \frac{I/c}{s\left(s+\frac{2}{Rc}\right)} = \frac{A}{s} + \frac{B}{s+\frac{2}{Rc}}$$

$$V_c(s) = \frac{IR}{2s} - \frac{IR}{2\left(s+\frac{2}{Rc}\right)}$$

$$v_c(t) = \frac{IR}{2}\left[1 - e\frac{-2t}{Rc}\right] \tag{1}$$

$$I_c(s) = scV_c(s)$$

$$i_c(t) = c\frac{dv_c(t)}{dt}$$

$$i_c(t) = \frac{IRc}{2}.\left[0 + \frac{2}{Rc}e\frac{-2t}{Rc}\right]$$

$$i_c(t) = Ie^{-2t/Rc}$$

Example 31: Switch S is suddenly closed at $t = 0$ in Fig. 3.30. There is no charge on capacitor initially and no current flowing through the inductor at $t = 0$. Determine current $i_1(t)$ for $t > 0$.

SOLUTION:

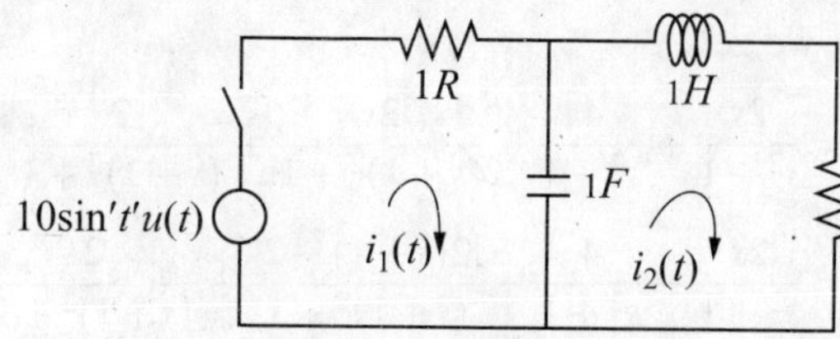

Fig. 3.30.

s-domain equivalent circuit of Fig. 3.30 is

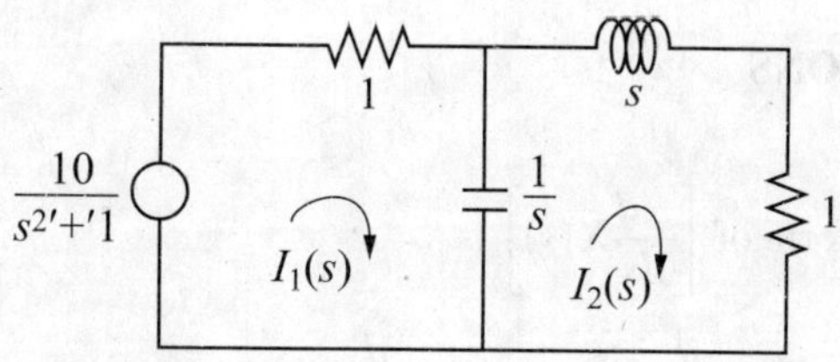

Applying KVL in loop (1)

$$I_1(s) + \frac{1}{s}\left(I_1(s) - I_2(s)\right) = \frac{10}{s^2+1}$$

$$I_1(s)\left(1+\frac{1}{s}\right) - \frac{1}{s}I_2(s) = \frac{10}{s^2+1} \tag{1}$$

KVL in loop (2)

$$I_2(s)(s+1) + \frac{1}{s}\left(I_2(s) - I_1(s)\right) = 0$$

$$-\frac{I_1(s)}{s} + I_2(s)\left(s+1+\frac{1}{s}\right) = 0$$

$$I_1(s) = \left(s^2+s+1\right) I_2(s) \tag{2}$$

$$I_2(s) = \frac{I_1(s)}{s^2+s+1} \tag{3}$$

On putting in (1), we get

$$I_1(s)(s+1) - \frac{I_1(s)}{s^2+s+1} = \frac{10s}{s^2+1}$$

$$I_1(s)\left\{s^3+s^2+s+s^2+s+1-1\right\} = \frac{10s(s^2+s+1)}{s^2+1}$$

$$I_1(s) = \frac{10s(s^2+s+1)}{(s^2+1)(s^3+2s^2+2s)} = \frac{10(s^2+s+1)}{(s^2+1)(s^2+2s+2)}$$

$$= \frac{10(s^2+s+1)}{(s^2+1)((s+1)^2+1)} = \frac{As+B}{s^2+1} + \frac{Cs+D}{(s+1)^2+1}$$

On solving using partial fraction, we get

$$= \frac{2s}{s^2+1} + \frac{4}{s^2+1} - \frac{2s}{(s+1)^2+1} + \frac{2}{(s+1)^2+1}$$

$$I_1(s) = \frac{2s}{s^2+1} + \frac{4}{s^2+1} - \frac{[2(s+1)-2]}{(s+1)^2+1} + \frac{2}{(s+1)^2+1}$$

$$= \frac{2s}{s^2+1} + \frac{4}{s^2+1} - \frac{2(s+1)}{(s+1)^2+1} + \frac{4}{(s+1)^2+1}$$

On taking ILT

$$X_1(t) = 2\cos t + 4\sin t - 2e^{-t}\cos t + 4e^{-t}\sin t$$

OBJECTIVE TYPE QUESTIONS

Q1: The inverse Laplace transform of $\left[\frac{-d}{ds}X(s)\right]$ is

(a) $-t\,x(t)$ (b) $t\,x(t)$ (c) $\frac{d}{dt}x(t)$ (d) $\frac{-d}{dt}x(t)$

Q2: The one-sided Laplace transform of $x(t) = \begin{cases} A & 0 \le t < T \\ 0 & \text{otherwise} \end{cases}$

(a) $\frac{A(1-e^{-sT})}{s}$ (b) $\frac{Ae^{-sT}}{s}$ (c) $\frac{A(1+e^{-sT})}{s}$ (d) $\frac{A(1-e^{sT})}{s}$

Q3: The ROC of the Laplace transform of $s(t)$ is

(a) jw axis (b) $\text{Re}(s) > 0$ (c) $\text{Re}(s) < 0$ (d) entire s-plane

Q4: One-sided Laplace transform of $x(t) = e^{-3t}\cos 4t$ is

(a) $X(s) = \frac{s+3}{s^2+6s+25}$ (b) $X(s) = \frac{s}{s^2+6s+25}$ (c) $X(s) = \frac{s+2}{s^2+8s+9}$ (d) $X(s) = \frac{s}{s^2+8s+9}$

Q5: If $X(s) = \frac{s}{s^2+2s+5}$ $x(t)$ is equal to

(a) $e^{-2t}\left[\sin 2t - \frac{1}{2}\cos 2t\right]$ (b) $e^{-t}\left[\sin 2t - \frac{1}{2}\cos 2t\right]$

(c) $e^{-t}\left[\cos 2t - \frac{1}{2}\sin 2t\right]$ (d) $e^{-2t}\left[\cos t - \frac{1}{2}\sin t\right]$

Q6: The ROC of $x(t) = u(t)$ is
(a) jw axis (b) Re(s) > 0 (c) Re(s) < 0 (d) entire s-place

Q7: If $X(s) = \dfrac{s-4}{s^2-2s}$, then $x(0)$ is equal to
(a) ∞ (b) 0 (c) 1 (d) 2

Q8: Given $X(s) = \dfrac{s+8}{s^2+6s+13}$, then $x(0)$ is
(a) 0 (b) ∞ (c) 1 (d) 2

Q9: The ROC of $x(t) = r(t)$ is
(a) entire s-plane (b) $\text{Re}(s) < 0$ (c) $\text{Re}(s) > 0$ (d) jw axis

Q10: The inverse Laplace transform of $X(s) = \dfrac{-1000}{s^2-100}$ for ROC $-10 < \text{Re}(s) < 10$
(a) $x(t) = 100\, e^{-10|t|}$ (b) $x(t) = 50\, e^{-10|t|}$ (c) $x(t) = 50e^{-10t}$
(d) $x(t) = 100\, e^{-10}t$

Answers
(1) b (2) a (3) d (4) a (5) c (6) b (7) c (8) c (9) c (10) b

UNSOLVED PROBLEMS

Q1: Find the Laplace transform of the following standard functions
(a) $x(t) = u(t)$ (b) $x(t) = u(t-a)$ (c) $x(t) = t\, u(t)$

Q2: Find the Laplace transform of the following
(a) $\tau(t) = S(t)$ (b) The unit doublet function $S^1(t)$

Q3: Find the Laplace transform of the following functions
(a) $x(t) = \sin wt$ (b) $x(t) = e^{-at} \sin wt$
(c) $x(t) = te^{-at} u(t)$

Q4: Find the Laplace transform of given waveforms

a)

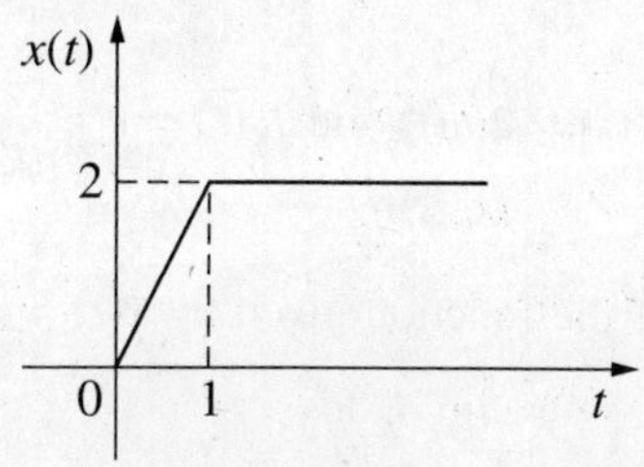

Fig. 3.1 P.

b)

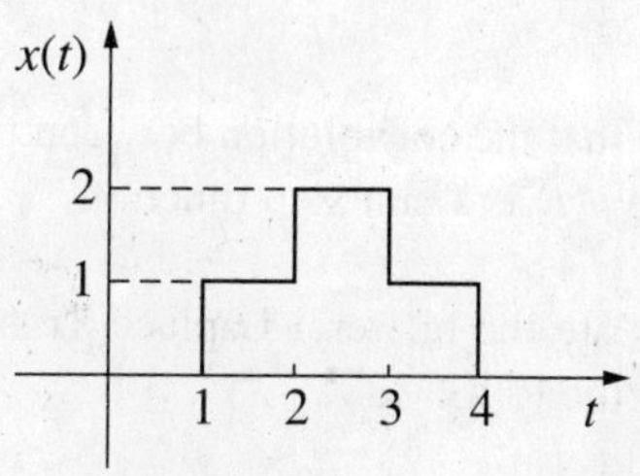

Fig. 3.2 P.

Q5: Find the L.T. of periodic waveform as given in figure.

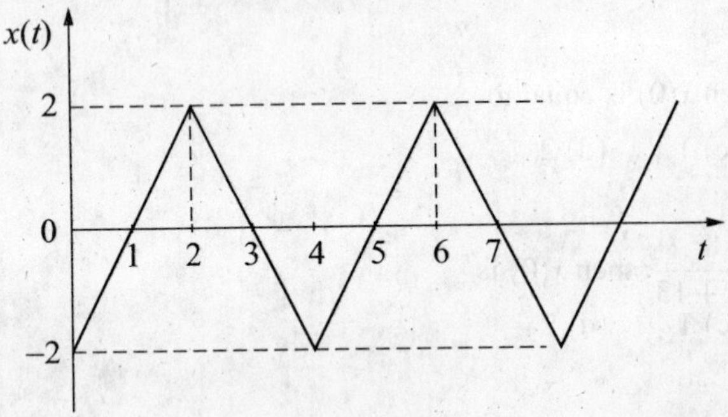

Fig. 3.3 P.

Q6: Solve the differential equation

$$\frac{d^2y(t)}{dt^2} + 3\frac{dy(t)}{dt} + 2y(t) = 0,\ y(0) = 2,\ \frac{dy(0)}{dt} = -3$$

Q7: Find the inverse L.T. of

$$X(s) = \frac{1}{(s+1)(s+5)} \text{ if the ROC is } -5 < \text{Re}(s) < -1$$

Q8: Find the signal $x(t)$ whose L.T. are as given below:

$X(s) = \dfrac{2s}{(s+1)(s+3)}$ if the ROC is (a) $-1 < \text{Re}(s) < -3$ (b) Re(s)>-1 (c) $\text{Re}(s) <$-3

Q9: Obtain the transfer function of the system when

$y(t) = e^{-t} - 2e^{-2t} + e^{-3t}$ and $x(t) = e^{-0.5t}$

Q10: Find the response of the system with transfer function

$H(s) = \dfrac{1}{(s+1)(s+0.5)}$ for a unit impulse input

QUESTION BANK

Q1: Verify that the convolution between two functions $f_1(t) = 2\,u(t)$ and $f_2(t) = e^{-3t}u(t)$ is $2/3(1 - e^{-3t})$; where $u(t)$ is a unit step function.

Q2: Calculate the bilateral Laplace Transform (L.T.) of the function given by $x(t) = e^{at}u(t) + e^{bt}u(-t)$, Draw the ROC.

Q3: Find the response $y(t)$ of the system, with the transfer function $H(s) = \frac{1}{s+4}; \text{Re}(s) > -4$ to the input $x(t) = e^{-t}u(t) + e^{-3t}u(-t)$.

Q4: Determine the signal $x(t)$, whose first derivative is as shown in figure and find $X(s)$;

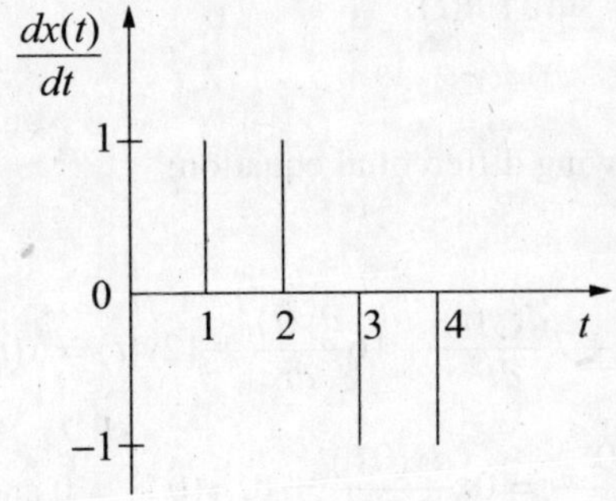

Fig. 3.4 P.

Q5: Determine the unilateral L.T. of the following functions:

(a) $x(t) = t^3 + 3t^2 - 6t + 4$ (b) $x(t) = \cos^3 3t$
(c) $x(t) = \sin(at)\cos(bt)$ (d) $x(t) = t\sin(at)$

Q6: In the circuit of following figure, the switch S is closed at $t = 0$ and there is no initial charge on either of the capacitors. Find the resulting current $i(t)$. Also determine the initial and final values of current.

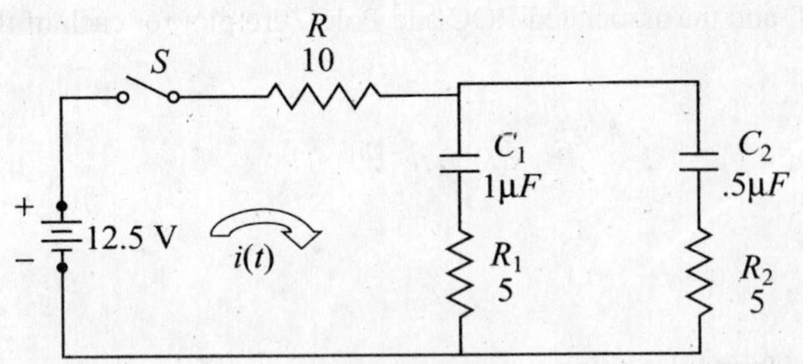

Fig. 3.5 P.

Q7: Find the inverse L.T. of the following:

$$\text{(a) } X(s) = \frac{3s^2 + 8s + 6}{(s+2)(s^2 + 2s + 1)}$$

$$\text{(b) } X(s) = \frac{2}{(s+4)(s-1)} \text{ ROC}: (1) -4 \; \text{Re}(s) < 1$$

(2) $\text{Re}(s) > 1$

(3) $\text{Re}(s) < -4$

Q8: Find the inverse L.T. using convolution integral.

$$X(s) = \frac{e^{-2s}}{s(s+3)}$$

Q9: Use the convolution theorem of L.T. to find $y(t) = x_1(t) * x_2(t)$

$x_1(t) = \cos 4\,t\;u(t) \qquad x_2(t) = \sin 2\,t\;u(t)$

Q10: By using L.T., solve the following differential equation:

(1)

$$\frac{d^3y(t)}{dt^3} + 7\frac{d^2y(t)}{dt^2} + 16\frac{dy(t)}{dt} + 12y(t) = x(t)$$

$$\text{if } \frac{d^2y(0^-)}{dt^2} = 0, \frac{dy(0^-)}{dt} = 0, \; y(0^-) = 0 \text{ and } x(t) = \delta(t)$$

(2)

$$\frac{d^2y(t)}{dt^2} + 3\frac{dy(t)}{dt} + 2y(t) = \frac{dx(t)}{dt}$$

$$\text{if } \frac{dy(0^-)}{dt} = 1, \; y(0^-) = 2 \text{ and } x(t) = e^{-1}u(t)$$

Q11: Determine the L.T. and the associated ROC and Pole-Zero plot for each of the following functions of time:

(a) $x(t) = e^{-2t}u(t) + e^{3t}u(t)$ (b) $x(t) = te^{-2|t|}$

(c) $x(t) = \delta(3t) + u(3t)$

Q12: Find the L.T. of the triangular pulse.

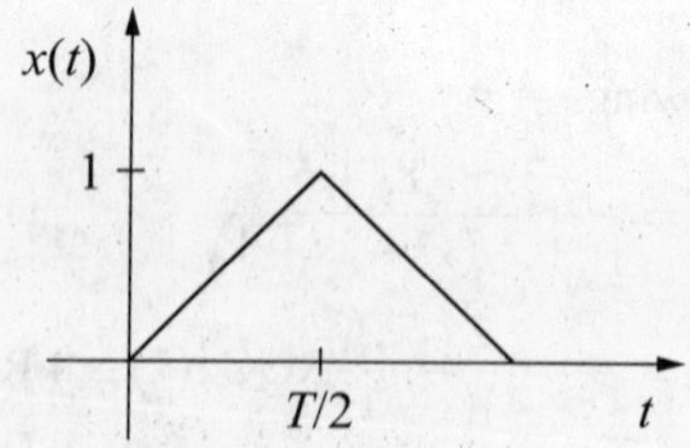

Fig. 3.6 P.

Q13: Find the inverse Laplace transform of $G(s) = \dfrac{10s^2e^{-s}}{(s+1)(s+3)}$

Q14: (a) Solve the differential equation

$$\frac{d^2x(t)}{dt^2}+7\frac{dx(t)}{dt}+12x(t)=0, \text{ for time } t>0 \text{ subject to the initial condition}$$

$$\frac{dx(t)}{dt}t=0^-=-4,\ x(0^-)=2$$

Q15: A signal $x(t)$ has the Laplace transform $X(s)=\dfrac{s+2}{s^2+4s+5}$.

Find the Laplace of the following signal: $y(t)=x(2t-1)u(2t-1)$.

SOLUTIONS OF QUESTION BANK

Q1:

$$f_1(t)=2u(t)$$
$$f_2(t)=e^{-3t}u(t)$$
$$F_1(s)=\frac{2}{s}\ F_2(s)=\frac{1}{s+3}$$
$$f(t)=f_1(t)^*f_2(t)$$
$$F(s)=F_1(s)F_2(s)=\frac{2}{s(s+3)}$$

Q2:

$$L\left[e^{at}u(t)\right]=\frac{1}{s-a}\ \text{Re}(s)>a$$
$$L\left[-e^{bt}u(-t)\right]=\frac{1}{s-a}\ \text{Re}(s)>b$$

Q3:

$$H(s)=\frac{1}{s+4}\ \text{Re}(s)>-4$$
$$x(t)=e^{-t}u(t)+e^{-3t}u(-t)$$
$$X(s)=\frac{1}{s+1}+\frac{-1}{s+3}$$

ROC:$(\text{Re}(s)>-1)$ $(\text{Re}(s)<-3)$

In this case, the ROC of $X(s)$ does not exist as there is no region for which $\text{Re}(s)<-3$ and $\text{Re}(s)>-1$. Therefore, we must determine the response of the system separately to each of the two inputs as

$$x_1(t)=e^{-t}u(t) \text{ and } x_2(t)=e^{-3t}u(-t)$$

If $y_1(t)$ and $y_2(t)$ are the responses of the system to the inputs $x_1(t)$ and $x_2(t)$ respectively, then

$$Y_1(s)=\frac{1}{(s+1)(s+4)}=\frac{1}{3(s+1)}=-\frac{1}{3(s+4)}$$
$$y_1(t)=\frac{1}{3}\left(e^{-t}-e^{-4t}\right)u(t)$$

$$Y_2(s) = \frac{-1}{(s+3)(s+4)} \quad -4 < \text{Re}(s) < -3$$
$$= -\frac{1}{s+3} + \frac{1}{s+4}$$
$$y_2(t) = e^{-3t}u(-t) + e^{-4t}u(t)$$
$$y(t) = y_1(t) + y_2(t) = \frac{1}{3}(e^{-t} - e^{-4t})u(t) + e^{-3t}u(-t) + e^{-4t}u(t)$$
$$= e^{-3t}u(-t) + \left(\frac{1}{3}e^{-t} + \frac{2}{3}e^{-4t}\right)u(t)$$

Q4:

$$\frac{dx(t)}{dt} = s(t-1) + s(t-2) - s(t-3) - s(t-4)$$
$$L\left(\frac{dx(t)}{dt}\right) = L(s(t-1) + s(t-2) - s(t-3) - s(t-4))$$
$$X(s) = \frac{e^{-s}}{s} + \frac{e^{-2s}}{s} - \frac{e^{-3s}}{s} - \frac{e^{-4s}}{s}$$

ILT

$$x(t) = u(t-1) + u(t-2) - u(t-3) - u(t-4)$$

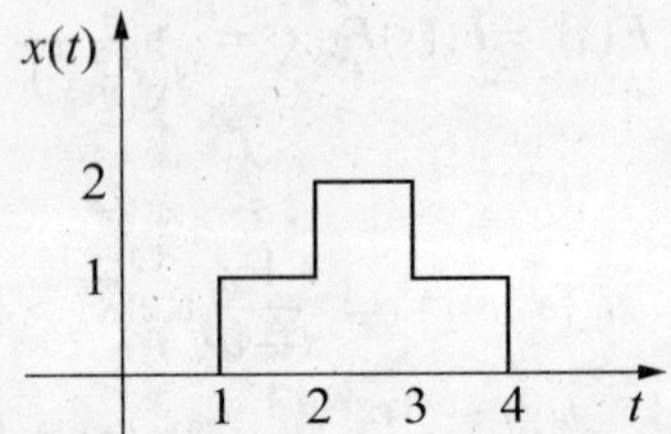

Q5:

(a)
$$L\left[\frac{t^{n-1}}{\lfloor n-1}x(t)\right] = \frac{d^n}{ds^n}X(s)$$
$$\frac{t^{n-1}}{\lfloor n-1}u(t) = \frac{1}{s^n}$$
$$X(s) = \frac{6}{s^4} + \frac{6}{s^3} - \frac{6}{s^2} + \frac{4}{s}$$
$$\text{For V.L.T } \frac{t^3}{\lfloor 3} = \frac{1}{s^4} \Rightarrow L(t^3) = \frac{6}{s^4}$$

(b) $x(t) = \cos^3 3t$

We know that $\cos 3A = 4\cos^3 A - 3\cos A$

$$x(t) = \cos^3 3t = \frac{1}{4}[\cos 9t + 3\cos 3t]$$
$$x(s) = \frac{1}{4}\left[\frac{s}{s^2+s1} + \frac{3s}{s^2+9}\right]$$

(c)
$$x(t) = \sin at bt = \frac{1}{2} 2 \sin at \cos bt = \frac{1}{2} (\sin (a+b)t + \sin (a-b)t)$$
$$X(s) = \frac{1}{2}\left[\frac{a+b}{s^2+(a+b)^2} + \frac{a-b}{s^2+(a-b)^2}\right]$$

(d)
$$x(t) = t \sin at$$
$$L[x(t)] = L[t \sin at] = -\frac{d}{ds} L[\sin at] = -\frac{d}{ds}\left[\frac{a}{s^2+a^2}\right] = \frac{2s}{(s^2+a^2)^2}$$

Q6:

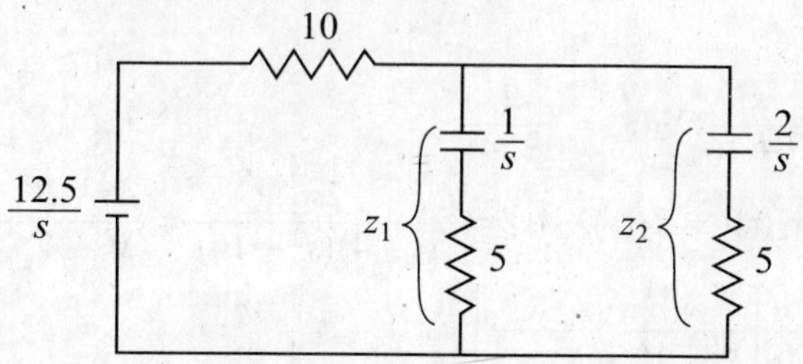

$$2(s) = 10 + (Z_1 \; Z_2)$$
$$Z(s) = 10 + \frac{\left(5+\frac{1}{s}\right)\left(5+\frac{2}{s}\right)}{\left(5+\frac{1}{s}+5+\frac{2}{s}\right)} = \frac{125s^2+45s+2}{s(10s+3)}$$
$$I(s) = \frac{V(s)}{Z(s)} = \frac{12.5}{s} \cdot \frac{s(10s+3)}{125s^2+45s+2} = \frac{s+0.3}{(s+.30s)(s+.052)}$$
$$I(s) = \frac{\frac{1}{32}}{s+.30s} + \frac{\frac{31}{32}}{s+0.052}$$
$$x(t) = \frac{1}{32}e^{-.30st} + \frac{31}{32}e^{-.052t}$$

Initial value
$$s(0) = \lim_{s\to\infty} sI(s) = 1$$

Final value $s(\infty) = \lim_{s\to 0} sI(s) = 0$

Q7:

(a) $x(t) = 2e^{-2t}u(t) + e^{-t}u(t) + te^{-t}u(t)$

(b) (1) $x(t) = -\frac{2}{5}e^{-ut}u(t) - \frac{2}{5}e^{t}u(-t)$

(2) $x(t) = -\frac{2}{5}e^{-4t}u(t) + \frac{2}{5}e^{t}u(t)$

(3) $x(t) = \frac{2}{5}e^{-4t}u(-t) - \frac{2}{5}e^{t}u(-t)$

Q8:

$$X(s) = \frac{e^{-Ls}}{s(s+3)} = X_1(s)X_2(s)$$

$$X_1(s) = \frac{e^{-2s}}{s} X_2(s) = \frac{1}{s+3}$$

$$x_1(t) = u(t-2) \; x_2(t) = e^{-3t}u(t)$$

$$x(t) = x_1(t) * x_2(t) = -\int_2^t e^{-3(t-\tau)} d\tau = e^{-3t} \int_2^t e^{3\tau} d\tau$$

$$= \frac{1}{3} e^{-3t} \left[e^{3t} - e^{6} \right] = \left[\frac{1}{3} - \frac{1}{3} e^{-3(t-2)} \right]$$

Q9:

$$X_1(s) = \frac{s}{s^2+16} \; X_2(s) = \frac{2}{s^2+4}$$

$$L[x_1(t) * x_2(t)] = X_1(s)X_2(s) = \frac{2s}{(s^2-4)(s^2+16)} = \frac{As+B}{s^2+4} + \frac{(s+1)}{s^2+16}$$

$$X(s) = -\frac{1}{6} \left[\frac{s}{s^2+16} - \frac{s}{s^2-4} \right]$$

$$x(t) = -\frac{1}{6} [\cos 4t - \cos 2t] \, u(t)$$

Q10:

(1) $y(t) = -e^{-2t}u(t) + te^{-2t}u(t) + e^{-3t}u(t)$

(2) $y(t) = 7e^{-t}u(t) - te^{-t}u(t) - 5e^{-2t}u(t)$

Q11:

(a)

$$x(t) = e^{-2t}u(t) + e^{-3t}u(t) \; X(s) = \frac{1}{s+2} + \frac{1}{s+3} = \frac{2s+5}{(s+2)(s+3)}$$

Pole at −2, −3 zero at −2.5 ROC: Re(s) > −2

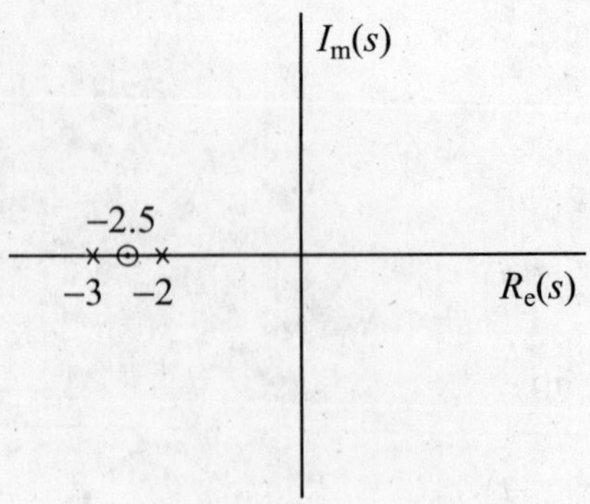

(b) $$x(t) = te^{-2(t)} = \begin{cases} te^{-2t} & t > 0 \\ te^{2t} & t < 0 \end{cases}$$

$$X(s) = \frac{1}{(s+2)2} + \frac{1}{(s-2)2} = \frac{2(s^2+4)}{(s^2-4)2} \qquad \text{ROC: Re}(s) > 2$$

pole : ± 2

zero : $\pm 2j$

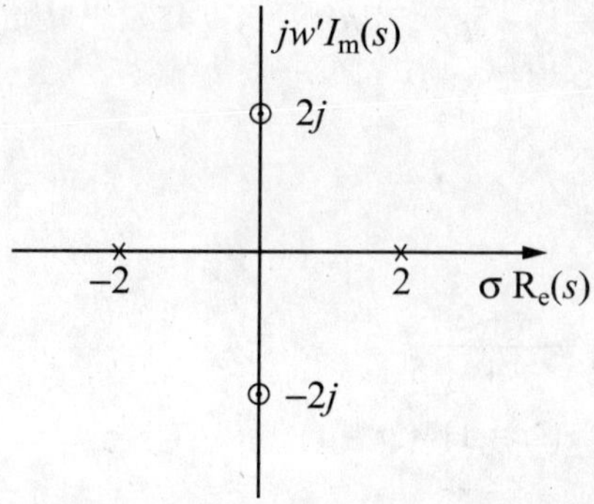

(c) $$x(t) = \delta(3t) + u(3t)\ X(s) = \frac{1}{3}.1 + \frac{1}{3}\cdot\frac{1}{\frac{s}{3}} = \frac{1}{3} + \frac{1}{s} = \frac{s+3}{3s}$$

pole : $s = 0$

zero : $s = -3$

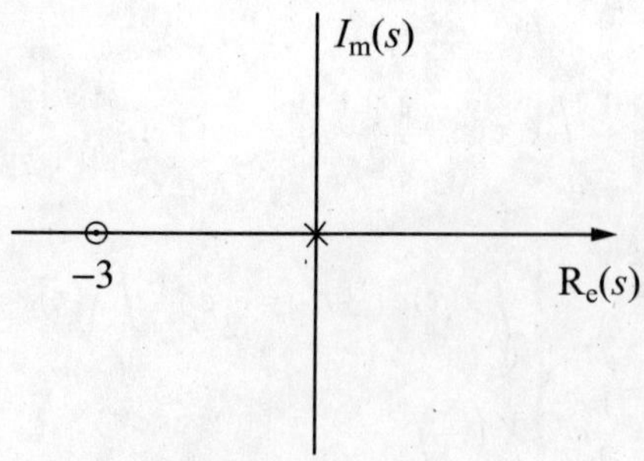

Q12:

$$x(t) = \frac{2t}{T}\left[u(t) - u\left(t - \frac{T}{2}\right)\right] + \left(\frac{-2t}{T} + 2\right)\left[u\left(t - \frac{T}{2}\right) - u(t-T)\right]$$

On solving

$$x(t) = \frac{2}{T}r(t) - \frac{4}{T}r\left(t - \frac{T}{2}\right) + \frac{2}{T}r(t-T)$$

$$X(s) = \frac{2}{Ts^2} - \frac{4}{T}\frac{e^{-s\frac{T}{2}}}{s^2} + \frac{2e^{-sT}}{Ts^2}$$

Q13:

$$G(s) = \frac{10s^2 e^{-2}}{(s+1)(s+3)} = e^{-s}G_1(s)$$

$$G_1(s) = \frac{10s^2}{(s+1)(s+3)} = 10\left[1 - \frac{(4s+3)}{(s+1)(s+3)}\right] = 10\left[1 + \frac{5}{s+1} - \frac{45}{s+3}\right]$$

$$g_1(t) = 10\left[\delta(t) + 5e^{-t}u(t) - 45e^{-3t}u(t)\right]$$

$$L^{-1}[G(s)] = L^{-1}[e^{-s}G_1(s)] = g_1(t-1)$$

$$g(t) = 10\left[\delta(t-1) + 5e^{-(t-1)}u(t-1) - 45e^{-3(t-1)}u(t-1)\right]$$

Q15:

$$X(s) = \frac{s+2}{s^2+4s+5}$$

$$y(t) = x(2t-1)4(2t-1)$$

$$Y(s) = \int_{-\infty}^{\infty} x(2t-1)u(2t-1)e^{-st}\,dt$$

$$= \int_{\frac{1}{2}}^{\infty} x(2t-1)e^{-st}\,dt$$

$$2t - 1 = \tau$$

$$= \frac{1}{2}\int_{0}^{\infty} x(\tau)e^{-s}\frac{\tau+1}{2}\,d_\tau$$

$$= \frac{1}{2}e^{-\frac{s}{2}}\int_{0}^{\infty} x(\tau)e^{-s\frac{\tau}{2}}\,d_\tau = \frac{1}{2}e^{-\frac{s}{2}}\int_{0}^{\infty} x(\tau)e^{-\left(\frac{s}{2}\right)\tau}\,d_\tau$$

$$Y(s) = \frac{1}{2}e^{-\frac{s}{2}}X\left(\frac{s}{2}\right)$$

$$Y(s) = \frac{1}{2}e^{-\frac{s}{2}}\frac{\frac{s}{2}+2}{\left(\frac{s}{2}\right)^2 + 4\cdot\frac{s}{2} + 5} = \frac{e^{-\frac{s}{2}}}{2}\frac{s+4}{2\left(\frac{s^2}{4} + 2s + 5\right)}$$

$$= e^{\frac{-s}{2}}\frac{(s+4)}{s^2+5s+20} = \frac{(s+4)e^{\frac{-s}{2}}}{s^2+5s+20}$$

4

State Variable Analysis

4.1 INTRODUCTION

The conventional approach used to study the behaviour of the systems, using the time domain or frequency domain methods. These methods require that the physical system be modelled in the form of transfer function, which is the ratio of Laplace transform of output to input, neglecting the initial conditions. Thus, the conventional approach faces all the limitations associated with the transfer function approach.

Some of its limitations can be stated as

(1) The method is insufficient and troublesome to give complete solution of higher order system.
(2) It is not very much convenient for the analysis of multiple input and multiple output.
(3) It gives analysis of the system for some specific types of inputs like step, ramp, impulse, etc.
(4) The classical methods like root locus, bode plot, etc., are basically trial and error procedures which fail to give the optimal solution required.
(5) It is only applicable to linear time invariant system.

Hence, it is necessary to use a method of analyzing systems which overcomes most of the above said difficulties. In this chapter state variable analysis is discussed which uses the concept of total internal state of the system considering all initial conditions. It is a time domain approach but it has number of advantages compared to conventional method of analysis.

4.2 ADVANTAGES OF STATE VARIABLE ANALYSIS

Advantages of state variable analysis are:

(1) The method takes into account the effect of all initial conditions.
(2) It can be applied to non-linear as well as time varying system.
(3) It can be applied to multiple input multiple output system.
(4) Any type of input can be considered for designing of the system.
(5) As the method involves matrix, algebra can be conveniently adopted for digital computers.
(6) The state variable selected need not necessarily be the physical quantities of the system.
(7) The vector matrix notation greatly simplifies the mathematical representation of the system.

4.3 IMPORTANT DEFINITIONS

State: The state of a system at any time 't_0'is the minimum set of numbers $x_1, x_2, x_3 \ldots, x_n$, which along with the input to the system for $t \geq t_0$ is sufficient to determine the behaviour of the system for all $t \geq t_0$.

In other words, the state of a system represents the minimum amount of information that we need to know about a system at 't_0' such that its future behaviour can be determined without reference to the input before 't_0'. The state can also be defined as the information of the system at time t_0, that together with input

$u\ (t_0, \infty)$ determine the unique behaviour of the system, we mean all responses, including the state of the system. If the system is a network we mean the voltage and current of every branch of the network.

Consider the network shown in Fig. 4.3. If the initial current through the inductor and initial voltage across the capacitor are known then for any driving voltage the behaviour of the network can be determined. Hence, the inductor current and capacitor voltage can be considered the state of the network.

State variable: The state of a dynamic system is the smallest set of variables such that the knowledge of these variables at $t = t_0$ with the knowledge of the input for $t \geq t_0$ completely determines the behaviour of the system for any time $t \geq t_0$. The set of variables is called state variables.

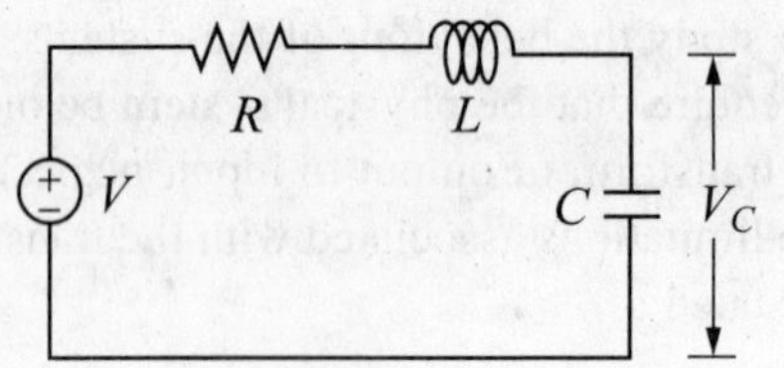

Fig. 4.1.

State vector: If we need n-variables to completely describe the behaviour of a given system, then these n-state variables may be considered n-component of a vector 'x'. Such a vector is called state vector. A state vector is thus a vector which determines uniquely the system $x(t)$ for any time $t \geq t_0$, once the state at $t = t_0$ is given and the input $u(t)$ for $t \geq t_0$ is specified.

State space: The n-dimensional space co-ordinates axes consist of the x_1 axis, x_2 axis,..., x_n axis called state space. Any state can be represented by a point in state space.

4.4 STATE SPACE REPRESENTATION

Consider a system which has n state variables, m inputs, p outputs. The state variables are represented as $x_1(t), x_2(t), \ldots$, input vector $u_1(t), u_2(t), \ldots$ and output to vector $y_1(t), y_2(t), \ldots$

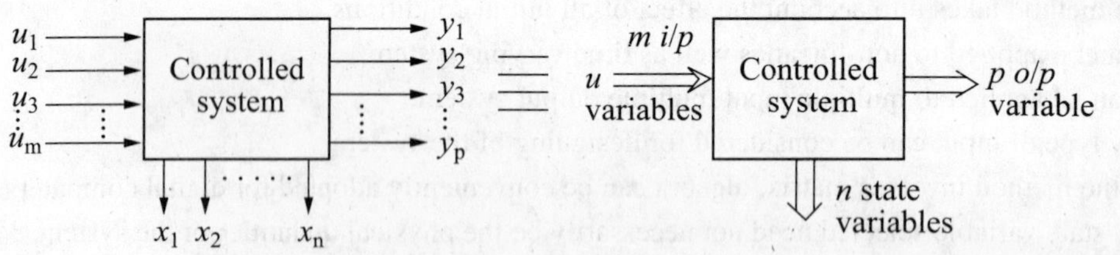

Fig. 4.2. Structure of general control system

Consider a single i/p single o/p system (SISO) which may be written by the following nth order differential equation.

$$\frac{dy^n(t)}{dt^n} + a_1\frac{d^{n-1}y(t)}{dt^{n-1}} + a_2\frac{d^{n-2}(y(t))}{dt^{n-2}} + a_3\frac{d^{n-3}y(t)}{dt^{n-3}} + \cdots + a_{n-1}\frac{d}{dt}y(t) + a_n\, y(t) = u(t) \quad (1)$$

Hence, $y(0), \frac{d}{dt}y(0), \frac{d^2y(0)}{dt^2}, \cdots \frac{d^ny(0)}{dt^n}$ and $u(t)$ are given.

Let us define $y = x_1$

$$\frac{dy}{dt} = \dot{x}_1 = x_2$$
$$\frac{d^2y}{dt^2} = \dot{x}_2 = x_3$$
$$\vdots$$
$$\frac{d^{n+y}}{dt^{n-1}} = \dot{x}_{n-1} = x_n$$

Then from eqn. (1)

$$\dot{x}_n = -a_n x_1 - a_{n1} x_1 - a_{n-2} x_2 \cdots - a_{-1} x_n + u \quad (2)$$

We can write the above equation in matrix form as

$$\dot{x} = Ax + Bu \quad (3)$$

Output $y = CX$

$$\text{where } x = \begin{bmatrix} x_1 \\ x_2 \\ \vdots \\ x_{n-1} \\ x_n \end{bmatrix} \quad A = \begin{bmatrix} 0 & 1 & 0 & \cdots & 0 \\ 0 & 0 & 1 & \cdots & 0 \\ \cdots & \cdots & \cdots & \cdots & \cdots \\ 0 & 0 & 0 & \cdots & 0 \\ 0 & 0 & 0 & \cdots & 1 \\ (-a_n) & (-a_{n-1}) & (-a_{n-2}) & \cdots & (-a_1) \end{bmatrix} \quad B = \begin{bmatrix} 0 \\ 0 \\ 0 \\ \vdots \\ 1 \end{bmatrix}$$

and $C = [1\ 0\ 0\ \cdots\ 0]$

where $x(t)$ = state vector matrix of order $n \times 1$

$u(t)$ = input vector matrix of order $m \times 1$

A = system or evolution matrix of order $m \times n$

B = input or control matrix of order $n \times m$

$y(t)$ = output vector matrix of order $p \times 1$

Equation (3) is called the state equation of the system. Hence, in general, the dynamics of nth order multivariable system having m inputs and p outputs can be written by vector matrix differential equation as $\dot{x} = Ax + Bu$ and $o/p\ y = Cx + Du$

$$\text{where } x = \begin{bmatrix} x_1 \\ x_2 \\ x_3 \\ \vdots \\ x_n \end{bmatrix} \quad u = \begin{bmatrix} u_1 \\ u_2 \\ u_3 \\ \vdots \\ u_m \end{bmatrix} \quad y = \begin{bmatrix} y_1 \\ y_2 \\ y_3 \\ \vdots \\ y_p \end{bmatrix}$$

$$A = \begin{bmatrix} a_{11} & a_{12} & a_{13} & \cdots & a_{1n} \\ a_{21} & a_{22} & \cdots & \cdots & a_{2n} \\ \vdots & & & & \vdots \\ a_{n1} & a_{n2} & \cdots & \cdots & a_{nn} \end{bmatrix}_{n\times n} = \text{Evolution matrix}$$

$$B = \begin{bmatrix} b_{11} & b_{12} & \cdots & \cdots & b_{1m} \\ b_{21} & b_{22} & \cdots & \cdots & b_{2m} \\ \vdots & & & & \vdots \\ b_{n1} & b_{n2} & \cdots & \cdots & b_{nm} \end{bmatrix}_{n\times m} = \text{Control matrix}$$

$$C = \begin{bmatrix} c_{11} & c_{12} & \cdots & \cdots & c_{1n} \\ c_{21} & c_{22} & \cdots & \cdots & c_{2n} \\ \vdots & & & & \vdots \\ c_{p1} & c_{p2} & \cdots & \cdots & c_{pn} \end{bmatrix}_{p\times n} = \text{Observation matrix}$$

$$D = \begin{bmatrix} d_{11} & d_{12} & \cdots & \cdots & d_{1m} \\ d_{21} & d_{22} & \cdots & \cdots & d_{2m} \\ \vdots & \vdots & & & \vdots \\ d_{p1} & d_{p2} & \cdots & \cdots & d_{pm} \end{bmatrix}_{p\times m} = \text{Transmission matrix}$$

Example 1: Obtain the state model of the given electrical system

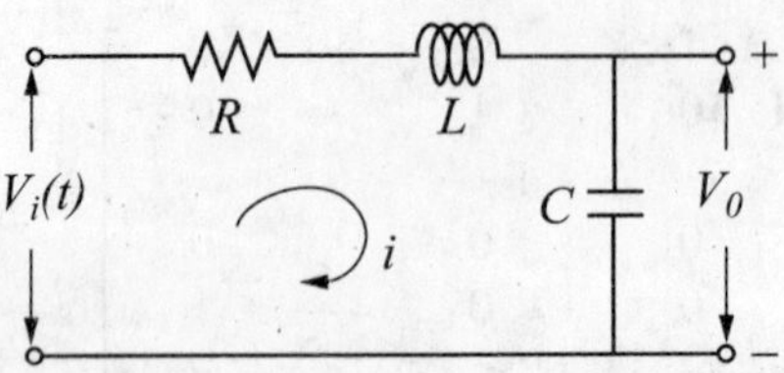

Fig. 4.3.

SOLUTION:

Consider two state variables $i(t)$ and o/p voltage $V_0(t)$.

$X_1(t) = i(t)$ $\quad X_2(t) = V_0(t)$ $\quad u(t) = V_i(t)$

Applying KVL to the loop

$$V_i(t) = i(t)R + L\frac{di}{dt} + V_0(t)$$

$$\frac{di(t)}{dt} = -i(t)\frac{R}{L} - \frac{V_0(t)}{L} + \frac{V_i(t)}{L}$$

$$\dot{X}_1(t) = -X(t)\frac{R}{L} - \frac{X_2(t)}{L} + \frac{u(t)}{L} \tag{1}$$

$$\because \quad V_0 = \frac{1}{C}\int idt$$

$$\therefore \quad v_0(t) = \frac{1}{C}\int i(t)dt$$

$$\frac{dv_0(t)}{dt} = \frac{1}{C}i(t)$$

$$\dot{X}_2(t) = \frac{1}{C}x(t) \tag{2}$$

Hence, eqns. (1) and (2) are state equations

$$\begin{bmatrix} \dot{X}_1(t) \\ \dot{X}_2(t) \end{bmatrix} = \begin{bmatrix} \frac{-R}{L} & -\frac{1}{L} \\ \frac{1}{C} & 0 \end{bmatrix} \begin{bmatrix} X_1(t) \\ X_2(t) \end{bmatrix} + \begin{bmatrix} \frac{1}{L} \\ 0 \end{bmatrix} u(t)$$

$$\text{i.e.} \quad \dot{X}(t) = A\,X(t) + B\,U$$

while the output variable $Y(t) = V_0(t) = X_2(t)$

$$Y(t) = [0\ 1] \begin{bmatrix} X_1(t) \\ X_2(t) \end{bmatrix} + 0u(t)$$

$$Y(t) = CX(t) \quad D = (0)$$

Example 2: A system is described by the following differential equation. Represent the system in state space

$$\frac{d^2x}{dt^2} + 3\frac{d^2x}{dt^2} + 4\frac{dx}{dt} + 4x = u_1 + 3u_2 + 4u_3$$

and output

$$y_1 = 4\frac{dx}{dt} + 3u_1$$

$$y_2 = \frac{d^2x}{dt^2} + 4u_2 + u_3$$

SOLUTION:

Select the state variables as

$$x_1 = x$$

$$\dot{x}_1 = \dot{x} = x_2$$

$$\dot{x}_2 = \ddot{x}_1 = \ddot{x} = x_3$$

$$\dot{x}_3 = u_1 + 3u_2 + 4u_3 - 3\dot{x}_3 - 4\dot{x}_2 - 4x_1$$

$$\begin{bmatrix} \dot{x}_1 \\ \dot{x}_2 \\ \dot{x}_3 \end{bmatrix} = \begin{bmatrix} 0 & 1 & 0 \\ 0 & 0 & 1 \\ -4 & -4 & -3 \end{bmatrix} \begin{bmatrix} x_1 \\ x_2 \\ x_3 \end{bmatrix} + \begin{bmatrix} 1 \\ 3 \\ 4 \end{bmatrix} \begin{bmatrix} u_1 \\ u_2 \\ u_3 \end{bmatrix}$$

and output $\rightarrow y_1 = 4\dot{x}_2 + 3u_1,\ y_2 = \dot{x}_3 + 4u_2 + u_3$

$$\text{Ans.} \quad \begin{bmatrix} y_1 \\ y_2 \end{bmatrix} = \begin{bmatrix} 0 & 4 & 0 \\ 0 & 0 & 4 \end{bmatrix} \begin{bmatrix} x_1 \\ x_2 \\ x_3 \end{bmatrix} = \begin{bmatrix} 3 & 0 & 0 \\ 0 & 4 & 1 \end{bmatrix} \begin{bmatrix} u_1 \\ u_2 \\ u_3 \end{bmatrix}$$

4.5 STATE SPACE REPRESENTATION USING PHASE VARIABLES

Let us study how to obtain state space model using phase variable. The phase variables are those state variables which are obtained by assuming one of the system variables as a state variable and other state variables are derivatives of the selected system variable. Most of the time the system variable used is the output variable which is used to select the state variable.

Such sets of phase variable is easily obtained if the differential equation is known.

State model from differential equation

Consider a linear continuous-time system represented by nth order differential equation as

$$\frac{d^n y(t)}{dt^n} + a_{n-1}\frac{d^{n-1}y(t)}{dt^{n-1}} + a_{n-2}\frac{d^{n-2}y(t)}{dt^{n-2}} + \cdots + \cdots + a_1\frac{dy(t)}{dt} + a_0 y(t)$$

$$= b_0 u(t) + b_1\frac{du(t)}{dt} + b_2\frac{d^2u(t)}{dt^2} + \cdots + \ldots b_m\frac{d^m u(t)}{dt^m} \quad (1)$$

For time-variant system, the coefficients $a_{n-1}, a_{n-2}, a_{n-3}, \ldots a_0, b_0, b_1 \ldots b_m$ are constants.

For the system $y(t) =$ output variable

$u(t) =$ input variable

Consider a simple case when $\dot{u}(t) = \ddot{u}(t) = \cdots = 0$

then

$$\frac{d^n y(t)}{dt^n} + a_{n-1}\frac{d^{n-1}y(t)}{dt^{n-1}} + a_{n-2}\frac{d^{n-2}y(t)}{dt^{n-2}} + \cdots + a_0\, dy(t) = u(t)\, b_0$$

- As choice of state variable is generally output variable $y(t)$ itself and another variables are derivatives of the selected state variable $y(t)$.

  Consider $X_1(t) = y(t)$

$$X_2(t) = \dot{X}_1(t) = \dot{y}(t)$$
$$X_3(t) = \ddot{X}_1(t) = \dot{X}_2(t) = \ddot{y}(t)$$
$$\vdots$$
$$X_{n-1}(t) = \dot{X}_{n-2}(t)$$
$$X_n(t) = \dot{X}_{n-1}(t)$$
$$\dot{X}_n(t) = ?$$

Here $\dot{X}n(t)$ can be obtained by substituting the selected state variables in the original differential eqn. (2)

$$\dot{X}_n(t) + a_{n-1}X_n(t) + a_{n-2}X_{n-1}(t) + \cdots + a_1 X_2(t) + a_0 X_1(t) = b_0\, u(t)$$
$$\dot{X}_n(t) = b_0 u(t) - a_{n-1}X_n(t) - a_{n-2}X_{n-1}(t) - a_{n-3}X_{n-3}(t) - \cdots - a_1 X_2(t) - a_0 X_1(t)$$

and we have

$$y(t) = X_1(t) \quad \dot{y}(t) = X_2(t) \quad \ddot{y}(t) = X_3(t) \ldots$$

Hence, all the above equations can be expressed in vector matrix form as

$$\begin{bmatrix} \dot{X}_1 \\ \dot{X}_2 \\ \dot{X}_3 \\ \dot{X}_4 \\ \vdots \\ \dot{X}_n \end{bmatrix} = \begin{bmatrix} 0 & 1 & 0 & \ldots & 0 \\ 0 & 0 & 1 & \ldots & 0 \\ \vdots & \vdots & \vdots & & \vdots \\ \vdots & \vdots & \vdots & & \vdots \\ \vdots & \vdots & \vdots & & \vdots \\ -a_0 & -a_1 & -a_2 & \ldots & -a_{n-1} \end{bmatrix} \begin{bmatrix} X_1 \\ X_2 \\ \vdots \\ \vdots \\ \vdots \\ X_n \end{bmatrix} + \begin{bmatrix} 0 \\ 0 \\ \vdots \\ \vdots \\ \vdots \\ b_0 \end{bmatrix} u(t)$$

Hence, $\dot{X} = AX(t) + Bu(t)$

Such set of state variables is called set of phase variables. The matrix A is called matrix in phase variable form.

4.6 STATE MODEL

4.6.1 Derivation of transfer function from state model

Consider a standard state model derived for linear time-invariant system as –

$$\dot{X}(t) = AX(t) + BU(t)$$
$$Y(t) = CX(t) + DU(t)$$

Taking Laplace transform on both sides –

$$SX(S) - X(0) = AX(S) + BU(S)$$
$$\text{and } X(S) = CX(S) + DU(S)$$

Note that as the system is time invariant so the coefficients of matrix A, B, C, D are constants. While the definition of transfer function is based on the assumptions of zero initial condition, i.e. $X(0) = 0$

$$SX(S) = AX(S) + BU(S)$$
$$SX(S) - AX(S) = BU(S)$$
$$[SI - A]X(S) = BU(S)$$

Pre-multiplying both sides by $(SI - A)^{-1}$

$$[SI - A]^{-1}[SI - A]X(S) = [SI - A]^{-1}BU(S)$$

Now $[SI - A]^{-1}[SI - A] = I$

$$\text{So } X(S) = [SI - A]^{-1}BU(S)$$
$$Y(S) = C[SI - A]^{-1}BU(S)DU(S)$$

Transfer function $= \frac{Y(s)}{U(s)}$

$$\text{So T. F.} = C[SI - A]^{-1}B + D$$
$$[SI - A]^{-1} = \frac{\text{Adj}[SI - A]}{|SI - A|}$$
$$\text{T.F} = T(s) = C\frac{\text{Adj}[SI - A]}{|SI - A|} + D$$

* Here the state model of any system is not unique but the transfer function of any system obtained from any state model is unique.

4.6.2 Characteristic equation

It is seen from the expression of transfer function that the denominator is $|SI - A|$. Now the equation obtained by equating denominator of transfer function to zero is called characteristic equation. The roots of this equation are the closed loop poles of the system. Thus, the characteristic equation of the system is $|SI - A| = 0$ characteristic equation.

The stability of the system depends on the roots of the characteristic equation. In matrix algebra, the roots of the equation $|SI - A| = 0$ are called eigenvalues of matrix A and these are generally denoted by λ.

Example 3: Consider a system having state model –

$$\begin{bmatrix} \dot{X}_1 \\ \dot{X}_2 \end{bmatrix} = \begin{bmatrix} -2 & -3 \\ 4 & 2 \end{bmatrix} \begin{bmatrix} X_1 \\ X_2 \end{bmatrix} + \begin{bmatrix} 3 \\ 5 \end{bmatrix} u$$

$$Y = [1\ 1] \begin{bmatrix} X_1 \\ X_2 \end{bmatrix}$$

With $D = 0$, obtain transfer function T. F.

SOLUTION:

$$\text{T.F.} = C[SI - A]^{-1} B$$

$$[SI - A]^{-1} = \frac{\text{Adj}[SI - A]}{|SI - A|}$$

$$[SI - A] = S \begin{bmatrix} 1 & 0 \\ 0 & 1 \end{bmatrix} - \begin{bmatrix} -2 & -3 \\ 4 & 2 \end{bmatrix} = \begin{bmatrix} S & 0 \\ 0 & S \end{bmatrix} - \begin{bmatrix} -2 & -3 \\ 4 & 2 \end{bmatrix}$$

$$[SI - A] = \begin{bmatrix} S+2 & 3 \\ -4 & S-2 \end{bmatrix}$$

$$|SI - A| = (S+2)(S-2) - (-4)(3)$$

$$= S^2 - 4 + 12 = S^2 + 8$$

$$[SI - A]^{-1} = \frac{\text{Adj}[SI - A]}{|SI - A|} = \frac{\text{Adj} \begin{bmatrix} S+2 & 3 \\ -4 & S-2 \end{bmatrix}}{S^2 + 8}$$

$$\text{Adj}[SI - A] = \begin{bmatrix} S-2 & -3 \\ 4 & S+2 \end{bmatrix}$$

$$[SI - A]^{-1} = \frac{\begin{bmatrix} S-2 & -3 \\ 4 & S+2 \end{bmatrix}}{s^2 + 8}$$

$$\text{T.F.} = \frac{[1\ 1] \begin{bmatrix} S-2 & -3 \\ 4 & S+2 \end{bmatrix} \begin{bmatrix} 3 \\ 5 \end{bmatrix}}{s^2 + 8}$$

$$\text{T.F.} = \frac{[1\ 1] \begin{bmatrix} 3S - 21 \\ 5S + 22 \end{bmatrix}}{s^2 + 8} = \frac{8S + 1}{s^2 + 8}$$

Example 4: Determine the transfer matrix for MIMO system given by –

$$\begin{bmatrix} \dot{X}_1 \\ \dot{X}_2 \end{bmatrix} = \begin{bmatrix} 0 & 3 \\ -2 & -5 \end{bmatrix} \begin{bmatrix} X_1 \\ X_2 \end{bmatrix} + \begin{bmatrix} 1 & 1 \\ 1 & 1 \end{bmatrix} \begin{bmatrix} u_1 \\ u_2 \end{bmatrix}$$

$$\begin{bmatrix} Y_1 \\ Y_2 \end{bmatrix} = \begin{bmatrix} 2 & 1 \\ 1 & 0 \end{bmatrix} \begin{bmatrix} X_1 \\ X_2 \end{bmatrix}$$

SOLUTION:

From the given state model –

$$A = \begin{bmatrix} 0 & 3 \\ -2 & -5 \end{bmatrix} \qquad B = \begin{bmatrix} 1 & 1 \\ 1 & 1 \end{bmatrix}$$

$$C = \begin{bmatrix} 2 & 1 \\ 1 & 0 \end{bmatrix} \qquad D = (0)$$

$$\text{T. F.} = T(S) = C[SI - A]^{-1}B + D$$

$$[SI - A] = S\begin{bmatrix} 1 & 0 \\ 0 & 1 \end{bmatrix} - \begin{bmatrix} 0 & 3 \\ -2 & -5 \end{bmatrix} = \begin{bmatrix} S & -3 \\ 2 & S+5 \end{bmatrix}$$

$$[SI - A]^{-1} = \frac{\text{adj}[SI - A]}{|SI - A|}$$

$$\text{adj}[SI - A] = \text{adj}\begin{bmatrix} S & -3 \\ 2 & S+5 \end{bmatrix}$$

$$= \begin{bmatrix} S+5 & 3 \\ -2 & S \end{bmatrix}$$

$$|SI - A| = (S+5)S - (-2)(3) = S^2 + 5S + 6$$

$$\text{Transform function} = C(SI - A)^{-1}B + D$$

$$= \frac{\begin{bmatrix} 2 & 1 \\ 1 & 0 \end{bmatrix}\begin{bmatrix} S+5 & 3 \\ -2 & S \end{bmatrix}\begin{bmatrix} 1 & 1 \\ 1 & 1 \end{bmatrix} + 0}{(S^2 + 5S + 6)}$$

$$= \frac{\begin{bmatrix} 2 & 1 \\ 1 & 0 \end{bmatrix}\begin{bmatrix} S+8 & S+8 \\ (S-2) & (S-2) \end{bmatrix}}{(S^2 + 5S + 6)}$$

$$= \frac{\begin{bmatrix} 2(S+8) + (S-2) & 2(S+8) + S - 2 \\ S+8+0 & S+8+0 \end{bmatrix}}{(S^2 + 5S + 6)}$$

$$\text{T.F.} = \frac{\begin{bmatrix} 3S+14 & 3S+14 \\ S+8 & S+8 \end{bmatrix}}{(S^2 + 5S + 6)}$$

$$\text{T.F.} = \begin{bmatrix} \frac{3S+14}{(S+2)(S+3)} & \frac{3S+14}{(S+2)(S+3)} \\ \frac{S+8}{(S+2)(S+3)} & \frac{S+8}{(S+2)(S+3)} \end{bmatrix}$$

The above example is of MIMO (multiple input and multiple output system) and for this system a single transfer function does not exist. There exists a mathematical relationship between each output and all the inputs. Hence, for such systems there exists a **transfer matrix** rather than transfer function.

4.7 EIGENVALUES

Consider an equation $AX = y$ which indicates the transform of $(n \times 1)$ vector matrix X into $(n \times 1)$ vector matrix y by $(n \times n)$ matrix operator A.

If there exists such a vector 'X' such that A transforms it to a vector λX, then X is called the solution of equation.

$$AX = \lambda X$$

$$AX - \lambda X = 0$$

$$[\lambda I - A]X = 0 \tag{1}$$

The set of homogeneous eqn. (1) has a non-trivial solution only under the condition,

$$|\lambda I - A| = 0 \tag{2}$$

The determinant $|\lambda I - A| = 0$ is called characteristic polynomial while the eqn. (2) is called the characteristic equation.

After expanding, we get

$$|\lambda I - A| = \lambda^n + a_1\lambda^{n-1} + a_2\lambda^{n-2} + \cdots + a_n = 0 \tag{3}$$

The 'n' roots of the eqn. (3) i.e. the value of λ satisfying the above eqn. (3) are called **eigenvalues** of the matrix A. Eigenvalues are also known as closed loop poles of the system.

4.8 EIGENVECTORS

Any non-zero vector X_i such that $AX_i = \lambda_i X_i$ is said to be eigenvector associated with the eigenvalue λ_i. Thus, let $\lambda = \lambda_i$ satisfy the equation, $(\lambda_i I - A)X = 0$.

The solution of this equation is called eigenvector of A associated with eigenvalue λ_i and is denoted as M_i.

Mathematicaly, eigenvector can be calculated by taking cofactors of matrix $(\lambda_i I - A)$ along any row.

$$\text{Thus, } M_i = \text{Eigenvector for } \lambda_i = \begin{bmatrix} C_{k1} \\ C_{k2} \\ C_{k3} \\ \vdots \\ C_{kn} \end{bmatrix} \quad \text{where } k = 1, 2, 3 \ldots n$$

$$\begin{aligned} AM &= A[M_1 : M_2 : M_3 : \ldots M_n] \\ &= [\lambda_1 M_1 : \lambda_2 M_2 : \lambda_3 M_3 : \ldots \lambda_n M_n] \\ &= M\Lambda \end{aligned} \tag{1}$$

where $\Lambda =$ diagonal matrix

$$= \begin{bmatrix} \lambda_1 & 0 & 0 & \ldots & 0 \\ 0 & \lambda_2 & 0 & \ldots & 0 \\ \vdots & \vdots & \vdots & & \vdots \\ 0 & \ldots & 0 & 0 \ldots & \lambda_n \end{bmatrix}$$

Thus, premultiplying eqn. (1) by M^{-1}

$$\begin{aligned} M^{-1}AM &= M^{-1}M\Lambda \\ &= \Lambda \text{ (Diagonal matrix)} \end{aligned}$$

It can be noted that

(1) Both A and Λ have same characteristic equation.
(2) Both A and Λ have same eigenvalues.

Hence, due to transformation $M^{-1}AM$, eigenvalues of matrix A remain unchanged and matrix A gets converted to diagonal form, where Ck_i is cofactor of matrix $(\lambda_i I - A)$ of k_{th} row.

* If the cofactors along any row give null solution, i.e. all elements of corresponding eigenvectors are zero then cofactor along any other row must be obtained. Otherwise inverse of modal matrix M can not exist.

4.9 MODAL MATRIX

Let $\lambda_1, \lambda_2, \lambda_3 \ldots \lambda_n$ are the n eigenvalues of matrix A while $M_1, M_2 \ldots M_n$ are the eigenvectors corresponding to eigenvalues $\lambda_1, \lambda_2, \lambda_3 \ldots \lambda_n$ respectively. Each eigenvector is of $n \times 1$ and placing all eigenvectors along the columns of another matrix, $n \times n$ matrix can be obtained. Such a matrix obtained by placing all the eigenvectors is called a **modal matrix** or **diagonalising matrix** of matrix A

$$M = \text{Modal Matrix} = [M_1 : M_2 : M_3 : \cdots M_n]$$

The important property of modal matrix is that, it is a diagonal matrix with eigenvalues of A as its diagonal element. In fact, Λ could be written directly without the need to compute $[M^{-1}AM]$.

4.10 LINEAR OR SIMILARITY TRANSFORMATION

For a given system, choice of state is not unique, i.e. a system may have different state models, but transfer function and hence characteristic equation is always unique. Suppose that there exists a set of state variables $x = [x_1, x_2, x_3, \ldots, x_n]^T$. Then we may choose another set of variables

$$Z = [Z_1 Z_2 Z_3 \ldots Z_n]^T$$

so that a linear or similarity transformation exists.

$$\text{Let} \quad x = MZ \text{ or } Z = M^{-1}x \tag{1}$$

where M is a non-singular matrix.

Now differentiating eqn. (1)

$$\dot{Z} = M^{-1}\dot{x}$$

$$= M^{-1}(Ax + Bu) \quad \text{Since } \dot{x} = Ax + Bu$$

$$\dot{Z} = M^{-1}Ax + M^{-1}Bu$$

$$= M^{-1}AMZ + M^{-1}Bu \quad \because \quad x = Mz$$

$$\dot{Z} = \tilde{A}z + B\tilde{u} \tag{2}$$

$$\tilde{A} = M^{-1}AM \quad \tilde{B} = M^{-1}B$$

$$\text{and output } y = Cx = CMZ$$

$$\tilde{C} = CM \tag{3}$$

where $\tilde{A} = M^{-1}AM$ is a diagonal matrix.

Eqns. (2) and (3) represent the transformed system in the vector matrix differential form (using similarity transformation)

(1) Matrix A and $\tilde{A}$ have same characteristic equation and same eigenvalues.

(2) The transfer function of the system remains invariant under similarity transformation.

Example 5: A system is described by the state model—

$$\begin{bmatrix} \dot{x}_1 \\ \dot{x}_2 \\ \dot{x}_3 \end{bmatrix} = \begin{bmatrix} 0 & 1 & 0 \\ 0 & 0 & 1 \\ -6 & -11 & -6 \end{bmatrix} \begin{bmatrix} x_1 \\ x_2 \\ x_3 \end{bmatrix} + \begin{bmatrix} 0 \\ 0 \\ 6 \end{bmatrix} u$$

$$\text{and the output } y = [1 \quad 0 \quad 0] \begin{bmatrix} x_1 \\ x_2 \\ x_3 \end{bmatrix}$$

Transform it into diagonal form of representation and check for invariance of eigenvalues.

SOLUTION:

The characteristic equation of system

$$|SI - A| = 0$$

$$\begin{bmatrix} S & -1 & 0 \\ 0 & S & -1 \\ 6 & 11 & S+6 \end{bmatrix} = 0$$

$$S[S(S+6)+11]+1[0+6]+0=0$$

$$S(S^2+6S+11)+6=0$$

$$S^3+6S^2+11S+6=0$$

$$(S+1)(S+2)(S+3)=0$$

$$S=-1,-2,-3$$

or we can say $|\lambda I - A| = 0$

then $\lambda = -1, -2, -3$

The similarity transformation matrix

$$M = \begin{bmatrix} 1 & 1 & 1 \\ \lambda_1 & \lambda_2 & \lambda_3 \\ \lambda_1^2 & \lambda_2^2 & \lambda_3^2 \end{bmatrix} = \begin{bmatrix} 1 & 1 & 1 \\ -1 & -2 & -3 \\ 1 & 4 & 9 \end{bmatrix}$$

$$|M| = 1(-18+12) - 1(-9+3) + 1(-4+2)$$

$$|M| = -6+6-2 = -2$$

As $|M| \neq 0$, the matrix M is non-singular matrix and hence M^{-1} exists.

$$M^{-1} = \frac{1}{|M|} \begin{bmatrix} \begin{vmatrix} -2 & -3 \\ 4 & 9 \end{vmatrix} & -\begin{vmatrix} -1 & -3 \\ 1 & 9 \end{vmatrix} & \begin{vmatrix} -1 & -2 \\ 1 & 4 \end{vmatrix} \\ -\begin{vmatrix} 1 & 1 \\ 4 & 9 \end{vmatrix} & \begin{vmatrix} 1 & 1 \\ 1 & 9 \end{vmatrix} & -\begin{vmatrix} 1 & 1 \\ 1 & 4 \end{vmatrix} \\ \begin{vmatrix} 1 & 1 \\ -2 & -3 \end{vmatrix} & -\begin{vmatrix} 1 & 1 \\ -1 & -3 \end{vmatrix} & \begin{vmatrix} 1 & 1 \\ -1 & -2 \end{vmatrix} \end{bmatrix}^T$$

$$= \frac{1}{-2} \begin{bmatrix} -6 & 6 & -2 \\ -5 & 8 & -3 \\ -1 & 2 & -1 \end{bmatrix}^T$$

$$M^{-1} = \frac{-1}{2} \begin{bmatrix} -6 & -5 & -1 \\ 6 & 8 & 2 \\ -2 & -3 & -1 \end{bmatrix}$$

Hence, $\tilde{A} = M^{-1}AM$

$$= \frac{-1}{2} \begin{bmatrix} -6 & -5 & -1 \\ 6 & 8 & 2 \\ -2 & -3 & -1 \end{bmatrix} \begin{bmatrix} 0 & 1 & 0 \\ 0 & 0 & 1 \\ -6 & -11 & -6 \end{bmatrix} \begin{bmatrix} 1 & 1 & 1 \\ -1 & -2 & -3 \\ 1 & 4 & 9 \end{bmatrix}$$

$$= \frac{-1}{2}\begin{bmatrix} 6 & 5 & 1 \\ -12 & -16 & -4 \\ 6 & 9 & 3 \end{bmatrix}\begin{bmatrix} 1 & 1 & 1 \\ -1 & -2 & -3 \\ 1 & 4 & 9 \end{bmatrix}$$

$$= \frac{-1}{2}\begin{bmatrix} 2 & 0 & 0 \\ 0 & 4 & 0 \\ 0 & 0 & 6 \end{bmatrix}$$

$$= \begin{bmatrix} -1 & 0 & 0 \\ 0 & -2 & 0 \\ 0 & 0 & -3 \end{bmatrix}$$

Example 6: Consider a matrix A given below:

$$A = \begin{bmatrix} 0 & 1 & 0 \\ 3 & 0 & 2 \\ -12 & -7 & -6 \end{bmatrix}$$

Corresponding to this matrix, the characteristic equation is

$$|\lambda I - A| = \begin{bmatrix} \lambda & -1 & 6 \\ -3 & \lambda & -2 \\ 12 & 7 & \lambda+6 \end{bmatrix} = 0$$

$$= \lambda[\lambda(\lambda+6)+14] - (-1)[-3(\lambda+6)+2\times 12] + 6[-21-12\lambda]$$

$$= (\lambda+1)(\lambda+2)(\lambda+3) = 0$$

Therefore, the eigenvalues of matrix A are $\lambda_1 = -1 \quad \lambda_2 = -2 \quad \lambda_3 = -3$

The eigenvector M_i corresponding to $\lambda_1 = -1$ is obtained from cofactors of matrix

$$(\lambda_1 I - A) = \begin{bmatrix} -1 & -1 & 0 \\ -3 & -1 & -2 \\ 12 & 7 & 5 \end{bmatrix}$$

The result is

$$C_{11} = \begin{vmatrix} -1 & -2 \\ 7 & 5 \end{vmatrix}$$

$$= (-1)(5) - (-2)(7)$$

$$= -5 + 14 = 9$$

$$C_{12} = -\begin{vmatrix} -3 & -2 \\ 12 & 5 \end{vmatrix} = -[(-3)(5) - (-2)(12)]$$

$$= -[-15 + 24]$$

$$= -9$$

$$C_{13} = \begin{vmatrix} -3 & -1 \\ 12 & 7 \end{vmatrix} = (-3)(7) - (-1)(12)$$

$$= -21 + 12$$

$$= -9$$

$$[M_1] = \begin{bmatrix} C_{11} \\ C_{12} \\ C_{13} \end{bmatrix} = \begin{bmatrix} 9 \\ -9 \\ -9 \end{bmatrix} \text{ OR } \begin{bmatrix} 1 \\ -1 \\ -1 \end{bmatrix}$$

Similarly, eigenvectors M_2 and M_3 associated with $\lambda_2 = -2$ and $\lambda_3 = -3$ respectively are

$$[M_2] = \begin{bmatrix} 2 \\ -4 \\ 1 \end{bmatrix} \qquad [M_3] = \begin{bmatrix} 1 \\ -3 \\ 3 \end{bmatrix}$$

Hence, the modal matrix

$$[M] = [M_1\ M_2\ M_3]$$

$$= \begin{bmatrix} 1 & 2 & 1 \\ -1 & -4 & -3 \\ -1 & 1 & 3 \end{bmatrix}$$

It can also be verified that

$$\Lambda = [M^{-1}AM]$$

$$= \begin{bmatrix} -1 & 0 & 0 \\ 0 & -2 & 0 \\ 0 & 0 & -3 \end{bmatrix}$$

$$\hat{B} = M^{-1}B = \frac{-1}{2} \begin{bmatrix} -6 & -5 & -1 \\ 6 & 8 & 2 \\ -2 & -3 & -1 \end{bmatrix} \begin{bmatrix} 0 \\ 0 \\ 6 \end{bmatrix}$$

$$= \frac{-1}{2} \begin{bmatrix} -6 \\ +12 \\ -6 \end{bmatrix} = \begin{bmatrix} 3 \\ -6 \\ 3 \end{bmatrix}$$

$$\tilde{C} = CM = [1 \quad 0 \quad 0] \begin{bmatrix} 1 & 1 & 1 \\ -1 & -2 & -3 \\ 1 & 4 & 9 \end{bmatrix}.$$

$$= [1 \quad 1 \quad 1]$$

Therefore, the vector matrix differential equation of transformed system can be written as

$$\begin{bmatrix} \dot{Z}_1 \\ \dot{Z}_2 \\ \dot{Z}_3 \end{bmatrix} = \begin{bmatrix} -1 & 0 & 0 \\ 0 & -2 & 0 \\ 0 & 0 & -3 \end{bmatrix} \begin{bmatrix} Z_1 \\ Z_2 \\ Z_3 \end{bmatrix} + \begin{bmatrix} 3 \\ -6 \\ +3 \end{bmatrix} u$$

$$Y = \tilde{C}Z = [1 \quad 1 \quad 1] \begin{bmatrix} Z_1 \\ Z_2 \\ Z_3 \end{bmatrix}$$

4.11 SOLUTION OF STATE EQUATIONS

It is observed that output response depends upon the state variable and their values. Hence, it is necessary to obtain the state vector $x(t)$ which satisfies the state equation $\dot{X}(t) = AX(t) + BU(t)$ at any time t. This is called solution of state equations, which helps to obtain response of a system. Consider a state equation of linear time invariant system as

$$\dot{X}(t) = AX(t) + BU(t)$$

Here the matrices A and B are constant. This state equation can be of two types;

(1) Homogeneous
(2) Non-homogeneous

4.11.1 Homogeneous equation

If A is a constant matrix and input control forces are zero then the equation takes the form

$$\dot{X}(t) = AX(t) \tag{1}$$

Such a equation is called homogeneous equation. In such systems, the driving force is provided by the initial conditions of the system to produce the output.

4.11.2 Non-homogeneous equation

If A is a constant matrix and $U(t)$ is a non-zero vector, i.e. input control forces are applied to the system, then the equation takes normal form as

$$\dot{X}(t) = AX(t) + BU(t) \tag{2}$$

Such an equation is called non-homogeneous equation. Most of the practical systems require inputs to drive them. Such systems are non-homogenous systems.

4.12 STATE TRANSITION MATRIX (STM)

Once the state equations of LTI systems are known, the next step is to find the solution of state equations, given the initial state vector $X(t_0)$, the input vector $V(t)$ for $t =\geq t_0$.

$$\text{and } \dot{X}(t) = AX(t) + BU(t)$$

The state transition matrix is defined as a matrix that satisfies the linear homogeneous state equation.

$$\dot{X}(t) = AX(t) \tag{1}$$

Suppose $\phi(t)$ be $n \times n$ matrix that represents STM, then it must satisfy the above eqn. (1)

$$\dot{\phi}(t) = A\phi(t) \tag{2}$$

Furthermore, $X(0)$ denotes the initial state at $t = 0$, then $\phi(t)$ is also defined by the matrix equation.

$$X(t) = \phi(t)x(0) \tag{3}$$

which is a solution of homogeneous state equation for $t \geq 0$

Take Laplace transform on both sides of eqn. (1)

$$SX(S) - X(0) = AX(S)$$

$$X(S)[SI - A] = X(0)$$

$$X(S) = [SI - A]^{-1}X(0) \tag{4}$$

where it is assumed that $[SI - A]$ is non-singular.

Taking inverse Laplace transform on both sides of eqn. (4)

$$X(t) = L^{-1}\left[[SI - A]^{-1}X(0)\right] \tag{5}$$

By comparing eqns. (3) and (5), STM is given by

$$\phi(t) = L^{-1}\left[[SI - A]^{-1}\right]$$

$$= L^{-1}\left[S^{-1}\left[I - \frac{A}{S}\right]^{-1}\right]$$

$$= L^{-1}\left[\frac{1}{S}\left(I + \frac{A}{S} + \frac{A^2}{S^2} + \frac{A^3}{S^3} + \cdots + \ldots\right)\right]$$

$$\phi(t) = L^{-1}\left[\frac{I}{S} + \frac{A}{S^2} + \frac{A^2}{S^3} + \frac{A^3}{S^4} + \ldots\right]$$

$$= \left[I + At + \frac{A^2}{L^{2!}}t^2 + \frac{A^3}{L^{3!}}t^3 + \ldots\right]$$

$$\phi(t) = e^{At}$$

This is the expression of STM.

4.12.1 Significance of STM

Since the STM satisfies the homogeneous state equation, it represents the free response of the system. STM represents the response that is excited by the initial condition only. STM is dependent only on the matrix A. The STM completely defines the transition of the states from the initial time $t = 0$ to any time t when the inputs are zero.

4.12.2 Properties of STM

The properties of STM are given below:

(1) $\phi(0) = e^{0A} = I$

(2) $\phi^{-1}(t) = (e^{At})^{-1} = e^{-At} = \phi(-t)$

(3) $\phi(t_1 + t_2) = e^{A(t_1+t_2)} = e^{At_1}e^{At_2}$
$\phi(t_1 + t_2) = \phi(t_1)\phi(t_2)$

(4) $[\phi(t)]^n = (e^{At})^n = e^{nAt} = \phi(nt)$

(5) $\phi(t_2 - t_1)\phi(t_1 - t_0) = e^{A(t_2-t_1)}e^{A(t_1-t_0)}$
$= e^{A(t_2-t_1+t_1-t_0)}$
$= e^{A(t_2-t_0)}$

4.13 SOLUTION OF NON-HOMOGENEOUS STATE EQUATION (FORCED SYSTEM)

Let us now determine the solution of the non-homogeneous state equation

$$\dot{X}(t) = AX(t) + BU(t), \quad X(t)|_{at\,t=0} = X(0) \tag{6}$$

$$\dot{X}(t) - AX(t) = BU(t)$$

Multiply both sides by e^{-At}, we have

$$e^{-At}\left[\dot{X}(t) - AX(t)\right] = e^{-At}BU(t)$$

$$\left[\dot{X}(t)e^{-At} - Ae^{-At}X(t)\right] = e^{-At}BU(t)$$

$$\frac{d}{dt}\left[e^{-At}X(t)\right] = e^{-At}BU(t)$$

Integrating both sides with respect to t between the limits 0 and t, we get

$$e^{-At}X(t)\Big|_0^t = \int_0^t BU(\tau)e^{-A(t-\tau)}d\tau$$

$$e^{-At}X(t) - X(0) = \int_0^t BU(\tau)e^{-A(t-\tau)}d\tau$$

$$e^{-At}X(t) = X(0) + \int_0^t BU(\tau)e^{-A(t-\tau)}d\tau$$

$$X(t) = e^{At}X(0) + \int_0^t e^{At}e^{-A(t-\tau)}BU(\tau)d\tau$$

$$= \underbrace{e^{At}X(0)}_{\substack{\text{Homogeneous}\\ \text{solution}}} + \underbrace{\int_0^t e^{+A\tau}BU(\tau)\,d\tau}_{\substack{\text{Forced}\\ \text{solution}}} \tag{7}$$

If the initial state is known at $t = t_0$ rather than $t = 0$, eqn. (7) becomes

$$X(t) = e^{A(t-t_0)}X(t_0) + \int_{t_0}^t e^{A(t-\tau)}BU(\tau)d\tau \tag{8}$$

In terms of STM, eqn. (8) will be

$$X(t) = \phi(t-t_0)X(t_0) + \int_{t_0}^t \phi(t-\tau)BU(\tau)d\tau \tag{9}$$

Here $X(t)$ is the total solution of the state vector, which is the sum of homogeneous (complementary solution) and forced solution (particular solution)

The solution of linear non-homogeneous state equation given by equations from (7) to (9) are known as state transition equations.

Example 7: Find the state transition matrix

$$A = \begin{bmatrix} 0 & -1 \\ 2 & -3 \end{bmatrix}$$

SOLUTION:

Using Laplace transform method

$$[SI - A] = S\begin{bmatrix} 1 & 0 \\ 0 & 1 \end{bmatrix} - \begin{bmatrix} 0 & -1 \\ 2 & -3 \end{bmatrix}$$

$$= \begin{bmatrix} S & 1 \\ S-2 & S+3 \end{bmatrix}$$

$$[SI - A]^{-1} = \frac{\text{adj}\begin{bmatrix} S & 1 \\ -2 & S+3 \end{bmatrix}}{|SI - A|}$$

$$|SI - A| = S(S+3) + 2$$

$$= S^2 + 3S + 2$$

$$= (S+1)(S+2)$$

$$[SI - A]^{-1} = \frac{\begin{bmatrix} S+3 & -1 \\ 2 & S \end{bmatrix}}{(S+2)(S+1)}$$

$$\phi(S) = \begin{bmatrix} \frac{S+3}{(S+2)(S+1)} & \frac{-1}{(S+2)(S+1)} \\ \frac{2}{(S+2)(S+1)} & \frac{S}{(S+2)(S+1)} \end{bmatrix}$$

$$= \begin{bmatrix} \frac{2}{S+1} - \frac{1}{S+2} & \frac{-1}{S+1} + \frac{1}{S+2} \\ \frac{2}{S+1} - \frac{2}{S+2} & \frac{-1}{S+1} + \frac{2}{S+2} \end{bmatrix}$$

Take Laplace transform on both sides of the above equation.

$$L^{-1}[\phi(S)] = L^{-1}\begin{bmatrix} \frac{2}{S+1} - \frac{1}{S+2} & \frac{-1}{S+1} + \frac{1}{S+2} \\ \frac{2}{S+1} - \frac{2}{S+2} & \frac{-1}{S+1} + \frac{2}{S+2} \end{bmatrix}$$

$$e^{At} = \phi(t) = \begin{bmatrix} (2e^{-t} - e^{-2t})u(t) & (-e^{-t} + e^{-2t})u(t) \\ 2(e^{-t} - e^{-2t})u(t) & (-e^{-t} + 2e^{-2t})u(t) \end{bmatrix}$$

This is the required state transition matrix.

Example 8: A LTI system is characterised by the homogeneous state equation

$$\begin{bmatrix} \dot{X}_1 \\ \dot{X}_2 \end{bmatrix} = \begin{bmatrix} 1 & 0 \\ 1 & 1 \end{bmatrix} \begin{bmatrix} X_1 \\ X_2 \end{bmatrix}$$

Compute the solution of homogeneous equation, assume the initial state vector.

$$X(0) = \begin{bmatrix} 1 \\ 0 \end{bmatrix}$$

SOLUTION:

From the given model

$$A = \begin{bmatrix} 1 & 0 \\ 1 & 1 \end{bmatrix}$$

$$[SI - A] = S\begin{bmatrix} 1 & 0 \\ 0 & 1 \end{bmatrix} - \begin{bmatrix} 1 & 0 \\ 1 & 1 \end{bmatrix} = \begin{bmatrix} S-1 & 0 \\ -1 & S-1 \end{bmatrix}$$

$$|SI - A| = (S-1)^2 - 0 = (S-1)^2$$

$$\text{adj } [SI - A] = \begin{bmatrix} S-1 & 0 \\ 1 & S-1 \end{bmatrix}$$

$$\therefore \quad [SI - A]^{-1} = \frac{\text{adj}[SI - A]}{|SI - A|}$$

$$= \frac{\begin{bmatrix} S-1 & 0 \\ 1 & S-1 \end{bmatrix}}{(S-1)^2}$$

$$= \begin{bmatrix} \frac{1}{S-1} & 0 \\ \frac{1}{(S-1)^2} & \frac{1}{S-1} \end{bmatrix}$$

Take Laplace transform on both sides

$$e^{At} = L^{-1}\left[[SI - A]^{-1}\right]$$

$$= L^{-1}\begin{bmatrix} \frac{1}{S-1} & 0 \\ \frac{1}{(S-1)^2} & \frac{1}{S-1} \end{bmatrix}$$

$$\phi(t) = \begin{bmatrix} e^t & 0 \\ te^t & e^t \end{bmatrix} u(t)$$

$$X(t) = \phi(t)X(0)$$

$$= \begin{bmatrix} e^t & 0 \\ te^t & e^t \end{bmatrix} \begin{bmatrix} 1 \\ 0 \end{bmatrix}$$

$$X(t) = \begin{bmatrix} e^t \\ te^t \end{bmatrix} u(t)$$

This is the required solution.

Example 9: A state model of the system is given by

$$\begin{bmatrix} \dot{x_1} \\ \dot{x_2} \end{bmatrix} = \begin{bmatrix} 0 & 1 \\ -2 & 3 \end{bmatrix} \begin{bmatrix} x_1 \\ x_2 \end{bmatrix} + \begin{bmatrix} 0 \\ 1 \end{bmatrix} u$$

$$y = [1\ 0] \begin{bmatrix} x_1 \\ x_2 \end{bmatrix}$$

Find

(a) Characteristic equation and eigenvalues of the system

(b) The transfer function $\frac{Y(S)}{U(S)}$

(c) The STM

(d) The state equation for a unit step input under zero initial conditions.

(e) The time response $y(t)$.

SOLUTION:

(a)

$$[SI - A] = \begin{bmatrix} S & -1 \\ 2 & S+3 \end{bmatrix}$$

$$|SI - A| = S^2 + 3S + 2$$

Characteristic equation $= (S+1)(S+2)$

Eigenvalues $\Rightarrow |SI - A| = 0$

$$(S+1)(S+2) = 0$$

$$S = -1, -2$$

(b) The transfer function $= \frac{Y(S)}{U(S)} = C[SI - A]^{-1}B$

$$[SI - A]^{-1} = \frac{\text{adj}[SI - A]}{|SI - A|}$$

$$= \frac{\text{adj}\begin{bmatrix} S & -1 \\ 2 & S+3 \end{bmatrix}}{(S+1)(S+2)}$$

$$= \frac{\begin{bmatrix} S+3 & 1 \\ -2 & S \end{bmatrix}}{(S+1)(S+2)}$$

$$[SI-A]^{-1} = \begin{bmatrix} \frac{S+3}{(S+1)(S+2)} & \frac{1}{(S+1)(S+2)} \\ \frac{-2}{(S+1)(S+2)} & \frac{S}{(S+1)(S+2)} \end{bmatrix}$$

Transfer function $= C[SI-A]^{-1}B$

$$= [1\ 0]\begin{bmatrix} \frac{S+3}{(S+1)(S+2)} & \frac{1}{(S+1)(S+2)} \\ \frac{-2}{(S+1)(S+2)} & \frac{S}{(S+1)(S+2)} \end{bmatrix}\begin{bmatrix} 0 \\ 1 \end{bmatrix}$$

$$= [1\ 0]\begin{bmatrix} \frac{1}{(S+1)(S+2)} \\ \frac{S}{(S+1)(S+2)} \end{bmatrix}$$

$$\frac{Y(S)}{U(S)} = \frac{1}{(S+1)(S+2)} \text{T F.}$$

(c) $\varphi(t) = L^{-1}\{[SI-A]^{-1}\}$

$$= L^{-1}\begin{bmatrix} \frac{S+3}{(S+2)(S+1)} & \frac{1}{(S+1)(S+2)} \\ \frac{-2}{(S+1)(S+2)} & \frac{S}{(S+1)(S+2)} \end{bmatrix}$$

$$\phi(S) \Rightarrow \begin{bmatrix} \frac{-1}{S+2} + \frac{2}{S+1} & \frac{-1}{S+1} - \frac{1}{S+2} \\ \frac{-2}{S+1} + \frac{2}{S+2} & \frac{-1}{S+1} + \frac{2}{S+2} \end{bmatrix}$$

Take inverse Laplace transform on both sides

$$\phi(t) = L^{-1}\begin{bmatrix} \frac{-1}{S+2} + \frac{2}{S+1} & \frac{-1}{S+1} - \frac{1}{S+2} \\ \frac{-2}{S+1} + \frac{2}{S+2} & \frac{-1}{S+1} + \frac{2}{S+2} \end{bmatrix}$$

$$\Rightarrow \begin{bmatrix} -e^{-2t} + 2e^{-t} & e^{-t} - e^{-2t} \\ -2e^{-t} + 2e^{-2t} & -e^{-t} + 2e^{-2t} \end{bmatrix}$$

(d) The state equation of forced system is given by

$$x(t) = \phi(t)x(0) + \int_0^t \phi(t-\tau)BU(\tau)d\tau$$

$$= 0 + \int_0^t \begin{bmatrix} 2e^{-(t-\tau)} - e^{-2(t-\tau)} & e^{-(t-\tau)} - e^{-2(t-\tau)} \\ -2e^{-(t-\tau)} + 2e^{-2(t-\tau)} & -e^{-(t-\tau)} + 2e^{-2(t-\tau)} \end{bmatrix}\begin{bmatrix} 0 \\ 1 \end{bmatrix} 1 d\tau$$

$$\begin{bmatrix} x_1(t) \\ x_2(t) \end{bmatrix} = \int_0^t \begin{bmatrix} e^{-(t-\tau)} - e^{-2(t-\tau)} \\ -e^{-(t-\tau)} + 2e^{-2(t-\tau)} \end{bmatrix} d\tau$$

$$\Rightarrow \begin{bmatrix} +e^{-(t-\tau)} - \frac{e}{2}^{-2(t-\tau)} \\ -e^{-(t-\tau)} + e^{-2(t-\tau)} \end{bmatrix}_0^t$$

$$\begin{bmatrix} x_1(t) \\ x_2(t) \end{bmatrix} = \begin{bmatrix} e^0 - \frac{e^0}{2} - e^{-t} + \frac{e}{2}^{-2t} \\ -1 + 1 + e^{-t} - e^{-2t} \end{bmatrix} = \begin{bmatrix} \frac{1}{2} - e^{-t} + \frac{e}{2}^{-2t} \\ e^{-t} - e^{-2t} \end{bmatrix}$$

4.14 ZERO INPUT RESPONSE

The solution of the homogeneous state equation is under the condition of zero input. Such a response is called zero input response and it is denoted as ZIR. Thus, the behaviour of $X(t)$ under the initial conditions without any input $U(t)$ is called the zero input response of the system. It is also called free, natural or unforced response of the system.

4.15 SOLUTION OF STATE EQUATIONS BY LAPLACE TRANSFORM

The Laplace transform method converts integrodifferential equations to simple algebraic equations. Consider a non-homogeneous equation as

$$\dot{X}(t) = AX(t) + BU(t) \tag{1}$$

Take Laplace transform

$$SX(S) - X(0) = AX(S) + BU(S)$$
$$SX(S) - AX(S) = X(0) + BU(S)$$
$$(SI - A)X(S) = X(0) + BU(S)$$
$$X(S) = [SI - A]^{-1}X(0) + [SI - A]^{-1}BU(S) \tag{2}$$
$$= Z1R + ZSR$$
$$\text{As, } \varphi(S) = [SI - A]^{-1}$$

$\$L^{-1}[\phi(s)] = e^{At} \rightarrow$ resolvant matrix of A

Taking Laplace inverse on eqn. (2)

$$X(t) = L^{-1}\left\{[SI - A]^{-1}X(0)\right\} + L^{-1}\left\{[SI - A]^{-1}BU(S)\right\}$$
$$= L^{-1}\{\phi(S)X(0)\} + L^{-1}\{\phi(S)BU(S)\}$$
$$= \phi(t)X(0) + \int_0^t \phi(t - \tau)BU(\tau)d\tau \tag{4}$$
$$= ZIR + ZSR$$

Thus, the second part of eqn. (4) is the zero state response of the system.

$$ZSR = \int_0^t \phi(t - \tau)Bu(\tau)d\tau$$

Example 10: Find the ZSR (zero state response) of the system

$$\dot{X} = \begin{bmatrix} 0 & 1 \\ -2 & -3 \end{bmatrix} X + \begin{bmatrix} 2 & 1 \\ 0 & 1 \end{bmatrix} u(t),\ X(0) = \begin{bmatrix} 0 \\ 0 \end{bmatrix}$$

and $Y(t) = \begin{pmatrix} 1 & 0 \\ 1 & 1 \end{pmatrix} X$ to the following input

$U(t) = \begin{pmatrix} U_1(t) \\ U_2(t) \end{pmatrix} = \begin{pmatrix} U(t) \\ e^{-3t}U(t) \end{pmatrix}$ where $u(t)$ is unit step function.

SOLUTION:

$$\text{ZIR} = e^{At}X(0)$$
$$= e^{At}\begin{pmatrix} 0 \\ 0 \end{pmatrix} = 0$$

Now ZSR= $L^{-1}[\phi(S)BU(S)]$

From the question it is given

$$A=\begin{bmatrix}0 & 1\\ -2 & -3\end{bmatrix} B=\begin{bmatrix}2 & 1\\ 0 & 1\end{bmatrix} C=\begin{bmatrix}1 & 0\\ 1 & 1\end{bmatrix}$$

$$U(t)=\begin{bmatrix}U_1(t)\\ e^{-3t}u(t)\end{bmatrix}\Rightarrow\begin{bmatrix}1/S\\ \frac{1}{S+3}\end{bmatrix}U(S)$$

$$[SI-A]=S\begin{bmatrix}1 & 0\\ 0 & 1\end{bmatrix}-\begin{bmatrix}0 & 1\\ -2 & -3\end{bmatrix}$$

$$=\begin{bmatrix}S & -1\\ 2 & S+3\end{bmatrix}$$

$$[SI-A]^{-1}=\frac{\text{adj}[SI-A]}{|SI-A|}$$

$$|SI-A|=S(S+3)-(-1)(2)$$

$$=S(S+3)+2=S^2+3S+2$$

$$=(S+1)(S+2)$$

$$\text{adj}[SI-A]=\begin{bmatrix}S+3 & 1\\ -2 & S\end{bmatrix}$$

$$[SI-A]^{-1}=\frac{\begin{bmatrix}S+3 & 1\\ -2 & S\end{bmatrix}}{(S+1)(S+2)}$$

$$\therefore\quad \text{ZSR}=L^{-1}\left\{\begin{bmatrix}\frac{S+3}{(S+1)(S+2)} & \frac{1}{(S+1)(S+2)}\\ \frac{-2}{(S+1)(S+2)} & \frac{S}{(S+1)(S+2)}\end{bmatrix}\begin{bmatrix}2 & 1\\ 0 & 1\end{bmatrix}\begin{bmatrix}\frac{1}{S}\\ \frac{1}{S+3}\end{bmatrix}\right\}$$

$$\Rightarrow L^{-1}\left\{\begin{bmatrix}\frac{S+3}{(S+1)(S+2)} & \frac{1}{(S+1)(S+2)}\\ \frac{-2}{(S+1)(S+2)} & \frac{S}{(S+1)(S+2)}\end{bmatrix}\begin{bmatrix}\frac{2}{S}+\frac{1}{S+3}\\ \frac{1}{S+3}\end{bmatrix}\right\}$$

$$=L^{-1}\left\{\begin{bmatrix}\frac{S+3}{(S+1)(S+2)} & \frac{1}{(S+1)(S+2)}\\ \frac{-2}{(S+1)(S+2)} & \frac{S}{(S+1)(S+2)}\end{bmatrix}\begin{bmatrix}\frac{3(S+2)}{S(S+3)}\\ \frac{1}{S+3}\end{bmatrix}\right\}$$

$$\Rightarrow L^{-1}\left\{\begin{bmatrix}\frac{3(S+2)(S+3)}{S(S+1)(S+2)(S+3)}+\frac{1}{(S+1)(S+2)(S+3)}\\ \frac{-6(S+2)}{S(S+3)(S+1)(S+2)}+\frac{S}{(S+1)(S+2)(S+3)}\end{bmatrix}\right\}$$

$$=L^{-1}\left\{\begin{bmatrix}\frac{3}{S}-\frac{3}{S+1}+\frac{1}{2(S+1)}-\frac{1}{S+2}+\frac{1}{2(S+3)}\\ \frac{-2}{S}+\frac{3}{S+1}-\frac{1}{(S+3)}-\frac{1}{2(S+1)}+\frac{2}{S+2}-\frac{3}{2(S+3)}\end{bmatrix}\right\}$$

$$\text{ZSR}=\begin{bmatrix}3-3e^{-t}+0.5e^{-t}-e^{-2t}+0.5e^{-3t}\\ -2+3e^{-t}-e^{-3t}-0.5e^{-t}+2e^{-2t}-\frac{3e^{-3t}}{2}\end{bmatrix}$$

$$X(t)=ZIR+ZSR=0+ZSR$$

$$\begin{bmatrix}X_1(t)\\ X_2(t)\end{bmatrix}=\begin{bmatrix}3-2.5e^{-t}-e^{-2t}+0.5e^{-3t}\\ -2+2.5e^{-t}+2e^{-2t}-2.5e^{-3t}\end{bmatrix}$$

4.16 CONTROLLABILITY AND OBSERVABILITY

4.16.1 Controllability

A system is said to be completely state controllable if it is possible to transfer the system state from any initial state $X(t_0)$ to any other desired state $X(t_f)$ in a specified finite time internal t_f, by a control vector $U(t)$.

The concept of controllability and observability is given by Kalman. Hence, Kalman's tests are used to find out whether the system is controllable and observable or not.

Kalman's Test for Controllability

Consider n^{th} order multiple input LTI system represented by its state equation as-$\dot{X}(t) = AX(t) + BU(t)$ where A has order $n \times n$ matrix $U(t)$ is $m\times$ 1 vector, i.e. there are m inputs $X(t)$ is $n\times$ 1 state vector.

The necessary and sufficient condition for the system to be completely state controllable is that rank of the composite matrix Q_c is n.

The composite matrix is given by

$$Q_c = [B : AB : A^2B : ---: A^{n-1}B]$$

In Q_c, B, AB, A^2B, A^3B are various columns.

Example 11: Find the controllability of the system

$$\dot{X} = \begin{bmatrix} 1 & 1 \\ 2 & -1 \end{bmatrix} \begin{bmatrix} X_1 \\ X_2 \end{bmatrix} + \begin{bmatrix} 0 \\ 1 \end{bmatrix} U(t)$$

$$n = 2$$

SOLUTION:

$$A = \begin{bmatrix} 1 & 1 \\ 2 & -1 \end{bmatrix} B = \begin{bmatrix} 0 \\ 1 \end{bmatrix}$$

$$AB = \begin{bmatrix} 1 & 1 \\ 2 & -1 \end{bmatrix} \begin{bmatrix} 0 \\ 1 \end{bmatrix}$$

$$AB = \begin{bmatrix} 1\times 0 + 1\times 1 \\ 2\times 0 + (-1)\times 1 \end{bmatrix} = \begin{bmatrix} 1 \\ -1 \end{bmatrix}$$

$$Q_c = [B : AB] = \begin{bmatrix} 0 & 1 \\ 1 & -1 \end{bmatrix}$$

Rank of $Q_c = 2$, as there are two non-zero rows.

Hence, the system is controllable.

4.16.2 Observability

A system is said to be completely observable if every state $X(t_0)$ can be completely identified by measurements of output $Y(t)$ over a finite time interval. If the system is not completely observable, it means that few of its state variables are not practically measurable and are shielded from the observation.

Kalman's Test for observability

Consider n^{th} order multiple input multiple output linear time-invariant system, represented by its state equation as

$$\dot{X} = AX(t) + BU(t)$$

and $$Y(t) = CX(t)$$

$$Y(t) = P \times 1 \text{ output vector}$$

and $$C = 1 \times n \text{ matrix}$$

The system is completely observable if and only if the rank of composite matrix Q_0 is n.

The composite matrix

$$Q_0 = [C^T : A^T C^T : A^{2T} C^T - -- : (A^T)^{n-1} C^T]$$

$$C^T \Rightarrow \text{Transpose of matrix } C$$

$$A^T \rightarrow \text{Transpose of matrix } A.$$

Thus, if rank of Q_0 is n, then the system is completely observable.

Example 12: Evaluate the observability of the system with

$$A = \begin{bmatrix} 0 & 1 & 0 \\ 0 & 0 & 1 \\ 0 & -2 & -3 \end{bmatrix} \quad B = \begin{bmatrix} 0 \\ 0 \\ 1 \end{bmatrix} \quad C = [3\ 4\ 1]$$

SOLUTION:

The order of the system = 3.

$$Q = [C^T:\ A^T C^T:\ (A_T)^2 C^T]$$

$$[A^T] = \begin{bmatrix} 0 & 0 & 0 \\ 1 & 0 & -2 \\ 0 & 1 & -3 \end{bmatrix}$$

$$C^T = \begin{bmatrix} 3 \\ 4 \\ 1 \end{bmatrix}$$

$$[A^T C^T] = \begin{bmatrix} 0 & 0 & 0 \\ 1 & 0 & -2 \\ 0 & 1 & -3 \end{bmatrix} \begin{bmatrix} 3 \\ 4 \\ 1 \end{bmatrix} = \begin{bmatrix} 0 \\ 1 \\ 1 \end{bmatrix}$$

$$[(A^T)^2 C^T] = \begin{bmatrix} 0 & 0 & 0 \\ 1 & 0 & -2 \\ 0 & 1 & -3 \end{bmatrix} \begin{bmatrix} 0 \\ 1 \\ 1 \end{bmatrix} = \begin{bmatrix} 0 \\ -2 \\ +2 \end{bmatrix}$$

$$Q_0 = \begin{bmatrix} 3 & 0 & 0 \\ 4 & 1 & -2 \\ 1 & 1 & -2 \end{bmatrix}$$

Consider the determinant $= \begin{vmatrix} 3 & 0 & 0 \\ 4 & 1 & -2 \\ 1 & 1 & -2 \end{vmatrix}$

$|Q_0| = 0$ and Rank 30
Hence, the system is not completely observable.

4.17 APPLICATION OF STATE VARIABLE TECHNIQUES TO THE ANALYSIS OF LINEAR SYSTEM

Example 13: Considering V_c and I_i as state variables and I_x as the output variable in the circuit shown below.

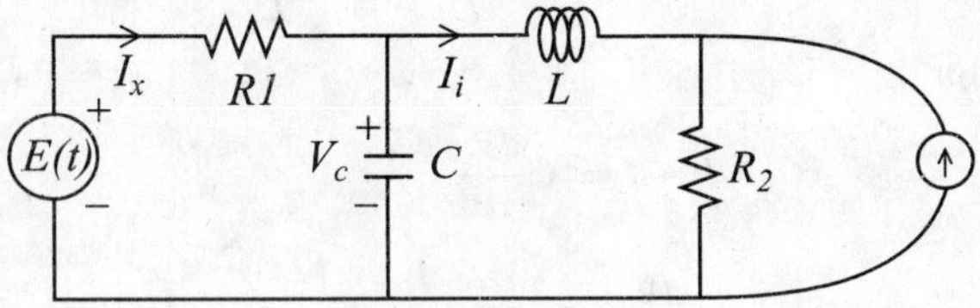

Fig. 4.4.

SOLUTION:
Convert the current source to voltage source as shown in figure.

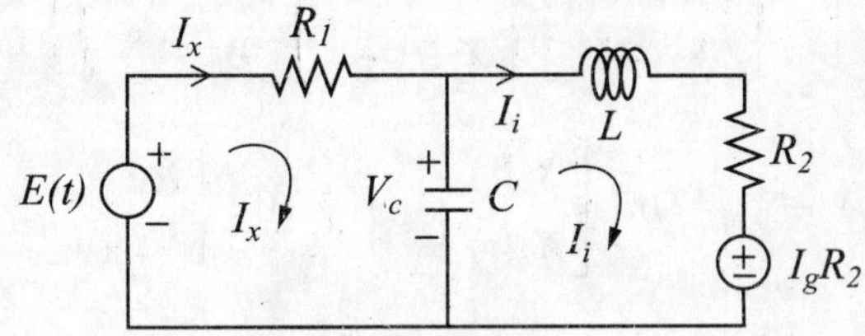

Fig. 4.5.

Two inputs, $E(t)$ and $I_g(t)$i.e. $U_1 = E(t)$
$U_2 = I_g(t)$
One output $I_x(t)$, i.e. $Y(t) = I_x(t)$

Variables $V_c(t) = X_1(t), I_i(t) = X_2(t)$
Let $I_x(t)$ and $I_i(t)$ be the loop currents.
Applying KVL to the two loops.

Loop 1

$$-I_xR_1 - V_c + E(t) = 0$$

$$I_xR_1 = E(t) - V_c(t)$$

$$I_x = \frac{E(t)}{R_1} - \frac{V_c(t)}{R_1}$$

$$Y(t) = \frac{U_1 - X_1(t)}{R_2} \cdots \text{output equation}$$

Loop 2

$$\frac{-LdI_i}{dt} - I_iR_2 - I_gR_2 + V_c = 0$$

$$\frac{dI_i}{dt} = \frac{1}{L}V_c(t) - \frac{R_2}{L}I_i - \frac{R_2}{L}I_g$$

$$\dot{X}_2(t) = \frac{1}{L}X_1(t) - \frac{R_2}{L}I_i$$

and current through the capacitor

$$I_x - I_i = C\frac{dV_c}{dt}$$

$$\frac{dV_c}{dt} = \frac{1}{C}[I_x - I_i].$$

Substituting I_x from output equation,

$$\dot{X}_1 = \frac{1}{CR_1}U_1 - \frac{1}{CR_1}X_1(t) - \frac{1}{C}X_2(t)$$

$$\begin{bmatrix} \dot{X}_1 \\ \dot{X}_2 \end{bmatrix} = \begin{bmatrix} \frac{-1}{R_1C} & \frac{-1}{C} \\ \frac{1}{L} & \frac{-R_2}{L} \end{bmatrix} \begin{bmatrix} X_1(t) \\ X_2(t) \end{bmatrix} + \begin{bmatrix} \frac{1}{CR_1} & 0 \\ 0 & \frac{-R_2}{L} \end{bmatrix} \begin{bmatrix} U_1 \\ U_2 \end{bmatrix}$$

$$Y = \begin{bmatrix} \frac{-1}{R_1} & 0 \end{bmatrix} \begin{bmatrix} X_1(t) \\ X_2(t) \end{bmatrix} + \begin{bmatrix} \frac{1}{R_1} & 0 \end{bmatrix} \begin{bmatrix} U_1 \\ U_2 \end{bmatrix}$$

Example 14: Consider the mechanical system shown in Fig. 4.6 for shown displacement and velocities. Obtain the state model in standard form. Assume velocity of M_2 as output.

SOLUTION:

Select the state variable as energy storing elements, i.e. displacements and velocities related to spring and friction.

$$\therefore \quad X_1(t) = Y_1(t)$$

$$X_2(t) = Y_2(t)$$

$$X_3(t) = V_1(t)$$

$$X_4(t) = V_2(t)$$

$$U(t) = F_1(t)$$

$$Y(t) = V_2(t)$$

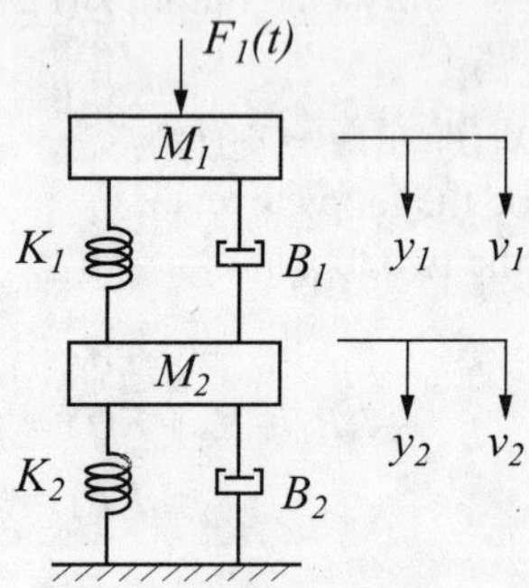

Fig. 4.6.

Equivalent diagram of above figure is

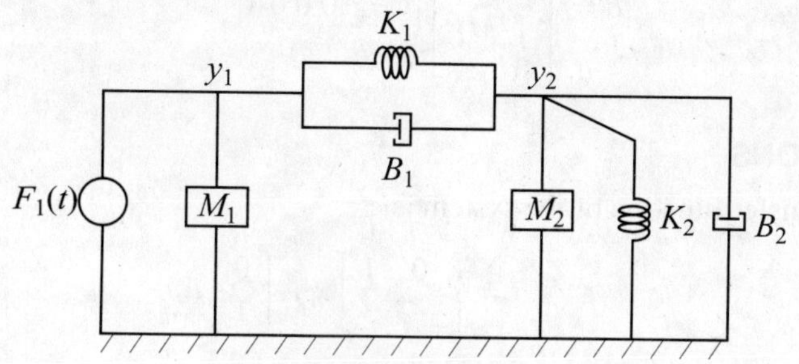

Fig. 4.7.

The spring force is proportional to net displacement in spring while friction force is proportional to velocity

At node y_1

$$F_1 = M_1\frac{d^2y_1}{dt^2} + K_1(y_1 - y_2) + B\frac{d}{dt}(y_1 - y_2) \tag{1}$$

At node y_2

$$O = M_2\frac{d^2y_2}{dt^2} + K_2y_2 + B_2\frac{dy_2}{dt} + K_1(y_2 - y_1) + B_1\frac{dy_2}{dt} - B_1\frac{dy_1}{dt} \tag{2}$$

Substituting all values in terms of state variables, we get

$$Y_1 = X_1, \frac{dY_1}{dt} = \dot{X}_1 = V_1 = X_3, \frac{d^2Y_1}{dt^2} = \dot{X}_3$$

$$Y_2 = X_2, \frac{dY_2}{dt} = \dot{X}_2 = V_2 = X_4, \frac{d^2Y_2}{dt^2} = \dot{X}_4$$

Substituting in eqns. (1) and (2)

$$U(t) = M_1\dot{X}_3 + K_1[X_1 - X_2] + B_1[X_3 - X_4] \tag{3}$$

$$O = M_2\dot{X}_4 + K_2X_2 + B_2X_4 + K_1[X_2 - X_1] + B_1[X_4 - X_3] \tag{4}$$

From eqns. (3) and (4), we can write

$$\dot{X}_3 = \frac{1}{M_1}[U(t) - K_1(X_1 - X_2) - B_1(X_3 - X_4)] \tag{5}$$

$$\dot{X}_4 = \frac{1}{M^2}[-K_2X_2 - B_2X_4 - K_1(X_2 - X_2) - B_1(X_4 - X_3)] \tag{6}$$

$$\text{and } \dot{X}_1 = X_3, \quad \dot{X}_2 = X_4, \quad Y(t) = V_2(t), \quad V_2(t) = X_4(t)$$

∴ State model can be constructed in the standard form

$$\dot{X} = AX + BU$$

$$Y = CX + DU$$

$$\text{where } A = \begin{bmatrix} 0 & 0 & 1 & 0 \\ 0 & 0 & 0 & 1 \\ \frac{-K_1}{M_1} & \frac{K_1}{M_1} & \frac{-B_1}{M_1} & \frac{B_1}{M_1} \\ \frac{K_1}{M_2} & \frac{-(K_1+K_2)}{M_2} & \frac{B_1}{M_2} & \frac{-(B_1+B_2)}{M_2} \end{bmatrix}$$

$$B = \begin{bmatrix} 0 \\ 0 \\ 1/M_1 \\ 0 \end{bmatrix} \quad C = [0\,0\,0\,1]$$

OBJECTIVE QUESTIONS

Q1: The equivalent transfer function of the system is

$$\dot{x} = \begin{bmatrix} 0 & 1 \\ -2 & 3 \end{bmatrix} x + \begin{bmatrix} 0 \\ 1 \end{bmatrix} u$$

$$\dot{y} = [1\ 0]x$$

(a) $\frac{1}{S^2+5S+2}$ (c) $\frac{3}{S^2+3S+2}$

(b) $\frac{1}{S^2+3S+2}$ (d) $\frac{3}{S^3+2S^2+2}$

Q2: Consider the following properties attributed to state model of a system:

(1) state model is unique

(2) state model can be derived from transfer function

(3) state model can be derived from time-invariant system.

(a) 1,2,3 are correct (c) 1,2 are correct

(b) 2,3 are correct (d) none

Q3: The transfer function of the system with the state space representation of

$X = Ax + Bu \qquad Y = Cx + Du$

(a) $C(SI-A)^{-1}B$ (b) $(SI-A)^{-1}B+D$

(c) $C(SI-A)^{-1}B+D$ (d) None

Q4: A system is described by the state equation

$$\begin{bmatrix} \dot{x}_1 \\ \dot{x}_2 \end{bmatrix} = \begin{bmatrix} 2 & 0 \\ 0 & 2 \end{bmatrix} \begin{bmatrix} x_1 \\ x_2 \end{bmatrix} + \begin{bmatrix} 1 \\ 1 \end{bmatrix} u$$

The STM of the system is

(a) $\begin{bmatrix} e^{2t} & 0 \\ 0 & e^{2t} \end{bmatrix}$ (b) $\begin{bmatrix} e^{-2t} & 0 \\ 0 & e^{-t} \end{bmatrix}$

(c) $\begin{bmatrix} e^{2t} & 1 \\ 1 & e^{2t} \end{bmatrix}$ (d) $\begin{bmatrix} e^{-2t} & 1 \\ 1 & e^{-t} \end{bmatrix}$

Q5: A linear time-invariant system is described by the state variable model

$$\begin{bmatrix} \dot{X}_1 \\ \dot{X}_2 \end{bmatrix} = \begin{bmatrix} -1 & 0 \\ 0 & -2 \end{bmatrix} \begin{bmatrix} X_1 \\ X_2 \end{bmatrix} + \begin{bmatrix} 0 \\ 1 \end{bmatrix} u$$

$$Y = [1\quad 2] \begin{bmatrix} X_1 \\ X_2 \end{bmatrix}$$

System is (a) completely controllable (b) observable

(c) controllable and observable (d) none

Q6: The transfer function $Y(S)/U(S)$ of a system described by state equation

$\dot{x}(t) = -2x(t) + 2u(t) \qquad y(t) = 0.5x(t)$

(a) $\frac{0.5}{S-2}$ (b) $\frac{1}{S-2}$

(c) $\frac{0.5}{S+2}$ (d) $\frac{1}{S+2}$

Hint Take Laplace on both equations, then find $\frac{Y(S)}{U(S)}$

Q7: The eigenvalues of the matrix $\Rightarrow \begin{bmatrix} a & 1 \\ a & 1 \end{bmatrix}$ are

(a) $(a+1)$, 0 (b) a, 0

(c) $(a-1)$, 0 (d) 0, 0

Q8: For a system $[\dot{X}] = \begin{bmatrix} 2 & 0 \\ 0 & 4 \end{bmatrix} X + \begin{bmatrix} 1 \\ 1 \end{bmatrix} u,$

$Y = [4 \quad 0]X$ with u as unit impulse and with zero initial state, the output y becomes

(a) $2\ e^{2t}$ (c) $4\ e^{2t}$

(b) $2\ e^{4t}$ (d) $4\ e^{4t}$

Q9: The eigenvalues of the given system are

$$\begin{bmatrix} \dot{x}_1 \\ \dot{x}_2 \\ \dot{x}_3 \end{bmatrix} = \begin{bmatrix} 0 & 1 & 0 \\ 0 & 0 & 1 \\ -6 & -11 & -6 \end{bmatrix} \begin{bmatrix} x_1 \\ x_2 \\ x_3 \end{bmatrix}$$

(a) $-1, -4, 5$ (c) $1, -2, 3$

(b) $-1, 4, 2$ (d) $-1, -2, -3$

Q10: What statement is not true for STM?

(a) $\varphi(0) = 1$ (c) $[\varphi(t)]^n = \varphi(t^n)$

(b) $\varphi(t_1 + t_2) = \varphi(t_1)\varphi(t_2)$ (d) None

Answers

(1) b (2) b (3) c (4) a (5) b (6) d (7) a (8) c (9) d (10) d

QUESTION BANK WITH ANSWERS

Q1: Use the Laplace transform to determine the output of a system represented by the differential equation

$$\ddot{y}(t) + 5\dot{y}(t) + 6y(t) = 2\dot{x}(t) + x(t)$$

in response to input $x(t) = u(t)$. Assume that all initial conditions are zero.

Ans. $\ddot{y}(t) + 5\dot{y}(t) + 6y(t) = 2\dot{x}(t) + x(t)$ and $x(t) = u(t)$

Take Laplace transform on both sides

$$S^2Y(S) + 5SY(S) + 6Y(S) = 2SX(S) + X(S)$$

$$\frac{Y(S)}{X(S)} = \frac{2S+1}{S^2+5S+6} = \frac{2S+1}{(S+2)(S+3)}$$

Here $X(S) = \frac{1}{S}$

$$Y(S) = \frac{(2S+1)}{S(S+2)(S+3)} = \frac{A}{S} + \frac{B}{S+2} + \frac{C}{S+3}$$

$$A = \frac{2S+1}{(S+2)(S+3)}\bigg|_{S=0} = \frac{1}{6}$$

$$B = \frac{2S+1}{S(S+3)}\bigg|_{S=-2} = \frac{-3}{-2\times 1} = \frac{3}{2}$$

$$C = \frac{2S+1}{(S+2)(S)}\bigg|_{S=-3} = \frac{-5}{-3\times -1} = \frac{-5}{3}$$

$$Y(S) = \frac{1/6}{S} + \frac{3/2}{S+2} + \frac{-5/3}{S+3}$$

$$y(t) = \left[\frac{1}{6} + \frac{3}{2}e^{-2t}\frac{-5}{3}e^{-3t}\right]u(t)$$

Q2: Obtain the state model for the electric circuit shown in figure.

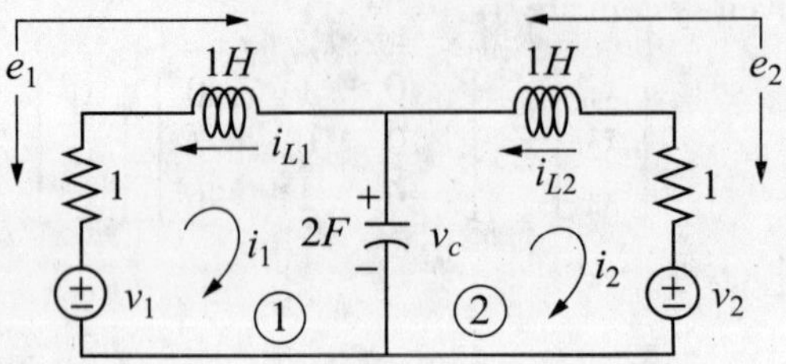

Fig. 4.1 P.

Ans. Apply KVL in loop (1)

$$V_1 = +i_1(1) + (1)\frac{di_1}{dt} + V_C \tag{1}$$

$$V_2 = -i_2 - (1)\frac{di_2}{dt} + V_C \tag{2}$$

Here $\Rightarrow i_1 = -i_{L1}$ and $i_2 = -i_{L2}$

and $i_1 - i_2 = 2\frac{dV_c}{dt}$ (3)

In terms of inductor's current eqns. (1), (2), (3) will be –

$$V_1 = -i_{L_1} - \frac{di_{L_1}}{dt} + V_C; \quad \frac{di_{L_1}}{dt} = -i_{L_1} + V_C + V_1$$

$$V_2 = -i_{L_2} + \frac{di_{L_2}}{dt} + V_C; \quad \frac{di_{L_2}}{dt} = -i_{L_2} + V_C + V_2$$

$$i_{L_2} - i_{L_1} = 2\frac{dV_C}{dt} \quad ; \frac{dV_C}{dt} = \frac{1}{2}i_{L_2} - \frac{i_{L_1}}{C}$$

Let $i_{L_1}; i_{L_2}$ and V_C are state variables x_1, x_2 and x_3 respectively and V_1, V_2 as forcing function u_1 and u_2.

Therefore, the state equation will be

$$\begin{bmatrix} \frac{di_{L_1}}{dt} \\ \frac{di_{L_2}}{dt} \\ \frac{dV_C}{dt} \end{bmatrix} = \begin{bmatrix} -1 & 0 & 1 \\ 0 & -1 & 1 \\ -\frac{1}{2} & \frac{1}{2} & 0 \end{bmatrix} \begin{bmatrix} i_{L_1} \\ i_{L_2} \\ V_C \end{bmatrix} + \begin{bmatrix} 1 & 0 \\ 0 & 1 \\ 0 & 0 \end{bmatrix} \begin{bmatrix} V_1 \\ V_2 \end{bmatrix} \tag{1}$$

If voltages e_1, e_2 and V_C are the required output y_1, y_2, y_3 respectively.

$$y_1 = e_1 = V_1 - V_C, \quad y_2 = V_2 - V_C, \quad y_3 = V_C$$

$$\begin{bmatrix} y_1 \\ y_2 \\ y_3 \end{bmatrix} = \begin{bmatrix} 0 & 0 & -1 \\ 0 & 0 & -1 \\ 0 & 0 & 1 \end{bmatrix} \begin{bmatrix} i_{L_1} \\ i_{L_2} \\ V_C \end{bmatrix} + \begin{bmatrix} 1 & 0 \\ 0 & 1 \\ 0 & 0 \end{bmatrix} \begin{bmatrix} V_1 \\ V_2 \end{bmatrix} \tag{2}$$

Equation (1) in addition to eqn. (2) represents the required state space model of the form

$$\dot{x} = A\,x + B\,u$$

$$y = C\,x + D\,u$$

Q3: Obtain the state variable representation of the system described by the following differential equation

(i) $\frac{d^3y}{dt^3} + \frac{4d^2y}{dt^2} + \frac{5dy}{dt} + 2y = V$

(ii) $\frac{d^3x}{dt^3} + \frac{3d^2x}{dt^2} + \frac{4dx}{dt} + 4x = u_1 + 3u_2 + 4u_3$ and output $y_1 = \frac{4dx}{dt} + 3u_1$, $y_2 = \frac{d^2x}{dt_2} + 4u_2 + u_3$

Ans. (i) $\frac{d^3y}{dt^3} + \frac{4d^2y}{dt^2} + \frac{5dy}{dt} + 2y(t) = V$

Suppose

$$x_1 = y \Rightarrow \dot{x}_1 = x_2$$

$$x_2 = \dot{y} \Rightarrow \dot{x}_2 = x_3$$

$$x_3 = \ddot{y}$$

$$\dot{x}_3 = -2x_1 - 5x_2 - 4x_3 + V$$

State space representation

$$\begin{bmatrix} \dot{x}_1 \\ \dot{x}_2 \\ \dot{x}_3 \end{bmatrix} = \begin{bmatrix} 0 & 1 & 0 \\ 0 & 0 & 1 \\ -2 & -5 & -4 \end{bmatrix} \begin{bmatrix} x_1 \\ x_2 \\ x_3 \end{bmatrix} + \begin{bmatrix} 0 \\ 0 \\ 1 \end{bmatrix} [V]$$

and output $y = x_1$

$$[y] = [1 \quad 0 \quad 0] \begin{bmatrix} x_1 \\ x_2 \\ x_3 \end{bmatrix}$$

(ii) $\frac{d^3x}{dt^3} + \frac{3d^2x}{dt^2} + \frac{4dx}{dt} + 4x = u_1 + 3u_2 + 4u_3$

and output, $y_1 = \frac{4dx}{dt} + 3u_1$

$$y_2 = \frac{d^2x}{dt^2} + 4u_2 + u_3$$

Suppose

$$\begin{aligned}\dot{x}_1 &= x\\ x_2 &= \dot{x}_1 = \dot{x}_1\\ x_3 &= \dot{x}_2 = \ddot{x}_1\\ \dot{x}_3 &= -3x_3 - 4x_2 - 4x_1\end{aligned}$$

State space representation

$$\begin{bmatrix}\dot{x}_1\\ \dot{x}_2\\ \dot{x}_3\end{bmatrix} = \begin{bmatrix}0 & 1 & 0\\ 0 & 0 & 1\\ -4 & -4 & -3\end{bmatrix}\begin{bmatrix}x_1\\ x_2\\ x_3\end{bmatrix} + \begin{bmatrix}0 & 0 & 0\\ 0 & 0 & 0\\ 1 & 3 & 4\end{bmatrix}\begin{bmatrix}u_1\\ u_2\\ u_3\end{bmatrix}$$

and outputs

$$y_1 = \frac{4dx}{dt} + 3u_1 = 4x_2 + 3u_1$$

$$y_2 = \frac{d^2x}{dt^2} + 4u_2 + u_3 = x_3 + 4u_2 + u_3$$

Hence, output matrix

$$\begin{bmatrix}y_1\\ y_2\end{bmatrix} = \begin{bmatrix}0 & 4 & 0\\ 0 & 0 & 1\end{bmatrix}\begin{bmatrix}x_1\\ x_2\\ x_3\end{bmatrix} + \begin{bmatrix}3 & 0 & 0\\ 0 & 4 & 1\end{bmatrix}\begin{bmatrix}u_1\\ u_2\\ u_3\end{bmatrix}$$

Q4: Obtain eigenvalues, eigenvec 'or and state model in canonical form for a system described by

$$\dot{x}(t) = \begin{bmatrix}0 & 1 & 0\\ 3 & 0 & 2\\ -12 & -7 & -6\end{bmatrix} x(t) + \begin{bmatrix}1\\ 0\\ 2\end{bmatrix} u(t)$$

$$y(t) = [1 \quad 0 \quad 0][x(t)]$$

SOLUTION:

(1) Eigenvalues

$$(\lambda I - A) = \begin{bmatrix}\lambda & 0 & 0\\ 0 & \lambda & 0\\ 0 & 0 & \lambda\end{bmatrix} - \begin{bmatrix}0 & 1 & 0\\ 3 & 0 & 2\\ -12 & -7 & -6\end{bmatrix}$$

$$= \begin{bmatrix}\lambda & -1 & 0\\ -3 & \lambda & -2\\ 12 & 7 & \lambda+6\end{bmatrix}$$

$$= \lambda\{\lambda(\lambda+6)\} - (-2)(7)\} - (-1)\{-3(\lambda+6)\} - (-2)12)\} = 0$$

$$= (\lambda+1)(\lambda+2)(\lambda+3) = 0$$

Eigenvectors are $-\lambda = -1,\ -2,\ -3$

(2) Eigenvectors

(a) Corresponding to $\lambda = -1$, the eigenvector is –

$$\begin{bmatrix} -1 & -1 & 0 \\ -3 & -1 & -2 \\ 12 & 7 & 5 \end{bmatrix} \begin{bmatrix} x_1 \\ x_2 \\ x_3 \end{bmatrix} = 0 \qquad \{\text{Place } \lambda = -1 \text{ in } 1^{\text{st}} \text{ row of } (\lambda I - A)\}$$

$$x_1 + x_2 = 0 \text{ or } x_2 = -x_1$$

$$3x_1 + x_2 + 2x_3 = 0 \text{ or } x_3 = -x_1$$

$$12x_1 + 7x_2 + 5x_3 = 0$$

If $x_1 = 1$ then x_2 and x_3 are -1. Therefore, eigenvector is $\begin{bmatrix} 1 \\ -1 \\ -1 \end{bmatrix}$

(b) Corresponding $\lambda = -2$

Place $\lambda = -2$ in 2nd row of $(\lambda I - A)$

$$\begin{bmatrix} -2 & -1 & 0 \\ -3 & -2 & -2 \\ 17 & 7 & 4 \end{bmatrix} \begin{bmatrix} x_1 \\ x_2 \\ x_3 \end{bmatrix} = 0$$

$$\text{i.e. } 2x_1 + x_2 = 0 \quad \text{or} \quad x_2 = -2x_1$$

$$3x_1 + 2x_2 + 2x_3 = 0 \quad \text{or} \quad x_3 = \frac{1}{2}x_1$$

If $x_1 = 2$ then

$x_2 = -4$ and $x_3 = 1$

Therefore, eigenvector is $\begin{bmatrix} 2 \\ -4 \\ 1 \end{bmatrix}$

(c) Corresponding to $\lambda = -3$, the eigenvector is

$$= \begin{bmatrix} 1 \\ -3 \\ 3 \end{bmatrix}$$

$$\text{Modal matrix} = \begin{bmatrix} 1 & 2 & 1 \\ -1 & -4 & -3 \\ -1 & 1 & 3 \end{bmatrix}$$

$$\text{adj}[M] = \begin{bmatrix} +(-12+3) & -(6-1) & +(-6+4) \\ -(-3-3) & +(3+1) & -(-3+1) \\ +(-1-4) & -(1+2) & +(-4+2) \end{bmatrix}$$

$$= \begin{bmatrix} -9 & -5 & -2 \\ 6 & 4 & 2 \\ -5 & -3 & -2 \end{bmatrix}$$

$$|M| = -2$$

$$[M^{-1}] = \frac{-1}{2} \begin{bmatrix} -9 & -5 & -2 \\ 6 & & 2 \\ -5 & -3 & --2 \end{bmatrix}$$

$$[M^{-1}AM] = -\frac{1}{2}\begin{bmatrix} -9 & -5 & -2 \\ 6 & 4 & 2 \\ -5 & -3 & -2 \end{bmatrix}\begin{bmatrix} 0 & 1 & 1 \\ 3 & 0 & 2 \\ -12 & -7 & -6 \end{bmatrix}\begin{bmatrix} 1 & 2 & 1 \\ -1 & -4 & -3 \\ -1 & 1 & 3 \end{bmatrix}$$

$$= \begin{bmatrix} -1 & 0 & 0 \\ 0 & -2 & 0 \\ 0 & 0 & -3 \end{bmatrix}$$

$$B = \begin{bmatrix} 1 \\ 0 \\ 2 \end{bmatrix}$$

$$[M^{-1}B] = \frac{-1}{2}\begin{bmatrix} -9 & -5 & -2 \\ 6 & 4 & 2 \\ -5 & -3 & -2 \end{bmatrix}\begin{bmatrix} 1 \\ 0 \\ 2 \end{bmatrix}$$

$$= \frac{-1}{2}\begin{bmatrix} -13 \\ -10 \\ -9 \end{bmatrix} = \begin{bmatrix} 13/2 \\ 5 \\ 9/2 \end{bmatrix}$$

Therefore, canonical representation is

$$\begin{bmatrix} \dot{Z}_1 \\ \dot{Z}_2 \\ \dot{Z}_3 \end{bmatrix} = \begin{bmatrix} -1 & 0 & 0 \\ 0 & -2 & 0 \\ 0 & 0 & -3 \end{bmatrix}\begin{bmatrix} Z_1 \\ Z_2 \\ Z_3 \end{bmatrix} + \begin{bmatrix} 13/2 \\ 5 \\ 9/2 \end{bmatrix} u$$

$$[y] = [1 \quad 1 \quad 1]\begin{bmatrix} Z_1 \\ Z_2 \\ Z_3 \end{bmatrix}$$

Q5: A linear time-invariant system is characterized by the homogeneous state equation

$$\begin{bmatrix} \dot{x}_1 \\ \dot{x}_2 \end{bmatrix} = \begin{bmatrix} 1 & 0 \\ 1 & 1 \end{bmatrix}\begin{bmatrix} x_1 \\ x_2 \end{bmatrix}$$

(a) Compute the solution of the homogeneous equation assuming that initial state vector

$$[x_0] = \begin{bmatrix} 1 \\ 0 \end{bmatrix}$$

(b) Consider now that the system has forcing function and is represented by the following non-homogeneous state equation.

$$\begin{bmatrix} x_1 \\ x_2 \end{bmatrix} = \begin{bmatrix} 1 & 0 \\ 1 & 1 \end{bmatrix}\begin{bmatrix} x_1 \\ x_2 \end{bmatrix} + \begin{bmatrix} 1 \\ 0 \end{bmatrix} u$$

where u is a unit step function. Compute the solution of this state equation assuming

$$[x_0] = \begin{bmatrix} 1 \\ 0 \end{bmatrix}$$

SOLUTION:

(a)
$$x(t) = \phi(t)x(0)$$
$$\phi(t) = LT^{-1}[SI - A]^{-1} = \begin{bmatrix} e^t & 0 \\ te^t & e^t \end{bmatrix}$$
$$\therefore \quad x(t) = \begin{bmatrix} e^t & 0 \\ te^t & e^t \end{bmatrix}\begin{bmatrix} 1 \\ 0 \end{bmatrix} = \begin{bmatrix} e^t \\ te^t \end{bmatrix} = \begin{bmatrix} x_1 \\ x_2 \end{bmatrix}$$

(b)
$$x(t) = e^{At}x(0) + \int_0^t e^{A(t-\tau)}Bu(\tau)d\tau$$
$$= e^{At}x(0) + \int_0^t e^{At}e^{-At}Bu(\tau)d\tau$$
$$= e^{At}\left[x(0) + \int_0^t e^{-A\tau}Bu(\tau)d\tau\right]$$
$$= \phi(t)\left[x(0) + \int_0^t \phi(-\tau)Bu(\tau)d\tau\right]$$
$$\therefore \quad \phi(-\tau)Bu(\tau) = \begin{bmatrix} e^{-\tau} & 0 \\ -\tau e^{-\tau} & e^{-\tau} \end{bmatrix}\begin{bmatrix} 0 \\ 1 \end{bmatrix}1$$
$$= \begin{bmatrix} 0 \\ e^{-\tau} \end{bmatrix}$$
$$\therefore \quad \int_0^t \phi(-\tau)Bu(\tau)d\tau = \int_0^t \begin{bmatrix} 0 \\ e^{-\tau} \end{bmatrix} d\tau$$
$$\Rightarrow \begin{bmatrix} 0 \\ 1 - e^{-\tau} \end{bmatrix}$$

Therefore,
$$x(t) = \begin{bmatrix} e^t & 0 \\ te^t & e^t \end{bmatrix}\left\{\begin{bmatrix} 1 \\ 0 \end{bmatrix} + \begin{bmatrix} 0 \\ 1 - e^{-t} \end{bmatrix}\right\}$$
$$= \begin{bmatrix} e^t & 0 \\ te^t & e^t \end{bmatrix}\begin{bmatrix} 1 \\ 1 - e^{-t} \end{bmatrix}$$
$$x(t) = \begin{bmatrix} e^t \\ te^t + e^t - 1 \end{bmatrix} = \begin{bmatrix} e^t \\ (t+1)e^t - 1 \end{bmatrix}$$

UNSOLVED QUESTION BANK

Q1: Explain the concept of state.

Q2: Define and explain the terms

(a) State variables

(b) State vector

(c) State space

Q3: Explain the advantages of state space method over conventional method.

Q4: Voltage across the capacitor is V_c with V_c and i_l as a set of state variables. Derive the state model

Q5: Derive the expression for the transfer function from the state model.

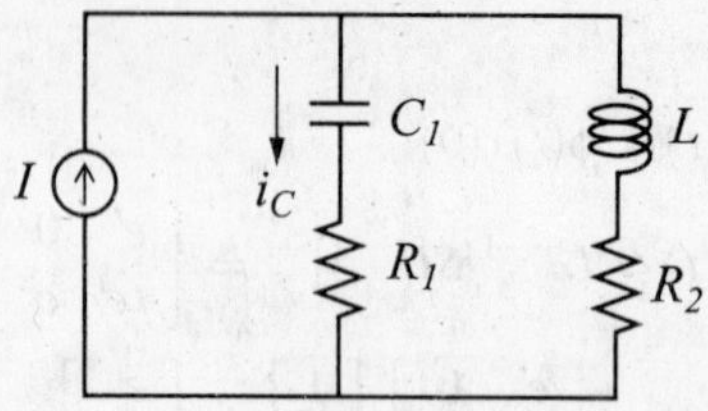

Fig. 4.2 P.

Q6: What is characteristic equation of a system matrix A.

Q7: Define eigenvalues and eigenvector of a matrix.

Q8: What is modal matrix? State its importance.

Q9: Obtain the transfer function of the system having state model

$$\dot{x}(t) = \begin{bmatrix} -5 & -1 \\ 3 & -1 \end{bmatrix}(t) + \begin{bmatrix} 2 \\ 5 \end{bmatrix} U(t)$$

$$Y(t) = (1 \quad 2)X(t)$$

Ans. $\frac{12S+59}{(S+2)(S+4)}$

Q10: What is homogeneous and non-homogeneous state equation.

Q11: Define state transition matrix (STM).

Q12: Obtain the complete solution of non-homogeneous state equation using time domain method.

Q13: State importance and various properties of state transition matrix (STM)

Q14: Define and explain the state controllability.

Q15: What is output controllability? How to evaluate for a system?

Q16: Define and explain concept of observability.

Q17: Obtain e^{At} for the following matrix

(a) $\begin{bmatrix} 0 & 1 \\ -2 & -3 \end{bmatrix}$ Ans. $\begin{bmatrix} 2e^{-t} - e^{-2t} & e^{-t} - e^{-2t} \\ -2e^{-t} + 2e^{-2t} & -e^{-t} + 2e^{-2t} \end{bmatrix}$

(b) $\begin{bmatrix} 0 & 1 \\ -2 & 0 \end{bmatrix}$ Ans. $\begin{bmatrix} \cos\sqrt{2}t & \frac{1}{\sqrt{2}}\sin\sqrt{2}t \\ -\sqrt{2}\sin\sqrt{2}t & \cos\sqrt{2}t \end{bmatrix}$

(c) $\begin{bmatrix} 1 & 0 \\ 1 & 1 \end{bmatrix}$ Ans. $\begin{bmatrix} e^t & 0 \\ te^t & e^t \end{bmatrix}$

Q18: Evaluate the controllability of the following system with

(a) $A = \begin{bmatrix} -2 & 0 \\ 0 & -1 \end{bmatrix} B = \begin{bmatrix} 3 \\ 1 \end{bmatrix} C^T = \begin{bmatrix} 1 \\ 0 \end{bmatrix}$

Ans. Controllable

(b) $A = \begin{bmatrix} -1 & 0 \\ 0 & -2 \end{bmatrix} B = \begin{bmatrix} 2 \\ 5 \end{bmatrix}$

Ans. Controllable

(c) $A = \begin{bmatrix} -1 & 1 & 0 \\ 0 & -1 & 0 \\ 0 & 0 & -2 \end{bmatrix} B = \begin{bmatrix} 4 & 2 \\ 0 & 0 \\ 3 & 0 \end{bmatrix}$ Ans. Not controllable

Q19: Evaluate the observability of the following system with

(a) $A = \begin{bmatrix} 2 & 1 & 0 \\ 0 & 2 & 1 \\ 0 & 0 & 0 \end{bmatrix}$ $C = \begin{bmatrix} 3 & 0 & 0 \\ 4 & 0 & 0 \end{bmatrix}$ Ans. Observable

(b) $A = \begin{bmatrix} -1 & 0 \\ 0 & -2 \end{bmatrix}$ $C = (0\ 1)$

Ans. Not observable

(c) $A = \begin{bmatrix} 0 & 1 & 0 \\ 0 & 0 & 1 \\ -6 & -11 & -6 \end{bmatrix}$ $C = (4\ 5\ 1)$ Ans. Not observable

Q20: Obtain the homogeneous solution of the equation $\dot{X} = AX$ where

$$A = \begin{bmatrix} -9 & 1 \\ -14 & 0 \end{bmatrix} \text{ and } X(0) = \begin{bmatrix} 1 \\ 1 \end{bmatrix}$$

Ans. $\begin{bmatrix} -\frac{1}{5}e^{-2t} + \frac{6}{5}e^{-7t} \\ \frac{-7}{5}e^{-2t} + \frac{15}{5}e^{-7t} \end{bmatrix}$

Q21: Consider a state model

$$\dot{X} = AX + BU$$

$$A = \begin{bmatrix} 0 & 1 & 0 \\ 0 & 0 & 1 \\ -40 & -34 & -10 \end{bmatrix} \quad B = \begin{bmatrix} 0 \\ 0 \\ 1 \end{bmatrix}$$

Show that the eigenvalues of A are -3 $\pm j, -4$.

Q22: Consider a state model

$$\begin{bmatrix} \dot{x}_1 \\ \dot{x}_2 \end{bmatrix} = \begin{bmatrix} 0 & 1 \\ -2 & --3 \end{bmatrix} \begin{bmatrix} x_1 \\ x_2 \end{bmatrix} + \begin{bmatrix} 0 \\ 2 \end{bmatrix} u$$

$$\begin{bmatrix} x_1(0) \\ x_2(0) \end{bmatrix} = \begin{bmatrix} 0 \\ 1 \end{bmatrix} \quad u \rightarrow \text{unit step.}$$

Compute the STM and then find the state response $x(t)$ for $t > 0$

$$\phi(t) = \begin{bmatrix} e^{-t} & 0 \\ 0 & e^{-2t} \end{bmatrix}$$

Ans. $x(t) = \begin{bmatrix} 1 - e^{-t} \\ e^{-t} \end{bmatrix}$

Q23: Determine the transfer matrix for MIMO system given by-

$$\begin{bmatrix} \dot{x}_1 \\ \ddot{x}_2 \end{bmatrix} = \begin{bmatrix} 0 & 3 \\ -2 & -5 \end{bmatrix} \begin{bmatrix} x_1 \\ x_2 \end{bmatrix} + \begin{bmatrix} 1 & 1 \\ 1 & 1 \end{bmatrix} \begin{bmatrix} u_1 \\ u_2 \end{bmatrix}$$

$$\begin{bmatrix} y_1 \\ y_2 \end{bmatrix} = \begin{bmatrix} 2 & 1 \\ 1 & 0 \end{bmatrix} \begin{bmatrix} x_1 \\ x_2 \end{bmatrix}$$

Ans. Hint $C(SI-A)^{-1}B = TM$ (Transfer matrix)

$$Y(S) = \text{TM}\, U(S)$$

$$= \begin{bmatrix} \dfrac{3S+14}{(S+2)(S+3)} & \dfrac{3S+14}{(S+2)(S+3)} \\ \dfrac{S+8}{(S+2)(S+3)} & \dfrac{S+8}{(S+2)(S+3)} \end{bmatrix} \begin{bmatrix} U_1(S) \\ U_2(S) \end{bmatrix}$$

Q24: Find the canonical form of representation and state transition matrix

$$\begin{bmatrix} x_1 \\ x_2 \\ x_3 \end{bmatrix} = \begin{bmatrix} 2 & -2 & 3 \\ 1 & 1 & 1 \\ 1 & 3 & -1 \end{bmatrix} \begin{bmatrix} x_1 \\ x_2 \\ x_3 \end{bmatrix} + \begin{bmatrix} 11 \\ 1 \\ -14 \end{bmatrix} u$$

$$y = [-3 \quad 5 \quad -2] \begin{bmatrix} x_1 \\ x_2 \\ x_3 \end{bmatrix}$$

$$\begin{bmatrix} \dot{Z}_1 \\ \dot{Z}_2 \\ \dot{Z}_3 \end{bmatrix} = \begin{bmatrix} 1 & 0 & 0 \\ 0 & -2 & 0 \\ 0 & 0 & 3 \end{bmatrix} \begin{bmatrix} Z_1 \\ Z_2 \\ Z_3 \end{bmatrix} + \begin{bmatrix} 0 \\ 1 \\ \frac{4}{5} \end{bmatrix} u$$

Ans. $Y = [1 \quad 1 \quad 1] \begin{bmatrix} Z_1 \\ Z_2 \\ Z_3 \end{bmatrix}$

Q25: Obtain the transfer function if

$$\begin{bmatrix} \dot{x}_1 \\ \dot{x}_2 \end{bmatrix} = \begin{bmatrix} -5 & -1 \\ 3 & -1 \end{bmatrix} \begin{bmatrix} x_1 \\ x_2 \end{bmatrix} + \begin{bmatrix} 2 \\ 5 \end{bmatrix} u$$

$$y = [1 \quad 2] \begin{bmatrix} x_1 \\ x_2 \end{bmatrix}$$

Ans. $\frac{12S+59}{(S+4)(S+2)}$

5 Z-Transform

The z-transform transforms differential equations into algebraic equations.

5.1 DEFINITION OF THE Z-TRANSFORM

The z-transform of a discrete-time signal $x(n)$ is defined as:

$$X(z) = \sum_{n=-\infty}^{\infty} x(n)z^{-n} \tag{5.1}$$

The inverse z-transform is defined as given below:

$$x(n) = \frac{1}{2\pi j}\int_{c} X(z)z^{n-1}dz \tag{5.2}$$

where z is a complex variable, c is a closed contour. In polar form z can be expressed as:

$$z = re^{j\omega} \tag{5.3}$$

where r is the radius of the circle.

On putting in eqn. (5.1)

$$X(re^{j\omega}) = \sum_{n=-\infty}^{\infty} x(n)r^{-n}e^{-j\omega n} \tag{5.4}$$

For $r = 1$, the above expression reduces to discrete-time Fourier transform of $x(n)$ as shown in following equation.

$$X(e^{j\omega}) = \sum_{n=-\infty}^{\infty} x(n)e^{-j\omega n} \tag{5.5}$$

Equation (5.1) represents the two-sided z-transform. If $x(n)$ is a causal sequence, i.e. $x(n) = 0$ for $n < 0$, then z-transform is

$$X(z) = \sum_{n=0}^{\infty} x(n)z^{-n} \tag{5.6}$$

Equation (5.1) represents bilateral z-transform and eqn. (5.6) represents unilateral or one-sided z-transform, since summation is from 0 to ∞.

5.2 ROC (REGION OF CONVERGENCE)

The Region of Convergence of $X(z)$ is the set of values of z for which $X(z)$ attains a finite value.

Example 1: Find the z-transform and ROC of the following sequences:

$$x(n) = \left\{ \begin{matrix} 1,1,2,-1,2 \\ \uparrow \qquad\qquad\quad \end{matrix} \right\} \qquad x(n) = \left\{ \begin{matrix} 1,1,2,-1,2 \\ \qquad\qquad\quad \uparrow \end{matrix} \right\}$$

$$x(n) = \left\{ \begin{matrix} 1,1,2,-1,2 \\ \uparrow \end{matrix} \right\}$$

SOLUTION:

The z-transform of $x(n)$ is given by

$$X(z) = \sum_{n=-\infty}^{\infty} x(n)z^{-n}$$

On expanding this equation, we get

$$X(z) = x(0) + x(1)z^{-1} + x(2)z^{-2} + x(3)z^{-3} \tag{5.7}$$

For right-hand sequence

$$x(n) = \left\{ \begin{matrix} 1,1,2,-1,2 \\ \uparrow \qquad\qquad\quad \end{matrix} \right\}$$

$$x(0) = 1;\ x(1) = 1;\ x(2) = 2;\ x(3) = -1;\ x(4) = 2$$

On putting in eqn. (5.6)

$$X(z) = 1 + z^{-1} + 2z^{-2} - z^{-3} + 2z^{-4}$$

$$X(z) = 1 + \frac{1}{z} + \frac{2}{z^2} - \frac{1}{z^3} + \frac{2}{z^4}$$

From the above equation, we can see that the $X(z)$ converges for all values except at $z = 0$

For left-hand sequence

$$x(n) = \left\{ \begin{matrix} 1, & 1, & 2, & -1, & 2 \\ & & & & \uparrow \\ -4 & -3 & -2 & -1 & 0 \end{matrix} \right\}$$

$$x(-4) = 1;\ x(-3) = 1;\ x(-2) = 2;\ x(-1) = -1;\ x(0) = 2$$

On putting these value in eqn. (5.1), we get

$$X(z) = 2 - z + 2z^2 + z^3 + z^4$$

From the above equation, we can see that $X(z)$ converges for all values except at $z = \infty$

For two-sided sequence

$$x(n) = \left\{ \begin{matrix} 1, & 1, & 2, & -1, & 2 \\ & & \uparrow & & \\ -2 & -1 & 0 & 1 & 2 \end{matrix} \right\}$$

$$x(-2) = 1;\ x(-1) = 1;\ x(0) = 2;\ x(1) = -1;\ x(2) = 2$$

On putting these values in eqn. (5.1), we get

$$X(z) = z^2 + z + 2 - z^{-1} + 2z^{-2}$$

From the above equation, we can see that $X(z)$ converges for all values except at $z = 0$ and $z = \infty$.

For Infinite duration sequence

Example 2: Determine the z-transform and ROC of the right-sided exponential sequence

$$x(n) = a^n u(n)$$

The z-transform of $x(n)$ is given by

$$X(z) = \sum_{n=-\infty}^{\infty} x(n) z^{-n}$$

Here $x(n) = a^n$ for $n \geq 0$ and hence the above equation becomes

$$X(z) = \sum_{n=0}^{\infty} x(n) z^{-n} = \sum_{n=0}^{\infty} a^n z^{-n} = \sum_{n=0}^{\infty} (az^{-1})^n = (az^{-1})^{-1} + (az^{-1})^{-2} + (az^{-1})^{-3} + \cdots$$

This is an infinite series, since we know that

$$1 + r + r + r + \cdots = \frac{1}{1-r} \text{if } |r| < 1$$

Therefore,

$$X(z) = \frac{1}{1 - az^{-1}} = \frac{z}{z-a}$$

$X(z)$ will converge for $|az^{-1}| < 1$, i.e. $|z| > |a|$

Hence, the required z-transform is $\frac{1}{1-az^{-1}}$ and ROC is $|z| > |a|$ shown in Fig. 5.1 as shaded portion.

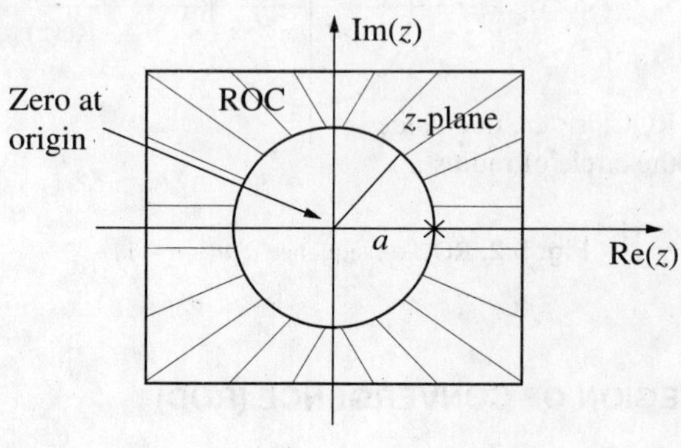

Fig. 5.1. ROC of sequence $a^n u(n)$

Example 3: Determine the z-transform and ROC of the left-sided exponential sequence

$$x(n) = -a^n u(-n-1)$$

The z-transform of $x(n)$ is given by

$$X(z) = \sum_{n=-\infty}^{\infty} x(n) z^{-n}$$

Here $x(n) = a^n$ for $n \leq 0$ and hence the above equation becomes

$$X(z) = -\sum_{n=-\infty}^{-1} x(n)z^{-n} = -\sum_{n=-\infty}^{-1} a^n z^{-n} = -\sum_{n=-\infty}^{-1} (az^{-1})^n$$

Let $m = -n$; on putting in above equation, we get

$$X(z) = -\sum_{m=\infty}^{1} (a^{-1}z)^m = -\{(a^{-1}z) + (a^{-1}z)^2 + (a^{-1}z)^3 + \cdots\}$$

$$X(z) = -(a^{-1}z)\{1 + (a^{-1}z) + (a^{-1}z)^2 + (a^{-1}z)^3 + \cdots\}$$

This is an infinite series, since we know that

$$1 + r + r + r + \cdots = \frac{1}{1-r} \text{ if } |r| < 1$$

Therefore,

$$X(z) = -\frac{a^{-1}z}{1 - a^{-1}z} = \frac{1}{1 - az^{-1}}$$

$X(z)$ will converge for $|a^{-1}z| < 1$, i.e. $|z| < |a|$

Hence, the required z-transform is $\frac{1}{1-az^{-1}}$ and ROC is $|z| < |a|$ shown in figure.

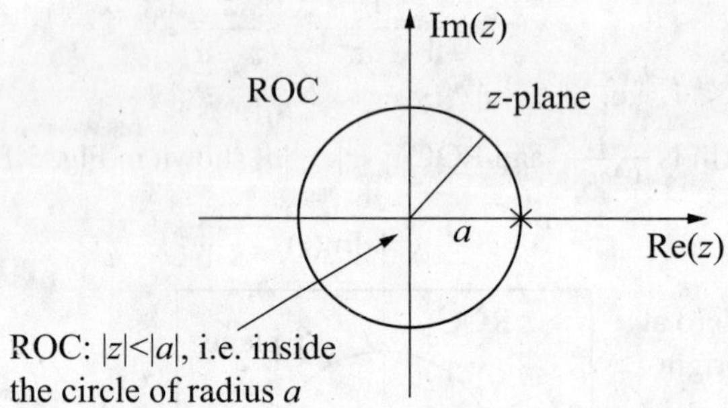

Fig. 5.2. ROC of sequence $a^n u(-n-1)$

5.3 PROPERTIES OF REGION OF CONVERGENCE (ROC)

- The ROC is a ring or disk in the z-place centred at the origin.
- The ROC cannot contain any poles.
- If $x(n)$ is a causal sequence then the ROC is the entire z-plane except at $z = 0$.
- If $x(n)$ is a non-causal sequence then the ROC is the entire z-plane except at $z = \infty$.
- If $x(n)$ is finite duration, two-sided sequence, then the ROC is the entire z-plane except at $z = 0$ and $z = \infty$.
- If $x(n)$ is infinite duration, two-sided sequence then the ROC will consist of a ring in the z-plane, bounded on the interior and exterior by a pole, not containing any poles.

Z-TRANSFORM OF SOME IMPORTANT SEQUENCES

1. Unit sample sequence

The unit sample sequence is given below:

$$\delta(n) = \begin{cases} 1 & \text{for } n = 0 \\ 0 & \text{elsewhere} \end{cases}$$

The z-transform of $x(n)$ is given by

$$X(z) = \sum_{n=-\infty}^{\infty} x(n) z^{-n}$$

$$X(z) = \sum_{n=0}^{\infty} \delta(n) z^{-n} = 1z^0 = 1$$

Therefore, the z-transform of unit sample sequence is 1 which is constant and does not depend on values of z. Hence, ROC is entire z-plane.

2. Unit step sequence

The unit step sequence is defined as:

$$u(n) = \begin{cases} 1 & \text{for } n \geq 0 \\ 0 & \text{elesewhere} \end{cases}$$

The z-transform of $x(n) = a^n u(n)$ is given as:

$$X(z) = \frac{1}{1 - az^{-1}}$$

If $a = 1$, $X(z)$ becomes

$$X(z) = \frac{1}{1 - z^{-1}}, \text{The ROC is } |z| > 1$$

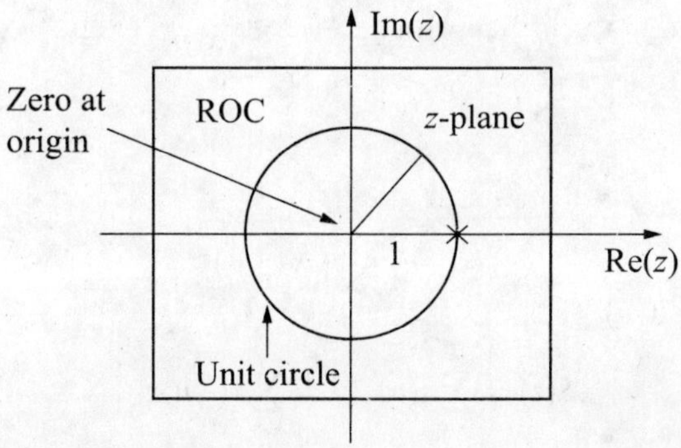

Fig. 5.3. ROC of unit step sequence

5.4 PROPERTIES OF Z-TRANSFORM

From the z-transform pair

$$x(n) \overset{z}{\longleftrightarrow} X(z)$$

5.4.1 Linearity

$$x_1(n) \overset{z}{\longleftrightarrow} X_1(z)$$

If

$$x_2(n) \overset{z}{\longleftrightarrow} X_2(z)$$

The linearity property states that

$$ax_1(n) + bx_2(n) \overset{z}{\longleftrightarrow} aX_1(z) + bX_2(z)$$

The new sequence is $s(n) = ax_1(n) + bx_2(n)$

$$Z\{s(n)\} = S(z) = \sum_{-\infty}^{\infty} s(n)z^{-n} = \sum_{-\infty}^{\infty} [ax_1(n) + bx_2(n)]\,z^{-n}$$

$$S(z) = \sum_{n=-\infty}^{\infty} ax_1(n)z^{-n} + \sum_{n=-\infty}^{\infty} bx_2(n)z^{-n} = a\sum_{n=-\infty}^{\infty} x_1(n)z^{-n} + b\sum_{n=-\infty}^{\infty} x_2(n)z^{-n} = aX_1(z) + bX_2(z)$$

5.4.2 Time shifting

$$x(n) \overset{z}{\longleftrightarrow} X(z)$$

Then $\quad x(n-m) \overset{z}{\longleftrightarrow} z^{-m}X(z)$

The z-transform of $x(n)$ is given by

$$X(z) = \sum_{n=-\infty}^{\infty} x(n)z^{-n}$$

The new sequence is $s(n) = x(n-m)$

$$S(z) = \sum_{-\infty}^{\infty} s(n)z^{-n} = \sum_{-\infty}^{\infty} x(n-m)z^{-n}$$

Let $n - m = p$

$$S(z) = \sum_{p=-\infty}^{\infty} x(p)z^{-p}z^{-m} = z^{-m}\sum_{p=-\infty}^{\infty} x(p)z^{-p} = z^{-m}X(z)$$

5.4.3 Time reversal

$$x(-n) \overset{z}{\longleftrightarrow} X(z^{-1})$$

The new sequence is $s(n) = a^n x(n)$

$$S(z) = \sum_{-\infty}^{\infty} s(n)z^{-n} = \sum_{-\infty}^{\infty} x(-n)z^{-n}$$

Let $p = -n$

$$S(z) = \sum_{p=-\infty}^{\infty} x(p)z^{p} = \sum_{p=-\infty}^{\infty} x(p)(z^{-1})^{-p} = X(z^{-1})$$

5.4.4 Multiplication by exponential sequence

$$a^n x(n) \overset{z}{\longleftrightarrow} X(a^{-1}z)$$

The new sequence is $s(n) = a^n X(n)$

$$S(z) = \sum_{-\infty}^{\infty} s(n)z^{-n} = \sum_{-\infty}^{\infty} a^n x(n)z^{-n}$$

$$S(z) = \sum_{n=-\infty}^{\infty} a^n x(n)z^{-n} = \sum_{n=-\infty}^{\infty} x(n)(az^{-1})^n = X(a^{-1}z)$$

5.4.5 Multiplication by *n*

$$nx(n) \overset{z}{\longleftrightarrow} -z\frac{d}{dz}X(z)$$

The new sequence is $s(n) = nx(n)$

$$S(z) = \sum_{-\infty}^{\infty} s(n)z^{-n} = \sum_{-\infty}^{\infty} nx(n)z^{-n}$$

On differentiating the eqn. (5.1)

$$\frac{d}{dz}X(z) = \sum_{n=-\infty}^{\infty} x(n)\frac{d}{dz}(z^{-n}) = \sum_{n=-\infty}^{\infty} (-n)x(n)(z^{-n-1})$$

$$\frac{d}{dz}X(z) = -z^{-1}\sum_{n=-\infty}^{\infty} \{nx(n)\}(z^{-n})$$

$$-z\frac{d}{dz}X(z) = \sum_{n=-\infty}^{\infty} \{nx(n)\}(z^{-n}) \overset{z}{\longleftrightarrow} nx(n)$$

5.4.6 Convolution

$$x(n) * h(n) \overset{z}{\longleftrightarrow} X(z)H(z)$$

where $s(n) = x(n) * h(n)$ is the convolution of sequences $x(n)$ and $h(n)$ defined as:

$$s(n) = x(n) * h(n) = \sum_{k=-\infty}^{\infty} x(k)h(n-k)$$

$$S(z) = \sum_{n=-\infty}^{\infty} s(n)z^{-n} = \sum_{n=-\infty}^{\infty}\left[\sum_{k=-\infty}^{\infty} x(k)h(n-k)\right]z^{-n} = \sum_{k=-\infty}^{\infty} x(k)\sum_{n=-\infty}^{\infty} h(n-k)z^{-n}$$

Let $n-k=p$

$$S(z) = \sum_{k=-\infty}^{\infty} x(k)\sum_{p=-\infty}^{\infty} h(p)z^{-(p+k)} = \sum_{k=-\infty}^{\infty} x(k)z^{-k}\sum_{p=-\infty}^{\infty} h(p)z^{-p} = X(z)H(z)$$

5.5 PROPERTIES OF UNILATERAL Z-TRANSFORM

The unilateral z-transform has almost all the properties similar to bilateral z-transform except time shift property:

5.5.1 Time shift property

$$x[n-m] \xleftrightarrow{z} z^{-m}X(z)+z^{-m+1}x(-1)+z^{-m+2}x(-2)+\cdots+x(-m)$$

let $s(n)=x(n-1)$

$$S(z)=\sum_{n=0}^{\infty}s(n)z^{-n}=\sum_{n=0}^{\infty}x(n-1)z^{-n}=x(-1)+\sum_{n=1}^{\infty}x(n-1)z^{-n}$$

Put $n-1=p$ in the above equation

$$S(z)=x(-1)+\sum_{p=0}^{\infty}x(p)z^{-(p+1)}=x(-1)+z^{-1}\sum_{p=0}^{\infty}x(p)z^{-p}=x(-1)+z^{-1}X(z)$$

$$\text{Similarly,}\quad x(n-2)=x(-2)+x(-1)z^{-1}+z^{-2}X(z)$$

Now we can generalize the time shifting property as

$$x[n-m] \xleftrightarrow{z} x(-m)+x(-m+1)z^{-1}+\cdots+z^{-m+2}x(-2)+z^{-m+1}x(-1)++z^{-m}X(z)$$

5.5.2 Initial value theorem

If

$$x(n) \xleftrightarrow{z} X(z) \quad \text{then}$$
$$x(0)=\lim_{z\to\infty}X(z)$$

Proof:

$$X(z)=\sum_{n=0}^{\infty}x(n)z^{-n}$$

On expanding this equation, we get

$$X(z)=x(0)+x(1)z^{-1}+x(2)z^{-2}+\ldots$$

$$\lim_{z\to\infty}X(z)=\lim_{z\to\infty}\left[x(0)+x(1)z^{-1}+x(2)z^{-2}+\ldots\right]$$

$$\lim_{z\to\infty}X(z)=\lim_{z\to\infty}x(0)+\lim_{z\to\infty}x(1)z^{-1}+\lim_{z\to\infty}x(2)z^{-2}+\cdots=\lim_{z\to\infty}x(0)+\lim_{z\to\infty}x(1)\frac{1}{z}+\lim_{z\to\infty}x(2)\frac{1}{z^2}+\ldots$$

$$\lim_{z\to\infty}X(z)=x(0)\qquad\left\{\text{since}\frac{1}{z}=\frac{1}{\infty}=0\right\}$$

5.5.3 Final value theorem

If

$$x(n) \xleftrightarrow{z} X(z) \qquad \text{then final value theorem states that}$$

$$\lim_{n\to\infty}x(n)=\lim_{z\to1}(z-1)X(z)$$

Proof: First we find the z-transform of $[x(n+1)-x(n)]$ and the proof for the final value theorem

$$x(n+1)-x(n) \stackrel{z}{\longleftrightarrow} \sum_{n=0}^{\infty}[x(n+1)-x(n)]\,z^{-n}$$

$$zX(z)-x(0)-X(z) \stackrel{z}{\longleftrightarrow} \lim_{k\to\infty}\sum_{n=0}^{k}[x(n+1)-x(n)]z^{-n}$$

From time shifting property of unilateral z-transform, we know that

$$x(n+m) \stackrel{z}{\longleftrightarrow} z^{m}X(z)-x(0)z^{m}-x(1)z^{m-1}-\cdots-x(m-1)z$$

$$x(n-m) \stackrel{z}{\longleftrightarrow} z^{-m}X(z)+x(-1)z^{-m+1}+x(-2)z^{-m+2}+\cdots+x(-m)$$

$$(z-1)X(z)-x(0)=\lim_{k\to\infty}\left\{[x(1)-x(0)]+[x(2)-x(1)]\,z^{-1}+\cdots+[x(k+1)-x(k)]\,z^{-k}\right\}$$

Now let $z\to 1$, then

$$\lim_{z\to 1}(z-1)X(z)-x(0)=x(1)-x(0)+x(2)-x(1)+x(3)-x(2)\ldots\left\{\text{since } (1)^{k}=\cdots=1\right\}$$

$$\lim_{z\to 1}(z-1)X(z)-x(0)=x(\infty)-x(0)$$

$$x(\infty)=\lim_{n\to\infty}x(n)=\lim_{z\to 1}(z-1)X(z)$$

5.6 INVERSE Z-TRANSFORM

The inverse z-transform can be obtained by

1. Long Division Method
2. Partial Fraction Method
3. Residue Method
4. Convolution Method

5.6.1 Long division method

If system is causal, i.e. $x(n)=0$ for $n<0$ then,

$$\sum_{n=0}^{\infty}x(n)z^{-n}=x(0)+\frac{x(1)}{z}+\frac{x(2)}{z^{2}}+\ldots \tag{5.8}$$

For anticausal system

$$\sum_{n=-\infty}^{0}x(n)z^{-n}=\cdots+x(-N)z^{N}+\ldots x(-3)z^{3}+x(-2)z^{2}+x(-1)z+x(0)\ldots \tag{5.9}$$

Example 4: Using long division method, determine inverse z-transform

$$X(z)=\frac{z+1}{z^{2}-3z+2}$$

(a) $x(n)$ is causal

(b) $x(n)$ is anticausal

SOLUTION:

(a)

$$
\begin{array}{r|l}
z^2-3z+2 & \overline{z+1}\ \ z^{-1}+4z^{-2}+10z^{-3}+22z^{-4} \\
 & \underline{z-3+2z^{-1}} \\
 & 4-2z^{-1} \\
 & \underline{4-12z^{-1}+8z^{-2}} \\
 & 10z^{-1}-8z^{-2} \\
 & \underline{10z^{-1}-30z^{-2}+20z^{-3}} \\
 & 22z^{-2}-20z^{-3} \\
 & \underline{22z^{-2}-66z^{-3}+44z^{-4}} \\
 & 46z^{-3}-44z^{-4}\ldots
\end{array}
$$

On comparing from eqn. (5.9), we get the values of $x(n)$

$$x(0)=0,\ x(1)=1,\ x(2)=4,\ x(3)=10,\ x(4)=22$$

Now the $x(n)$ can be represented as

$$x(n)=\left\{\begin{array}{l}0,1,4,10,22,\cdots \\ \uparrow\end{array}\right\}$$

(b) For anticausal system order of divider and dividend is changed as follows:

$$
\begin{array}{r|l}
2-3z+z^2 & \overline{1+z}\ \ 0.5+1.25z+1.625z^2 \\
 & \underline{1-1.5z+0.5z^2} \\
 & 2.5z-0.5z^2 \\
 & \underline{2.5z-3.75z^2+1.258z^3} \\
 & 3.25z^2-1.25z^3 \\
 & \underline{3.25z^2-4.875z^3+1.625z^4} \\
 & 3.625z^3-1.625z^4+\ldots
\end{array}
$$

On comparing with eqn. (5.10), we get the sequence $x(n)$ as

$$x(n)=\left\{\begin{array}{r}\ldots 1.625, 1.25, 0.5 \\ \uparrow\end{array}\right\}$$

5.6.2 Partial fraction method

Example 5: Find inverse z-transform using Partial Fraction Method

(a) $X(z)=\frac{\frac{1}{4}z^{-1}}{(1-\frac{1}{2}z^{-1})(1-\frac{1}{4}z^{-1})}$ ROC : $|z|>\frac{1}{2}$

(b) $X(z)=\frac{z^2}{(1-az)(z-a)}$ ROC : $a<|z|<\frac{1}{a}$

SOLUTION:

(a)
$$X(z) = \frac{\frac{1}{4}z^{-1}}{(1-\frac{1}{2}z^{-1})(1-\frac{1}{4}z^{-1})} = \frac{A}{(1-\frac{1}{2}z^{-1})} + \frac{B}{(1-\frac{1}{4}z^{-1})}$$

$$\left\{ \begin{array}{l} (1-\frac{1}{4}z^{-1})A + (1-\frac{1}{2}z^{-1})B = \frac{1}{4}z^{-1}, \text{on comparing the coefficients, we get} \\ A+B=0, -\frac{1}{4}A - \frac{1}{2}B = \frac{1}{4} \Rightarrow \text{on solving we get } A=1 \,\&\, B=-1 \end{array} \right\}$$

$$X(z) = \frac{1}{(1-\frac{1}{2}z^{-1})} - \frac{1}{(1-\frac{1}{4}z^{-1})} \quad \left\{ \begin{array}{l} \text{Since } a^n u(n) \overset{z}{\longleftrightarrow} \frac{1}{1-az^{-1}}, \text{ ROC } |z|>a \\ \text{and } -a^n u(-n-1) \overset{z}{\longleftrightarrow} \frac{1}{1-az^{-1}}, \text{ ROC } |z|<a \end{array} \right\}$$

On taking IZT

$$x(n) = \left(\frac{1}{2}\right)^n u(n) - \left(\frac{1}{4}\right)^n u(n)$$

Since ROC is $|z| > \frac{1}{2}$

(b)
$$X(z) = \frac{z^2}{(1-az)(z-a)} = \frac{-z^2}{a(z-\frac{1}{a})(z-a)} = \frac{A}{(z-\frac{1}{a})} + \frac{B}{(z-a)}$$

$$X(z) = \frac{-1}{a(1-a^2)}\frac{z}{(z-\frac{1}{a})} + \frac{a}{(1-a^2)}\frac{z}{(z-a)}$$

On taking IZT

$$x(n) = \frac{1}{a(1-a^2)}\left(\frac{1}{a}\right)^n u(-n-1) + \frac{a}{(1-a^2)}a^n u(n)$$

$$\left\{ \begin{array}{ll} \text{Since} & a^n u(n) \overset{z}{\longleftrightarrow} \frac{1}{1-az^{-1}}, \text{ ROC } |z|>a \\ \text{and} & -a^n u(-n-1) \overset{z}{\longleftrightarrow} \frac{1}{1-az^{-1}}, \text{ ROC } |z|<a \end{array} \right\}$$

5.6.3 Residue method

We know that inverse z-transform

$$x(n) = \frac{1}{2\pi j}\oint_c X(z)z^{n-1}dz \tag{5.10}$$

where c is a circle in the z-plane in the ROC of $X(z)$.

Equation 5.10 can be evaluated by finding sum of all residues of the poles that are inside the circle c. Equation 5.10 can be rewritten as

$$x(n) = \sum(\text{residue of } X(z)z^{n-1} \text{ at the poles inside } c)$$

$$x(n) = \sum_i (z-z_i)X(z)z^{n-1}\big|_{z=z_i} \tag{5.11}$$

if $X(z)z^{n-1}$ has no poles inside the contour c for one or more values of n, then $x(n) = 0$ for these values.

Example 6: By using residue method, find inverse z-transform

$$X(z) = \frac{1+3z^{-1}}{1+3z^{-1}+2z^{-2}} \qquad |z| > 2$$

SOLUTION:

Given

$$X(z) = \frac{1+3z^{-1}}{1+3z^{-1}+2z^{-2}} = \frac{z(z+3)}{z^2+3z+2}$$

$$X(z) = \frac{z(z+3)}{(z+1)(z+2)} \tag{1}$$

We know that $x(n) = \frac{1}{2\pi j}\oint_c X(z)z^{n-1}dz$

$$x(n) = \sum \text{residues of } X(z)z^{n-1} \text{ at poles of } X(z)z^{n-1} \text{ within } c$$

$$= \sum \text{residues of } \frac{z(z+3)}{(z+1)(z+2)}z^{n-1} \text{ at poles of same within } c$$

The closed contour c, being in the ROC $|z| > 2$ encloses the poles at $z = -1$ and $z = -2$.

$$x(n) = \sum \text{residues of} \frac{(z+3)z^n}{(z+1)(z+2)} \text{ at poles } z = -1 \text{ and } z = -2$$

$$= \text{residue of} \frac{(z+3)z^n}{(z+1)(z+2)} \text{ at } z = -1 + \text{ residue of } \frac{(z+3)z^n}{(z+1)(z+2)} \text{ at } z = -2$$

$$= \cancel{(z+1)}\frac{(2+3)z^n}{\cancel{(z+1)}(z+2)}\Bigg|_{z=-1} + \cancel{(z+2)}\frac{(z+3)z^n}{(z+1)\cancel{(z+2)}}\Bigg|_{z=-2}$$

$$= z(-1)^n - (-z)^n$$

5.6.4 Convolution method

Example 7: Find inverse z-transform using convolution method

$$X(z) = \frac{1+3z^{-1}}{1+3z^{-1}+2z^{-2}} \qquad |z| > 2$$

$$X(z) = \frac{z(z+3)}{z^2+3z+2} = \frac{z(z+3)}{(z+1)(z+2)} = X_1(z)X_2(z)$$

$$X_1(z) = \frac{z}{(z+1)} \quad X_2(z) = \frac{z+3}{z+2} = \frac{z}{z+2} + \frac{3}{z+2}$$

IZT of $X_1(z)$ and $X_2(z)$ is

$$x_1(n) = (-1)^n u(n), \; x_2(n) = (-2)^n u(n) + 3(-2)^{n-1}u(n-1)$$

$$\left\{\begin{array}{l}\text{From time shifting property } x(n-m) = z^{-m}X(z) \\ \frac{3}{z+2} \text{ can be represented as } \left[z^{-1}\frac{3z}{z+2}\right] \stackrel{IZT}{\longleftrightarrow} 3(-2)^{n-1}u(n-1)\end{array}\right\}$$

$$x_1(n) * x_2(n) \stackrel{z}{\longleftrightarrow} X_1(z)X_2(z)$$

where $s(n) = x_1(n) * x_2(n)$ is the convolution of sequences $x_1(n)$ and $x_2(n)$ defined as

$$s(n) = x_1(n) * x_2(n) = \sum_{k=-\infty}^{\infty} x_1(k)x_2(n-k)$$

$$x(n) = x_1(n) * x_2(n) = \sum_{k=-\infty}^{\infty} (-1)^k u(k)\left\{(-2)^{n-k}u(k) + 3(-2)^{n-k-1}u(n-k-1)\right\}$$

5.7 ANALYSIS AND CHARACTERIZATION OF LTI SYSTEM

Consider an output of the relaxed LTI system, which is given as

$$y(n) = x(n) * h(n) = \sum_{k=-\infty}^{\infty} x(k)h(n-k)$$

where $y(n) =$ Output of LTI system

$x(n) =$ Input to the system

$h(n) =$ Impulse response or unit sample response of the system

If $y(n) \stackrel{z}{\longleftrightarrow} Y(z)$, then

$Y(z) = X(z)H(z)$ From convolution property

$$H(z) = \frac{Y(z)}{X(z)} \tag{5.12}$$

$H(z) \rightarrow$ Transfer function or System function (z-transform of impulse response or unit sample response $\{h(n)\}$ of the system)

Thus, the system is characterized by $H(z)$ in z-domain.

Example 8: A difference equation of the system is given as:

$$y(n) = 2y(n-1) + x(n)$$

Determine

(a) Transfer function or system function

(b) Pole zero plot of the transfer function

(c) Impulse response of the system

SOLUTION:

To obtain the transfer function, we take the z-transform of the given difference equation and then we get a relation between $X(z)$ and $Y(z)$ from there we get $H(z)$ which is given as $H(z) = \frac{Y(z)}{X(z)}$ and then we take IZT.

(a)

$$Y(z) = 2z^{-1}Y(z) + X(z) \qquad \{\text{From time shifting property } x(n-m) \overset{z}{\longleftrightarrow} z^{-m}X(z)\}$$

$$Y(z)[1-2z^{-1}] = X(z)$$

$$\frac{Y(z)}{X(z)} = \frac{1}{1-2z^{-1}}$$

$$\text{System function } H(z) = \frac{Y(z)}{X(z)} = \frac{1}{1-2z^{-1}}$$

(b) $H(z) = \frac{1}{1-2z^{-1}} = \frac{z}{z-2}$

From the above equation, we can say that $H(z)$ has one zero $z = 0$, i.e. origin and one pole at $p = 2$

(c) Impulse response $h(n) = IZT[H(z)] = (2)^n u(n) \left\{ \text{Since } a^n u(n) \overset{z}{\longleftrightarrow} \frac{1}{(1-az^{-1})} \text{ ROC } |z| > a \right\}$

Example 9: For the following difference equation

$$y(n) + \frac{1}{4}y(n-1) = x(n) + \frac{1}{2}x(n-1)$$

(a) Find the system function $H(z)$ and obtain the corresponding ROC.
(b) Find the unit sample response of the system.
(c) Find the response of the system for input sequence $x(n) = \delta(n)$.
(d) Show that this system is a causal and stable system.
(e) Find the frequency response $H(e^{j\omega})$ and determine its magnitude and phase.

SOLUTION:

1. Methods for finding $H(z)$ and $h(n)$ are discussed in previous example.
2. For a LTI system to be causal $h(n) = 0$ for $n < 0$ (i.e. LTI system is causal if and only if the ROC of the system function is exterior of a circle of radius $r < \infty$.), for system to be stable $\sum_{n=-\infty}^{\infty} |h(n)| < \infty$ evaluated on unit circle (i.e. LTI system is BIBO stable if and only if the ROC of the system function includes the unit circle).
3. For finding frequency response $H(e^{j\omega})$, replace z by $e^{j\omega}$ and then find $|H(e^{j\omega})|$ & $H(e^{j\omega})$

(a) $y(n) + \frac{1}{4}y(n-1) = x(n) + \frac{1}{2}x(n-1)$

Taking the z-transform on both sides

$$Y(z) + \frac{1}{4}z^{-1}Y(z) = X(z) + \frac{1}{2}z^{-1}X(z)$$

$$H(z) = \frac{Y(z)}{X(z)} = \frac{(1+\frac{1}{2}z^{-1})}{(1+\frac{1}{4}z^{-1})} = \frac{(z+\frac{1}{2})}{(z+\frac{1}{4})} \qquad \text{ROC} : |z| > \frac{1}{4}$$

(b)

$$H(z)=\frac{(1+\frac{1}{2}z^{-1})}{(1+\frac{1}{4}z^{-1})}=\frac{(z+\frac{1}{2})}{(z+\frac{1}{4})}=\frac{z}{(z+\frac{1}{4})}+\frac{\frac{1}{2}}{(z+\frac{1}{4})}=\frac{z}{(z+\frac{1}{4})}+\frac{1}{2}z^{-1}\frac{z}{(z+\frac{1}{4})}$$

$$IZT[H(z)]=h(n)=(-\frac{1}{4})^n u(n)+\frac{1}{2}(-\frac{1}{4})^n u(n-1)$$

(c) Since $x(n)=\delta(n)$

On taking z-transform $X(z)=1$

$$H(z)=\frac{Y(z)}{X(z)}\Rightarrow H(z)=Y(z)\Rightarrow h(h)=y(n)$$

$$\text{so}\quad y(n)=(-\frac{1}{4})^n u(n)+\frac{1}{2}(-\frac{1}{4})^n u(n-1)$$

(d) For a LTI system to be causal $h(n)=0$ for $n<0$

$$h(n)=(-\frac{1}{4})^n u(n)+\frac{1}{2}(-\frac{1}{4})^n u(n-1)$$

In the above equation $h(n)=0$ for $n<0$ $\left\{\begin{array}{ll}\text{since } u(n) & =1 \, for\, n\geq 0\\ & =0\, for\, n<0\\ \&\quad u(n-1) & =1\, for\, n\geq 1\\ & =0\, for\, n<1\end{array}\right\}$

So the given LTI system is causal.

The pole $z=-\frac{1}{4}$ is inside the unit circle, therefore, the system is stable.

(e) For finding frequency response $H(e^{j\omega})$, replace z by $e^{j\omega}$ and then find $|H(e^{j\omega})|$ & $\underline{/H(e^{j\omega})}$

$$H(e^{j\omega})=H(e^{j\omega})|_{z=e^{j\omega}}=\frac{(e^{j\omega}+\frac{1}{2})}{(e^{j\omega}+\frac{1}{4})}=\frac{\frac{1}{2}+\cos\omega+j\sin\omega}{\frac{1}{4}+\cos\omega+j\sin\omega}$$

{ For finding the magnitude multiply the numerator and denominator by complex conjugate of the denominator and separate the real (A) and imaginary (B) parts. Then magnitude $=\sqrt{(A^2+B^2)}$ Phase $=\tan^{-1}\frac{B}{A}$. }

Magnitude of frequency response $H(e^{j\omega})=|H(e^{j\omega})|=\dfrac{\sqrt{(\frac{5}{4}+\cos\omega)}}{\sqrt{(\frac{17}{16}+\frac{1}{2}\cos\omega)}}$

Phase of frequency response $H(e^{j\omega})=\underline{/H(e^{j\omega})}=\tan^{-1}\dfrac{\sin\omega}{\frac{1}{2}+\cos\omega}-\tan^{-1}\dfrac{\sin\omega}{\frac{1}{4}+\cos\omega}$

Example 10: Determine the z-transform of $x(n)=a^n\sin(\frac{\pi}{2}n)$ for $n\geq 0$, where a is a real constant.

SOLUTION:

$$X(z)=\sum_{n=-\infty}^{\infty}x(n)z^{-n}$$

$$X(z)=\sum_{n=0}^{\infty}a^n\sin(\frac{\pi}{2}n)z^{-n}$$

$$X(z) = 0 + az^{-1} + 0 - a^3z^{-3} + 0 + a^5z^{-5} + 0 - \ldots$$

$$X(z) = az^{-1} - a^3z^{-3} + a^5z^{-5} - \ldots \begin{Bmatrix} \text{Since}\sin(0) = \sin(\pi) = \cdots = 0 \\ \&\sin\left[\frac{\pi}{2}, 5\frac{\pi}{2}\ldots\right] = 1, \sin\left[\frac{3\pi}{2}, \frac{7\pi}{2}\ldots\right] = -1 \end{Bmatrix}$$

$$X(z) = \frac{az^{-1}}{1 + a^2z^{-2}}, \text{ ROC } : |a^2z^{-2}| < 1 \Rightarrow |z| > |a|$$

Example 11: Determine the z-transform of $x(n) = [3(2)^n - 4(3)^n]u(n)$.

SOLUTION:

$$X(z) = \sum_{n=-\infty}^{\infty} x(n)z^{-n}$$

$$X(z) = \sum_{n=0}^{\infty}[3(2)^n - 4(3)^n]z^{-n} = \sum_{n=0}^{\infty}[3(2)^n]z^{-n} - \sum_{n=0}^{\infty}[4(3)^n]z^{-n}$$

$$\text{Let } x_1(n) = \sum_{n=0}^{\infty}[3(2)^n]z^{-n} = 3\sum_{n=0}^{\infty}(2)^n z^{-n}$$

$$\& \; x_2(n) = -\sum_{n=0}^{\infty}[4(3)^n]z^{-n} = -4\sum_{n=0}^{\infty}(3)^n z^{-n}$$

$$X_1(z) = 3\{1 + 2z^{-1} + 2^2z^{-2} + 2^3z^{-3} + \ldots\} = \frac{3}{1 - 2z^{-1}}, \text{ ROC } : |z| > 2$$

$$X_2(z) = -4\{1 + 3z^{-1} + 3^2z^{-2} + 3^3z^{-3} + \ldots\} = \frac{-4}{1 - 3z^{-1}}, \text{ ROC } : |z| > 3$$

The intersection of the ROC of $X_1(z)$ and $X_2(z)$ is $|z| > 3$, thus the overall z-transform of the given sequence is

$$X(z) = \frac{3}{1 - 2z^{-1}} - \frac{4}{1 - 3z^{-1}} = \frac{-1 - z^{-1}}{1 - 5z^{-1} + 6z^{-2}} = \frac{-(1 + z^1)}{1 - 5z^{-1} + 6z^{-2}}, \text{ ROC } : |z| > 3$$

Example 12: Find inverse z-transform using convolution method.

$$\text{Given } \; X(z) = \frac{1 + 3z^{-1}}{1 + 3z^{-1} + 2z^{-2}}, |z| > 2$$

$$X(z) = \frac{z(z+3)}{(z+1)(z+2)}$$

SOLUTION:

Let $X(z) = X_1(z)X_2(z)$

$$X_1(z) = \frac{z}{z+1} \qquad X_2(z) = \frac{z+3}{z+2}$$

$$X_1(z) = \frac{1}{1+z^{-1}} \qquad X_2(z) = \frac{z}{z+2} + \frac{3}{z+2}$$

$$X_1(z) = \frac{1}{1-(-1)z^{-1}} \qquad X_2(z) = \frac{1}{1+2z^{-1}} + \frac{3z^{-1}}{1+2z^{-1}}$$

$$\therefore \quad x_1(n) = (-1)^n u(n) \quad x_2(n) = (-2)^n u(n) + 3(-2)^{n-1}u(n-1)$$

$$x(n) = x_1(n) * x_2(n) = \sum_{k=-\infty}^{\infty} x_1(k)x_2(n-k) = \sum_{k=-\infty}^{\infty} x_1(n-k)x_2(k)$$

$$x(n) = \sum_{k=-\infty}^{\infty} (-1)^{n-k}u(n-k)\left\{(-2)^k u(k) + 3(-2)^{k-1}4(k-1)\right\}$$

$$x(n) = \sum_{k=-\infty}^{\infty} (-1)^{n-k}(-2)^k u(k)u(n-k) + 3\sum_{k=-\infty}^{\infty} (-1)^{n-k}(-2)^{k-1}u(k-1)u(n-k)$$

$$x(n) = \sum_{k=0}^{n}(-1)^n(-1)^{-k}(-2)^k + 3\sum_{k=1}^{n}(-1)^n(-1)^{-k}(-2)^{k-1}$$

$$x(n) = (-1)^n\sum_{k=0}^{n}(2)^k + 3(-1)^n \times (-2)^{-1}\sum_{k=1}^{n}(-1)^{-k}(-2)^k$$

$$= (-1)^n\sum_{k=0}^{n}(2)^k - \frac{3}{2}(-1)^n\sum_{k=1}^{n}(2)^k$$

$$u(k) = 1 \text{ for } k \geq 0$$

$$u(k-1) = 1 \text{ for } k \geq 1$$

$$u(n-k) = 1 \text{ for } n-k \geq 0$$

or

$$n \geq k$$

$$k \leq n$$

so $\quad u(k)u(n-k) = 1 \text{ for } 0 \leq k \leq n$

$$u(k-1)u(n-k) = 1 \text{ for } 1 \leq k \leq n$$

$$x(n) = (-1)^n\{1+2+2^2+2^3+\cdots 2^n\}\frac{-3}{2}(-1)^n\{2+2^2+\cdots 2^n\}$$

$$x(n) = (-1)^n\frac{\{1-2^{n+1}\}}{1-2} - 3\frac{(-1)^n\{1-2^n\}}{1-2}$$

$$x(n) = (-1)^{n-1}\{1-2^{n+1}\} - 3(-1)^{n-1}\{1-2^n\}$$

$$= (-1)^{n-1} + 2(-2)^n - 3(-1)^{n-1} + 3\times -1(-2)^n$$

$$= (-1)^{n-1} + 2(-2)^n - 3(-1)^{n-1} - 3(-2)^n$$

$$= -(-2)^n - 2(-1)^{n-1}$$

$$x(n) = [2(-1)^n - (-2)^n]u(n)$$

Example 13: Find the inverse z-transform using residue method.

$$\text{Given}\quad X(z) = \frac{1+3z^{-1}}{1+3z^{-1}+2z^{-2}}|z| > 2$$

SOLUTION:

Given

$$X(z) = \frac{1+3z^{-1}}{1+3z^{-1}+2z^{-2}} = \frac{z(z+3)}{z^2+3z+2}$$

$$X(z) = \frac{z(z+3)}{(z+1)(z+2)}$$

We know that $x(n) = \frac{1}{2\pi j}\oint_C X(z)z^{n-1}dz$

$x(n) = \Sigma$ residues of $X(z)z^{n-1}$ at poles of $X(z)z^{n-1}$ within C.

$x(n) = \Sigma$ residues of $\frac{z(z+3)}{(z+1)(z+2)}z^{n-1}$ at poles of same within C.

The closed contour C, being in the ROC$|z| > 2$ encloses the poles at $z = -1$ and $z = -2$

$x(n) = \Sigma$ residues of $\frac{(z+3)z^n}{(z+1)(z+2)}$ at poles $z = -1$ and $z = -2$

$x(n) =$ residue of $\frac{(z+3)z^n}{(z+1)(z+2)}$ at $z = -1+$ residue of $\frac{(z+3)z^n}{(z+1)(z+2)}$ at $z = -2$

$$= (z+1)\frac{(z+3)z^n}{(z+1)(z+2)}\bigg|_{z=-1} + (z+2)\frac{(z+3)z^n}{(z+1)(z+2)}\bigg|_{z=-2}$$

$$= 2(-1)^n - (-2)^n$$

Example 14: Find the inverse z- transform of $X(z) = \dfrac{z^2}{(1-az)(z-a)}$

SOLUTION:

$$X(z) = \frac{z^2}{(1-az)(z-a)}$$

$$\frac{X(z)}{z} = \frac{z}{(1-az)(z-a)} = \frac{z}{-3a\left(z-\frac{1}{a}\right)(z-a)}$$

$$\frac{X(z)}{z} = \frac{-z/a}{\left(z-\frac{1}{a}\right)(z-a)}$$

$$\frac{X(z)}{z} = \frac{A}{z-\frac{1}{a}} + \frac{B}{z-a} \tag{5.1}$$

A and B can be evaluated as

$$A = \left(z-\frac{1}{a}\right)\frac{X(z)}{z}\bigg|_{z=\frac{1}{a}} = \left(z-\frac{1}{a}\right)\frac{-z/a}{\left(z-\frac{1}{a}\right)(z-a)}\bigg|_{z=\frac{1}{a}}$$

$$= \frac{-1/a^2}{\left(\frac{1}{a}-a\right)} = \frac{-1/a^2}{\frac{1-a^2}{a}} = -\frac{1}{a(1-a^2)} \tag{5.2}$$

$$B = (z-a)\frac{X(z)}{z}\bigg|_{z=a} = (z-a)\,\frac{-z/a}{\left(z-\frac{1}{a}\right)(z-a)}\bigg|_{z=a}$$

$$= \frac{-1}{\left(a-\frac{1}{a}\right)} = \frac{-a}{(a^2-1)} = \frac{a}{1-a^2} \tag{5.3}$$

On putting in (1)

$$\frac{X(z)}{z} = \frac{-1}{a(1-a^2)} \cdot \frac{1}{\left(z-\frac{1}{a}\right)} + \frac{a}{1-a^2} \cdot \frac{1}{z-a}$$

$$X(z) = -\frac{1}{a(1-a^2)}\,\frac{z}{z-\frac{1}{a}} + \frac{a}{1-a^2}\,\frac{z}{z-a}$$

$$X(z) = -\frac{1}{a(1-a^2)}\,\frac{1}{1-\frac{1}{a}z^{-1}} + \frac{a}{1-a^2}\,\frac{1}{1-az^{-1}}$$

Given ROC is $a < |z| < \frac{1}{a}$

We know that

$$ZT\left[a^n u(n)\right] = \frac{1}{1-az^{-1}} \;\text{ROC} : |z| > a$$

$$ZT\left[-a^n u(-n-1)\right] = \frac{1}{1-az^{-1}} \;\text{ROC} : |z| < a$$

$$\text{so} \quad X(z) = \frac{1}{a(1-a^2)} \cdot \frac{-1}{1-\frac{1}{a}z^{-1}} + \frac{a}{1-a^2}\frac{1}{1-az^{-1}}$$

On taking inverse z-transform of the above equation

$$x(n) = \frac{1}{a(1-a^2)}\left(\frac{1}{a}\right)^n u(-n-1) + \frac{a}{1-a^2}a^n u(n)$$

$$x(n) = \frac{a}{1-a^2}\left\{\left(\frac{1}{a}\right)^{n+2} u(-n-1) + a^n u(n)\right\}$$

Example 15: Find the inverse z- transform of $X(z) = \dfrac{8z-9}{z^2-5z+6}$ ROC : $|Z| > 3$

SOLUTION:

$$X(z) = \frac{8z-9}{z^2-5z+6} \quad \text{ROC} : |z| > 3$$

$$\frac{X(z)}{z} = \frac{8z-9}{z(z^2-5z+6)} = \frac{8z-9}{z(z-2)(z-3)} \tag{5.1}$$

$$\frac{X(z)}{z} = \frac{A}{z} + \frac{B}{z-2} + \frac{C}{z-3} \tag{5.2}$$

$$A = z\frac{X(z)}{z}\bigg|_{z=0} = z\frac{8z-9}{z(z-2)(z-3)}\bigg|_{z=0} = \frac{-9}{6} = \frac{-3}{2}$$

$$B = (z-2)\frac{X(z)}{z}\bigg|_{z=2}$$

$$= \cancel{(z-2)}\frac{8z-9}{z\cancel{(z-2)}(z-3)}\bigg|_{z=2}$$

$$B = \frac{-7}{2}$$

$$C = (z-3)\frac{X(z)}{z}\bigg|_{z=3}$$

$$= \cancel{(z-3)}\frac{8z-9}{z(z-2)\cancel{(z-3)}}\bigg|_{z=3}$$

$$C = 5$$

On putting in (2)

$$\frac{X(z)}{z} = \frac{-3}{2z} - \frac{7}{2(z-2)} + \frac{5}{z-3}$$

$$X(z) = \frac{-3}{2} - \frac{7z}{2(z-2)} + \frac{5z}{z-3} = \frac{-3}{2}.1 - \frac{7}{2}\cdot\frac{1}{1-2z^{-1}} + \frac{5}{1-3z^{-1}}$$

On taking inverse transform

$$x(n) = -\frac{3}{2}\delta(n) - \frac{7}{2}(2)^n u(n) + 5(3)^n u(n)$$

Since we know that

$$ZT[\delta(n)] = 1$$

$$ZT[a^n u(n)] = \frac{1}{1 - az}$$

Example 16: Find the z- transform of $x(n) = a^{n-1}u(n-1)$

SOLUTION:

$x(n) = a^{n-1}u(n-1)$

The z- transform of the sequence

$$x_1(n) = a^n u(n) \text{ is given by}$$

$$X_1(z) = \frac{1}{1 - az^{-1}} \quad \text{ROC: } |z| > a$$

By using time shifting property of z- transform, we have

$$ZT\{x(n-k)\} = z^{-k}X(z)$$

$$\text{so} \quad ZT[a^{n-1}u(n-1)] = z^{-1} \cdot \frac{1}{1 - az^{-1}} = \frac{z^{-1}}{1 - az^{-1}}$$

$$X(z) = \frac{1}{z - a} \quad \text{ROC: } |Z| > a$$

Example 17: Find $x(\infty)$ when $X(z)$ is given by

(i) $X(z) = \dfrac{z+1}{3(z-1)(z+0.9)}$

(ii) $X(z) = \dfrac{3z+4}{(z-1)(z+1)}$

SOLUTION:

Final value theorem status that the steady-state value of $x(n)$ is given as

$$X(\infty) = \lim_{z \to 1}(z-1)X(z)$$

(i)

$$X(z) = \frac{z+1}{3(z-1)(z+0.9)}$$

$$(z-1)X(z) = (z-1)\frac{(z+1)}{3(z-1)(z+0.9)} = \frac{z+1}{3(z+0.9)}$$

$(z-1)X(z)$ has no poles on or outside the unit circle

So

$$x(\infty) = \lim_{z \to 1}(z-1)X(z) = \lim_{z \to 1} \frac{z+1}{3(z+0.9)}$$

$$x(\infty) = \frac{2}{3 \times 1.9} = 0.35$$

(ii)
$$X(z) = \frac{3z+4}{(z-1)(z+1)}$$
$$(z-1)X(z) = \frac{(z-1)(3z+4)}{(z-1)(z+1)} = \frac{3z+4}{z+1}$$
$$x(\infty) = \lim_{z\to 1}(z-1)X(z) = \lim_{z\to 1}\frac{3z+4}{z+1}$$

In this case $(z-1)X(z)$ has a pole on the unit circle so $x(\infty)$ is undefined.

Example 18: Find the z- transform of sequence $x(n) = 3\delta(n+2)+6\delta(n)+4\delta(n-3)$.

SOLUTION:

We know that

$$ZT[x(n\pm k)] = z^{\pm k}X(z) \quad \text{from time shifting property}$$
$$ZT[\delta(n)] = \sum_{n=-\infty}^{\infty}\delta(n)z^{-n}\left\{\begin{array}{ll}\text{since } \delta(n)=1 & \text{for } n=0\\ 0 & \text{for } n\neq 0\end{array}\right\}$$
$$ZT\delta(n) = 1$$
$$X(z) = ZT[x(n)]$$
$$= ZT[3\delta(n+2)+6\delta(n)+4\delta(n-3)]$$
$$X(z) = 3ZT[(\delta(n+2))+6ZT[\delta(n)]+4ZT(\delta(n-3))]$$
$$X(z) = 3z^2+6+4z^{-3}$$

Example 19: Determine the z-transform of the function

$$x(n) = \left(\frac{1}{5}\right)^n u(n) + \left(\frac{3}{4}\right)^n u(n-4) - 2u(n-8)$$

SOLUTION:

$$x(n) = x_1(n)+x_2(n)+x_3(n)$$
$$x_1(n) = \left(\frac{1}{5}\right)^n u(n), x_2(n) = \left(\frac{3}{4}\right)^n u(n-u)\, x_3(n) = -2u(n-8)$$
$$X_1(z) = \sum_{n=-\infty}^{\infty}\left(\frac{1}{5}\right)^n z^{-n}u(n) = \sum_{n=0}^{\infty}\left(\frac{1}{5}z^{-1}\right)^n$$
$$= \frac{1}{1-\frac{1}{5}z^{-1}} \quad \text{ROC}: |Z| > \frac{1}{5} \quad \text{Since } u(n) = \begin{cases}1 \text{ for } n\geq 0\\ 0 \text{ otherwise}\end{cases}$$
$$X_2(z) = ZT[x_2(n)] = \sum_{n=-\infty}^{\infty}\left(\frac{3}{4}\right)^n z^{-n}u(n-4)$$
$$= \sum_{n=4}^{\infty}\left(\frac{3}{4}z^{-1}\right)^n \left\{\begin{array}{l}u(n-4) = 1 \text{ for } n-4\geq 0\\ \text{or } n\geq 4\\ 0 \text{ otherwise}\end{array}\right.$$
$$= \left(\frac{3}{4}z^{-1}\right)^4 + \left(\frac{3}{4}z^{-1}\right)^5 + \left(\frac{3}{4}z^{-1}\right)^6 + \cdots$$

$$= \left(\frac{3}{4}z^{-1}\right)^4 \left(1 + \frac{3}{4}z^{-1} + \left(\frac{3}{4}z^{-1}\right)^2 + \cdots\right)$$

$$X_2(z) = \frac{\left(\frac{3}{4}z^{-1}\right)^4}{1 - \frac{3}{4}z^{-1}} \quad \text{ROC} : |z| > \frac{3}{4}$$

$$x_3(n) = -2u(n-8)$$

$$X_3(z) = -2 \sum_{n=-\infty}^{\infty} u(n-8)z^{-n}$$

$$= -2 \sum_{n=8}^{\infty} z^{-n} \qquad \text{Since } u(n-8) = \begin{cases} 1 \text{ for } n-8 \geq 0 \text{ or } n \geq 8 \\ 0 \text{ otherwise} \end{cases}$$

$$= -2\left(z^{-8} + z^{-9} + z^{-10} + z^{-15} + \cdots\right)$$

$$= -2z^{-8}\left(1 + z^{-1} + z^{-2} + z^{-3} + \cdots\right)$$

$$= \frac{-2z^{-8}}{1 - z^{-1}} \quad \text{ROC} : |z| > 1$$

$$X(z) = \frac{1}{1 - \frac{1}{5}z^{-1}} + \frac{81z^{-4}}{256\left(1 - \frac{3}{4}z^{-1}\right)} - \frac{2z^{-8}}{1 - z^{-1}}$$

ROC: $|z| > \frac{1}{5}, |z| > \frac{3}{4}$ and $|z| > 1$

Therefore, ROC is $|Z| > 1$

Example 20: Find the z-transform of the following

(i) $x(n) = na^n u(n)$ (ii) $x(n) = a^{n-1}u(n-1)$

SOLUTION:

We know that from differentiation in z-domain

$$ZT[nx(n)] = -z\frac{d}{dz}X(z)$$

$$x(n) = nx_1(n)$$

$$\text{So} \quad X_1(z) = ZT\left[a^n u(n)\right] = \sum_{n=-\infty}^{\infty} a^n u(n)z^{-n} = \sum_{n=0}^{\infty} \left(az^{-1}\right)^n$$

Example 21: Determine the z-transform and ROC of the given signal

$$x(n) = 3(2)^n u(n) - 4(3)^n u(n)$$

SOLUTION:

Given $x(n) = 3(2)^n u(n) - 4(3)^n u(n)$

Let $x_1(n) = 3(2)^n u(n) \quad x_2(n) = 4(3)^n u(n)$

z- Transform of $x(n)$ is defined as

$$X(z) = \sum_{n=-\infty}^{\infty} x(n)z^{-n}$$

$$X_1(z) = \sum_{n=-\infty}^{\infty} x_1(n)z^{-1} = \sum_{n=-\infty}^{\infty} 3(2)^n u(n)z^{-n}$$

$$= 3\sum_{n=0}^{\infty}(2)^n z^{-n} = 3\sum_{n=0}^{\infty}(2z^{-1})^n$$

$$[X_1(z) = 3\left[1+2z^{-1}+(2z^{-1})^2+(2z^{-1})^3+(2z^{-1})^4+\cdots\right]$$

$$= 3\left(\frac{1}{1-2z^{-1}}\right) \quad \text{ROC}: |2z^{-1}| < 1 \text{ or } |z| > 2$$

$$= \frac{3}{1-2z^{-1}} \tag{5.1}$$

Similarly, $X_2(z) = ZT\,[4(3)^n u(n)]$

$$X_2(z) = 4\sum_{n=0}^{\infty}(3z^{-1}) = \frac{4}{1-3z^{-1}} \quad \text{ROC}: |z| > 3$$

Since $X(z) = X_1(z) - X_2(z)$

$$X(z) = \frac{3}{1-2z^{-1}} - \frac{4}{1-3z^{-1}}$$

$$X(z) = \frac{-1-z^{-1}}{(1-2z^{-1})(1-3z^{-1})} \quad \text{ROC}: |z| > 3$$

OBJECTIVE TYPE QUESTIONS

Q1: Consider a system described by the following difference equation

$$y(n+3)+6y(n+2)+11y(n+1)+6y(n) = x(n+2)+9x(n+1)+20x(n)$$

where y is the output and x is the input. The transfer function of the system will be

(a) $\frac{2z^2+z+20}{3z^3+2z^2+z+6}$ (b) $\frac{z^2+9z+20}{z^3+6z^2+6z+11}$

(c) $\frac{z^3+6z^2+6z+11}{z^2+9z+20}$ (d) None of the above

Q2: If z–transform of $u(n)$ is $\frac{z}{z-1}$, then the z-transform of $a^n u(n)$ is

(a) $\frac{z}{z-a}$ (b) $\frac{z}{(z-a)^2}$

(c) $\frac{az}{(z-a)^2}$ (d) $\frac{z}{(z-\frac{1}{a})^2}$

Q3: The z-transform of $na^n u(n)$ is

(a) $\frac{z}{z-a}$ (b) $\frac{z^{-1}}{(z-a)}$

(c) $\frac{az}{(z-a)^2}$ (d) $\frac{z}{(z^{-1}-a)}$

Q4: If the function $H(z) = 1 + 1.5z$ and $H(z) = z^2 + 1.5z - 1$

(a) The poles and zeros of functions will be same

(b) The poles of the functions will be identical but not zeros

(c) The zeros of the functions will be identical but not poles

(d) Neither the poles nor the zeros of the two functions will be identical

Q5: The z-transform of $\{-a^n u(-n-1)\}$ is

(a) $\frac{az^{-1}}{(1-az^{-1})^2}$ ROC : $|z| > |a|$ (b) $\frac{1}{(1-az^{-1})}$ ROC : $|z| > |a|$

(c) $\frac{az^{-1}}{(1-az^{-1})}$ ROC : $|z| < |a|$ (d) $\frac{1}{(1-az^{-1})}$ ROC : $|z| < |a|$

Q6: If $u(n) \overset{z}{\longleftrightarrow} \frac{z}{z-1}$, then $Z[u(-n)]$

(a) $\frac{z}{1-z}$ (b) $\frac{1}{1-z}$

(c) $\frac{1}{z-1}$ (d) $\frac{z^2}{z-1}$

Q7: ROC of $Z[u(-n)]$ is

(a) $|z| > 1$ (b) $|z| > -6$

(c) $|z| < 1$ (d) $|z| > 6$

Q8: If $X(z) = \frac{0.5z^2}{(z-1)(z^2-0.95z+0.45)}$, the initial and final values of $x(n)$ are

(a) 0, 0 (b) 1, 0

(c) 0, 0.5 (d) 0, 1

Q9: If $H(z) = \dfrac{1+z^{-1}}{1-\frac{7}{12}z^{-1}+\frac{1}{12}z^{-2}}$, the poles of $H(z)$ are at

(a) $z = \frac{1}{3}$ and $z = \frac{1}{5}$ (b) $z = \frac{1}{3}$ and $z = -\frac{1}{4}$

(c) $z = \frac{1}{3}$ and $z = \frac{1}{4}$ (d) $z = \frac{1}{3}$ and $z = \pm\frac{1}{4}$

Q10: The z-transform of the time function $\sum_{k=0}^{\infty} \delta(n-k)$ is

(a) $\frac{z-1}{z}$ (b) $\frac{1}{1-z}$

(c) $\frac{z}{z-1}$ (d) $\frac{z^2}{z-1}$

Answers

1. (d) 2. (a) 3. (c) 4. (c) 5. (d)

6. (b) 7. (c) 8. (d) 9. (c) 10. (c)

QUESTION BANK

Q1: (a) Find the z-transform of te^{at}

(b) Find the inverse z-transform of $\frac{z}{(2z-1)^2}$

Q2: Find the z-transform of the following time functions

1. Ramp function
2. Impulse function
3. Unit step

Q3: Consider a system described by the following difference equation

$$y(n+3)+6y(n+2)+11y(n-1)+6y(n)=x(n+2)+9x(n+1)+20x(n)$$

where y is the output and x is the input. The transfer function of the system will be

Q4: If z-transform of $u(n)$ is $\frac{z}{z-1}$, then the z-transform of $a^n u(n)$ is

Q5: The z-transform of $na^n u(n)$ is

Q6: The z-transform of $\{a^n u(-n-1)\}$ is

Q7: Discuss the significance of the difference equation. Solve the following difference equation using the z-transform method.

$$C(k+2)-0.1C(k+1)+0.2C(k)=r(k+1)+r(k)$$

where $r(k)=1(k)$ for $k=0,1,2\ldots C(0)=0$ and $C(1)=0$

Q8: (a) Define initial and final value theorem for z-domain transfer function.

(b) Compute convolution $x(n)$ of the signals

$$x_1(n)=(1,-2,1)$$

$$x_2(n)=\begin{cases}1 \text{ for } 0\le n\le 5\\ 0, \text{ otherwise}\end{cases}$$

Q9: Explain region of convergence (ROC) and its properties in z-transform with suitable example. Explain the scaling and convolution property of z-transform.

Q10: Determine z-transform of the sequences $x(n)$ given as follows:

(i) $x(n)=\sin(\omega_0 n)u(n)$ $\left\{(\text{Ans.})\frac{z^{-1}\sin\omega_0}{1-2z^{-1}\cos\omega_0+z^{-2}},\ \text{ROC}: |z|>1\right\}$

(ii) $x(n)=\cos(\omega_0 n)u(n)$

(iii) $x(n)=\left(\frac{1}{2}\right)^n u[n+2]+(3)^n u[-n-1]$

Q11: Find the z-transform of the following sequence

(a) $x(n)=na^n u(n)$

(b) $x(n)=\delta(n+k)$

$$\left\{\begin{array}{l}(\text{Ans.})\ a)\ X(z)=\frac{az^{-1}}{(1-az^{-1})^2},\ \text{ROC}: |z|>|a|\\ b) X(z)=z^k,\ \text{ROC}: \text{entire z-plane except } z=\infty\end{array}\right\}$$

Q12: Determine the ROC of unit impulse signals $\delta(n), \delta(n-1)$ and $\delta(n+1)$ & $u(n), -u(-n-1)$.

$$\left\{ \begin{array}{l} \text{Entire z-plane, Entire z-plane except } z=0 \\ \text{Entire z-plane except } z=\infty \end{array} \right\}$$

Q13: A causal, stable discrete-time system is given by

$$y(n) = \frac{5}{6}y(n-1) - \frac{1}{6}y(n-2) + x(n) - 2x(n-1)$$

Determine system function $H(z)$ and magnitude response at zero frequency.

$$\left\{ \text{Ans. } H(z) = \frac{1-2z^{-1}}{1-\frac{5}{6}z^{-1}+\frac{1}{6}z^{-2}}, -3 \right\}$$

Q14: For a discrete-time system given in above Q. 13, find the impulse response of the system

$$\left\{ \text{Ans. } h(n) = \{10(\tfrac{1}{3})^n - 9(\tfrac{1}{2})^n\} u(n) \right\}$$

Q15: Determine the signal $x(n)$, for given z-transform

$$X(z) = \frac{5}{(1-\frac{1}{2}z^1)^2(1+\frac{1}{3}z^{-1})} |z| > \frac{1}{2} \qquad \left\{ \text{Ans. } x(n) = \left[\tfrac{4}{5}\left(\tfrac{1}{3}\right)^n + \tfrac{21}{5}\left(\tfrac{1}{2}\right)^n + 3n\left(\tfrac{1}{2}\right)^n\right] u(n) \right\}$$

Q16: A causal system has $H(z) = \frac{8}{z^2-6z^{-1}+8}$. Determine the pole locations. Is the system stable?

$$\left\{ \begin{array}{l} \text{Poles}: z = \frac{1}{2}, \frac{1}{4} \\ \text{Stable} \end{array} \right\}$$

Q17: Derive the expression for the z-transform with ROC of a sequence which is convolution of two sequences.

Q18: Using the z-transform method, solve the difference equation
$y(n+2) - \frac{3}{2}y(n+1) + \frac{1}{2}y(n) - (\frac{1}{4})$: for $n \geq 0$ with initial conditions $y(0) = 10$ and $y(1) = 4$

Q19: Determine the z-transforms and their ROC for the following discrete-time signals.
(i) $x(n) = \left[3\left(\frac{4}{5}\right)^n \left(\frac{2}{3}\right)^{2n}\right] u(n)$ (ii) $x(n) = [(2)^n u(n) - (3)^n u(-u)]$

Q20: Determine the system function $H(z)$ and the frequency response of the system whose impulse response is given as
$h(n) = \frac{1}{2}\left[\left(\frac{1}{2}\right)^n + \left(-\frac{1}{4}\right)^n\right] u(n)$ and locate zeros and poles in z-plane.

Q21: If $u(n) \overset{z}{\longleftrightarrow} \frac{z}{z-1}$, then $Z[u(-n)]$ also fine the ROC of $Z[u(-n)]$.

Q22: If $X(z) = \frac{0.5z^2}{(z-1)(z^2-0.95z+0.45)}$, the initial and final value of $x(n)$ are

Q23: If $H(z) = \frac{1+z^{-1}}{1-\frac{7}{12}z^{-1}+\frac{1}{12}z^{-2}}$, the poles of $H(z)$ are at

Q24: The z-transform of the time function $\sum_{k=0}^{\infty} \delta(n-k)$ is

Q25: Find the impulse response of the causal system and test its stability.

$$y(n) - y(n-1) = x(n) + x(n-1)$$

$$\{\text{Ans. } h(n) = u(n) + u(n-1), \text{Unstable}\}$$

Q26: Find the response of causal system $y(n) - y(n-1) = x(n) + x(n-1)$ to input $x(n) = 2^{-2}u$

$$\left\{ \text{Ans. } y(n) = \left[4 - 3\left(\tfrac{1}{2}\right)^n\right] u(n) \right\}$$

Q27: Determine z-transform of the signal $x(n) = 2(n-1)^2\left(\tfrac{1}{2}\right)^n u(n-1)$. Sketch its ROC and pole-zero plot.

$$\left\{ \begin{array}{l} \text{Hint: Let } n-1 = m \text{ and then apply time shifting property} \\ \text{Ans. } \frac{z^{-2}(1+\frac{1}{2}z^{-1})}{(1-\frac{1}{2}z^{-1})^3}, \text{ ROC} : |z| > \frac{1}{2}; \text{poles } z = \frac{1}{2}, \frac{1}{2}, \frac{1}{2}; \text{zeros} : z = -\frac{1}{2} \end{array} \right\}$$

Q28: Determine the z-transform of $x(n) = \frac{a^n}{n}$ for $n \geq 0$ $\quad \{\text{Ans. } X(z) = e^{az^{-1}}\}$

Q29: Determine z-transform of

(a) $x(n) = \cos(\omega_0 n)u(n)$

(b) $x(n) = a^n \sin(\omega_0 n)u(n)$

$$\left\{ \begin{array}{l} \text{Ans. } a)\ X(z) = \frac{1-z^{-1}\cos\omega_0}{1-2z^{-1}\cos\omega_0+z^{-2}}, \text{ ROC} : |z| > 1 \\ b)\ X(z) = \frac{\left(\frac{z}{a}\right)^{-1}\sin\omega_0}{1-2\left(\frac{z}{a}\right)^{-1}\cos\omega_0+\left(\frac{z}{a}\right)^{-2}}, \text{ ROC} : |z| > |a| \end{array} \right\}$$

UNSOLVED QUESTIONS

Q1: Determine the ROC of unit impulse signals $\delta(n), \delta(n-1)$ and $\delta(n+1)$.

$$\left\{ \begin{array}{l} \text{Entire } z\text{-plane, Entire } z\text{-plane except } z = 0 \\ \text{Entire } z\text{-plane except } z = \infty \end{array} \right\}$$

Q2: For the signal $x(n) = \left(\frac{1}{3}\right)^n \sin\left(\frac{\pi}{4}n\right) u(n)$. Determine ROC.

$$\left\{ \text{Ans. } |z| > \frac{1}{3} \right\}$$

Q3: Determine the transfer function of the system described by

$$y(n) - \frac{3}{4}y(n-1) + \frac{1}{8}y(n-2) = 2x(n)$$

$a)\ |z| > \frac{1}{2} \quad b)\ |z| < \frac{1}{4} \quad c)\ \frac{1}{4} < |z| < \frac{1}{2}$

Also determine the output $y(n)$ if the input $x(n) = 2^n u(n)$

$$\left\{ \begin{array}{l} \text{Ans. } a)\ h(n) = \left[4\left(\frac{1}{2}\right)^n - 2\left(\frac{1}{4}\right)^n\right] u(n),\ b)\ h(n) = \left[-4\left(\frac{1}{2}\right)^n + 2\left(\frac{1}{4}\right)^n\right] u(-n-1) \\ c)\ h(n) = \left[-4\left(\frac{1}{2}\right)^n u(-n-1) - 2\left(\frac{1}{4}\right)^n u(n)\right],\ y(n) = \left[-\frac{4}{3}\left(\frac{1}{2}\right)^n + \frac{2}{7}\left(\frac{1}{4}\right)^n + \frac{64}{21}2^n\right] u(n) \end{array} \right\}$$

Q4: Using long division method, determine the inverse z-transform

$$X(z) = \frac{2z^2}{z^2 - 3z + 1}, \text{ ROC} : |z| > 1 \qquad \left\{ \text{Ans. } x(n) = \frac{2^{n+1} - 1}{2^n} \text{ for } n \geq 0 \right\}$$

Q5: By using partial fraction method, determine the inverse z-transform

$$H(z) = \frac{8z - 4z^2}{8 + 6z + z^2} \qquad \{ \text{Ans. } h(n) = [-12(-4)^n + 8(-2)^n] u(n) \text{ for } n \geq 0 \}$$

Q6: Find the impulse response of the causal system and test its stability.

$$y(n) - y(n-1) = x(n) + x(n-1) \qquad \{ \text{Ans. } h(n) = u(n) + u(n-1), \text{ unstable} \}$$

Q7: Find the response of the causal system $y(n) - y(n-1) = x(n) + x(n-1)$ to input $x(n) = 2^{-n}u(n)$

$$\left\{ \text{Ans. } y(n) = \left[4 - 3\left(\frac{1}{2}\right)^n \right] u(n) \right\}$$

Q8: Determine z-transform of the signal $x(n) = 2(n-1)^2 \left(\frac{1}{2}\right)^n u(n-1)$. Sketch its ROC and pole-zero plot.

$$\left\{ \begin{array}{l} \text{Hint: Let } n - 1 = m \text{ and then apply time shifting property} \\ \text{Ans. } \dfrac{z^{-2}\left(1 + \frac{1}{2}z^{-1}\right)}{\left(1 - \frac{1}{2}z^{-1}\right)^3}, \text{ ROC} : |z| > \dfrac{1}{2}; \text{poles } z = \dfrac{1}{2}, \dfrac{1}{2}, \dfrac{1}{2}; \text{zeros} : z = -\dfrac{1}{2} \end{array} \right\}$$

Q9: Determine the z-transform of $x(n) = \dfrac{a^n}{\lfloor n}$ for $n \geq 0$ $\qquad \left\{ \text{Ans. } X(z) = e^{az^{-1}} \right\}$

Q10: Determine z-transform of

(a) $x(n) = \cos(\omega_0 n)u(n)$

(b) $x(n) = a^n \sin(\omega_0 n)u(n)$

$$\left\{ \begin{array}{l} \text{Ans.}(a)\ X(z) = \dfrac{1 - z^{-1}\cos\omega_0}{1 - 2z^{-1}\cos\omega_0 + z^{-2}}, \text{ ROC} : |z| > 1 \\ (b)\ X(z) = \dfrac{\left(\frac{z}{a}\right)^{-1} \sin\omega_0}{1 - 2\left(\frac{z}{a}\right)^{-1}\cos\omega_0 + \left(\frac{z}{a}\right)^{-2}}, \text{ ROC} : |z| > |a| \end{array} \right\}$$

Model Question Paper

BASIC SYSTEM ANALYSIS

Time - 3 Hrs **Total Marks -100**

$5 \times 4 = 20$

(1) Attempt any *four* parts of the following:

(a) Define signal? Define the types of signals and their presentation.

(b) Determine whether the systems are time invariant or variant.

(a) $y(t) = x(t-2) + x(2-t)$ (b) $y(t) = \sin t \ x(t+1)$

(c) Explain D'Alembert's principle.

(d) Explain force voltage or force current analogy.

(e) Obtain the nodal equations for the system shown in figure and draw its analogous electrical network.

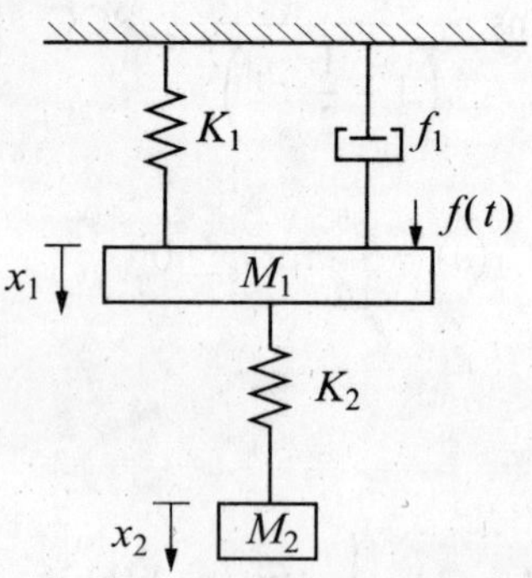

(f) Obtain the transfer function of the system shown in figure and draw its electrical analog.

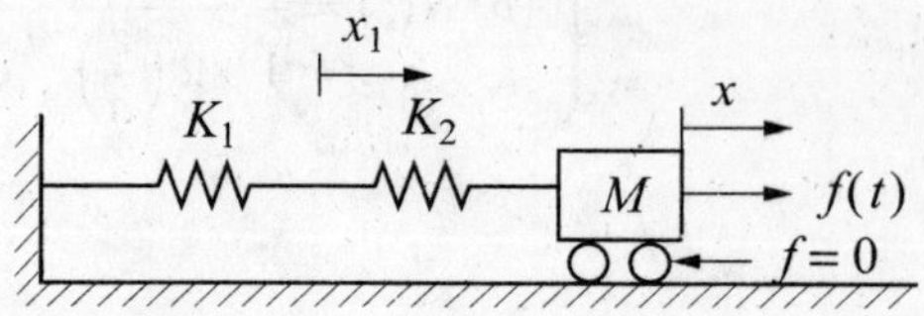

(2) Attempt any *four* parts of the following: $5 \times 4 = 20$

(a) What is Fourier series? Give its importance. What are the conditions required to be satisfied to expand a function $f(t)$ into fourier series.

(b) Explain Parseval's theorem.

(c) Determine the Fourier series coefficients for $g(t)$ as shown in figure.

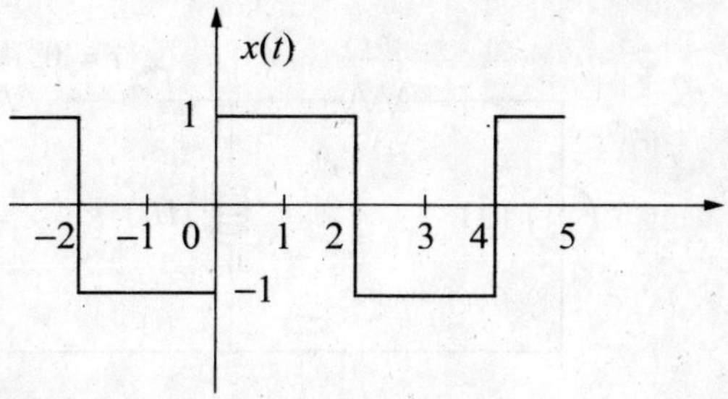

(d) Find the Fourier transform of $x(t)$ as shown in figure.

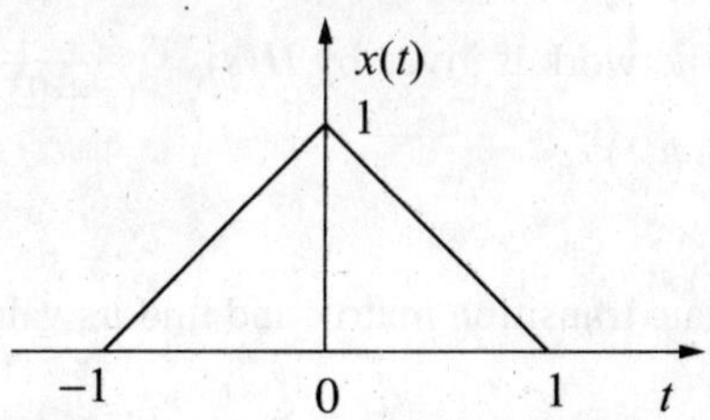

(e) Find $v_c(t)$ for the circuit shown in figure.

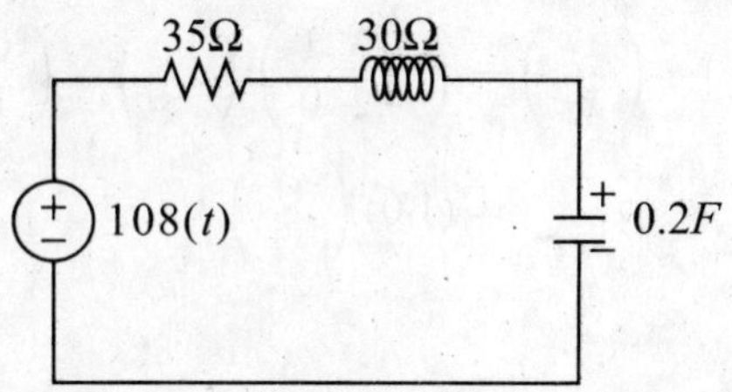

(3) Attempt any *four* questions: $4 \times 5 = 20$

(a) Give the relationship between Laplace transform and continuous-time Fourier transform.

(b) Determine the inverse Laplace of the following

$$\frac{s^3+1}{s(s+1)(s+2)}$$

(c) Determine the response of current in the network shown in Fig. 3(a) when a voltage having the waveform shown in Fig. 3(b) is applied to it by using the Laplace transform method.

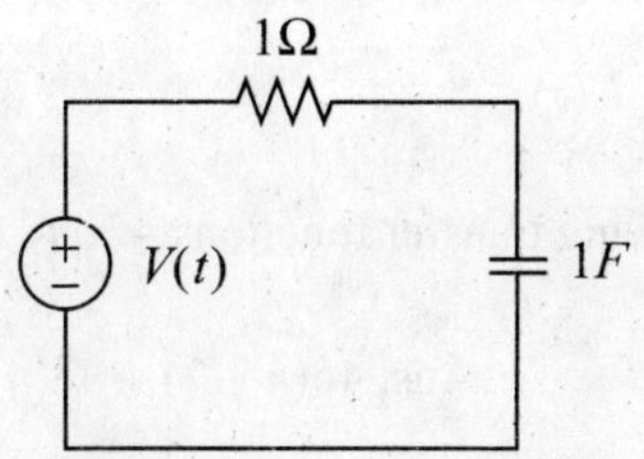

Fig. 3(a)

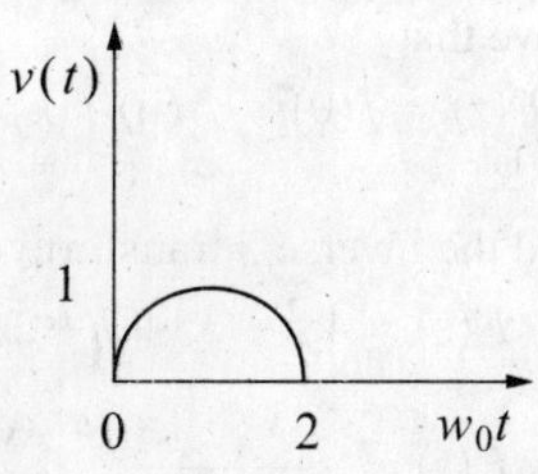

Fig. 3(b)

(d) Find $i(t)$ for $t > 0$ in the circuit shown in Fig. 3(c). Switch is opened at $t = 0$

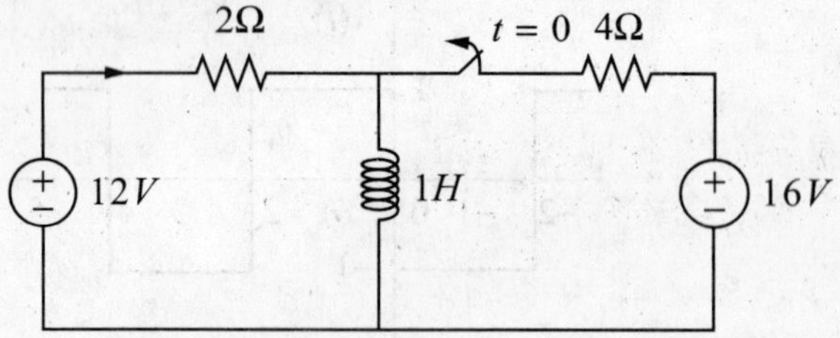

Ans. $u(t) = 2(3 + 2e^{-2t})u(t)$

(e) Show that the convolution integral of $f_1(t) = e^{-at}$ $f_2(t) = t$ is $\frac{1}{a^2}(at - 1 + e^{-at})$

(f) Laplace transform of the network is given by $H(s) = \dfrac{3s}{(s+2)(s^2+s+2)}$

Find its impulse response $h(t)$.

(4) Attempt any *two* questions: **$10 \times 2 = 20$**

(a) Write the properties of state transition matrix and find its value for the matrix

$$[A] = \begin{pmatrix} 0 & -1 \\ 2 & -3 \end{pmatrix}$$

Refer Question 6 of Unit 4.

(b) A network is characterised by the following state space equations.

$$\begin{pmatrix} \dot{x}_1 \\ \dot{x}_2 \end{pmatrix} = \begin{pmatrix} -3 & 1 \\ -2 & 0 \end{pmatrix}\begin{pmatrix} x_1 \\ x_2 \end{pmatrix} + \begin{pmatrix} 0 \\ 1 \end{pmatrix}(u)$$

$$y = (1\ 0)\begin{pmatrix} x_1 \\ x_2 \end{pmatrix}$$

Obtain

(i) Transfer function

(ii) State transition matrix

(iii) Eigenvalues

(c) What is homogeneous system? Explain the solution of state equations for non-homogeneous system.

(5) Attempt any *two* questions: **$10 \times 2 = 20$**

(a) Define region of convergence (ROC) and find the z-transform of the following function with its ROC.

(i) $\delta(k)$ (ii) $u(k)$ (iii) $e^{\pm\beta k}$ $K \geq 0$ (iv) $u(-k)$

(b) Prove that

(i) $\lim_{z\to\infty} F(z) = f(0)$ (ii) $\lim_{k\to\infty} f(k) = \left(\frac{z-1}{z}\right)_{z\to 1} F(z)$

(c) Find the inverse z-transform of the following transfer functions –

(i) $F(z) = 4z^2 + 3z + 7 + 2z^{-1}$ $0 < |z| < \infty$

Ans. $4\delta(n+2) + 3\delta(n+1) + 7\delta(n) + 2\delta(n-1)$

(ii) $F(z) = \frac{1}{z^2 - 1.2z + 0.2}$

Ans. $5\delta(n) + 1.25u(n) - 6.25(0.2)^n\, u(n)$

BASIC SYSTEM ANALYSIS

U.P.T.U. Examination 2007-08

Time: 3 Hours **Total Marks: 100**

Note: Attempt all questions.

Q1: Attempt any *four* parts of the following: **$(5\times 4=20)$**

(a) Calculate energies and powers of the following signals:
(i) $e^{-5t}u(t)$ (ii) $2\sin 2t+4\cos\pi t$

(b) Calculate the following integrals:
(i) $\int_{-\infty}^{5}(5+\cos t)\,\delta(t-10)\,dt$ (ii) $\int_{-\infty}^{\infty}(t-1)^2\,\delta(t-2)dt$
(iii) $\int_{-\infty}^{\infty} e^{-ut^2}\delta(t-10)dt$

(c) Differentiate between time-invariant and time-variant systems. Give a suitable example for both types of systems.

(d) What do you understand by linear systems? Show that the systems described by the following differential equation is linear

$$\dot{y}(t)+\,ty(t)=r(t)$$

(e) Sketch the waveforms of the following signals:

(i) $x(t)=u(t)-2u(t)+u(t-1)$

(ii) $x(t)=u(t+1)-2u(t)+u(t-1)$

(iii) $y(t)=r(t+1)-r(t)+r(t-2)$

where $u(t)$ and $r(t)$ are unit step and ramp signals respectively.

(f) Consider the R-L-C series circuit in Fig.1 which is closed at $t=0$. Write the loop equation for this circuit assuming zero initial condition. What are three different possible situations of the transient solution of above equations? Describe the corresponding relationship R, L and C for each of the situations.

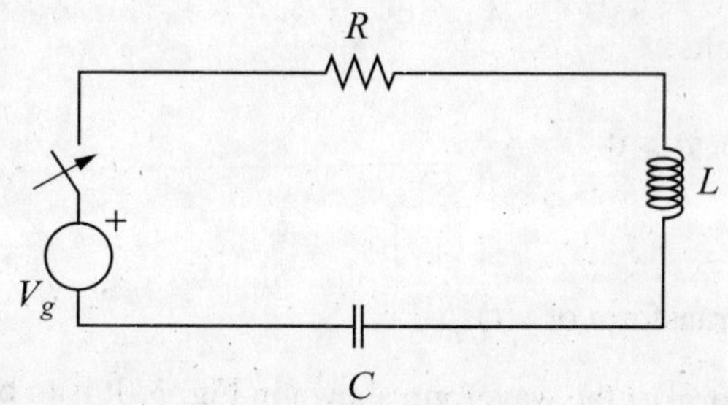

Fig. 1.

Q2: Attempt any *two* parts of the following: **$(10 \times 2 = 20)$**

(a) Determine the effect of each of the following symmetry conditions on the coefficients of the Fourier series expansion for $f(\theta)$ and obtain the formula for those coefficients which do not vanish.

(1) $f(\theta) = f(\pi - \theta)$ (2) $f(\theta) = -f(\pi - \theta)$

(b) What are Dirichlets's conditions?

(1) Find the Fourier transform of $x(t) = \delta(t)$.

(2) Find the inverse Fourier transform of $X(j\omega) = 2\pi\delta(\omega)$.

(c) Find the Fourier series of the signal shown in Fig. 2 using exponential form.

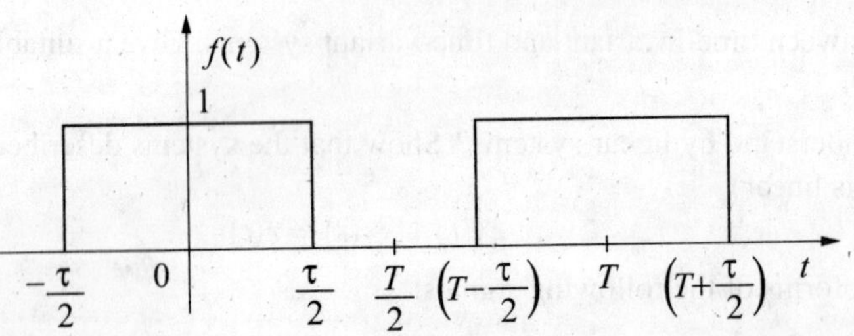

Fig. 2.

Q3: Attempt any *two* parts of the following: **$(2 \times 10 = 20)$**

(a) Use the Laplace transform to determine the output of a system represented by the different equation

$$\ddot{y}(t) + 5\dot{y}(t) + 6y(t) = 2\dot{x}(t) + x(t)$$

in response to the input $x(t) = u(t)$. Assume that all initial conditions are zero.

(b) Prove the following results:

(1) $L\left(y\left(\frac{t}{a}\right)\right) = a\,Y(s),\ a > 0$

(2) $L[ty(t)] = \frac{d}{ds}Y(s)$

where $Y(s)$ is Laplace transform of $y(t)$.

(c) Find the Laplace transform of the waveform shown in Fig. 3. It is to be noted that $v(t) = 0$ for $t > 2T$ and $t < 0$.

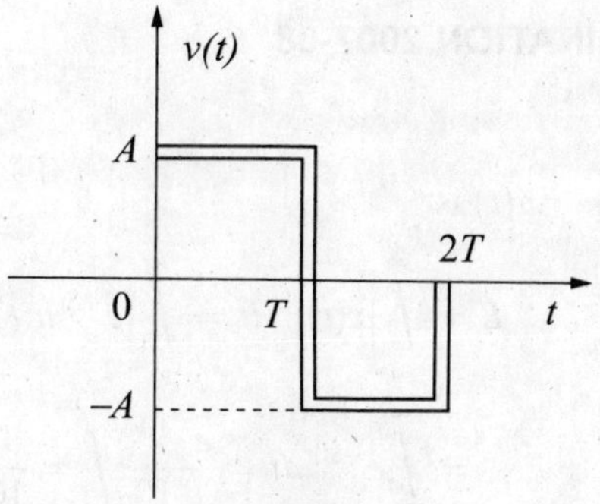

Fig. 3.

Q4: Attempt any *two* parts of the following: **(2 × 10 = 20)**

(a) Find the state transition matrix for

$$A = \begin{pmatrix} 0 & 1 \\ -6 & -5 \end{pmatrix}$$

Find the corresponding resolvant matrix also.

(b) Find the output response of the system described by the following state variable formulation.

$$\begin{pmatrix} \dot{x}_1 \\ \dot{x}_2 \end{pmatrix} = \begin{pmatrix} 0 & 1 \\ -2 & -3 \end{pmatrix} \begin{pmatrix} \dot{x}_1 \\ \dot{x}_2 \end{pmatrix} + \begin{pmatrix} 0 \\ 1 \end{pmatrix} u(t)$$

to unit step input.
It is given that

$$C = \begin{pmatrix} 1 & 0 \\ 1 & -1 \end{pmatrix} \text{ and } x^r(0) = [1\ 1].$$

(c) Consider the state variable model of a second-order system represented as

$$\begin{pmatrix} \dot{x}_1 \\ \dot{x}_2 \end{pmatrix} = \begin{pmatrix} 0 & 1 \\ -2 & -3 \end{pmatrix} \begin{pmatrix} \dot{x}_1 \\ \dot{x}_2 \end{pmatrix} + \begin{pmatrix} 0 \\ 2 \end{pmatrix} r$$

$$\begin{pmatrix} x_1(0) \\ x_2(0) \end{pmatrix} = \begin{pmatrix} 0 \\ 1 \end{pmatrix}; \ r = \text{unit step.}$$

Find the state response $x(t),\ t > 0$

Q5: Attempt any *two* parts of the following: **(10 × 2 = 20)**

(a) Find the z-transform of the sequences

(i) $\delta(k)$ (ii) $u(k)$ (iii) $e^{\pm\beta k}$, $k \geq 0$

(b) (i) Prove that $\lim\limits_{z\to\infty} F(z) = f(0)$

(ii) $\lim\limits_{k\to\infty} f(k) = \lim\limits_{z\to 1}\left[\left(\frac{z-1}{z}\right)F(z)\right]$

(c) Obtain the z-inverse of $F(z)$ for the following:

(i) $\frac{z}{(z-0.4)}, |z| > 0.4$ (ii) $\frac{z}{(z-0.4)}, |z| < 0.4$

SOLUTIONS OF U.P.T.U. EXAMINATION 2007-08

Q1:

(a) (i) Energy of signal $x(t) = e^{-5t}u(t)$ is

$$E = \int_{-\infty}^{\infty} |x(t)|^2 dt = \int_{-\infty}^{\infty} \left|e^{-5t}u(t)\right|^2 dt$$

$$= \int_{0}^{\infty} e^{-10t} dt = \frac{e^{-10t}}{-10}\Big|_{0}^{\infty} = \frac{1}{10}$$

Power of the signal $P = \lim_{T\to\infty} \frac{1}{2T} \int_{-T}^{T} |x(t)|^2 dt$

$$P = \lim_{T\to\infty} \frac{1}{2T} \int_{0}^{T} e^{-10t} dt = \lim_{T\to\infty} \frac{1}{2T} \frac{\left[e^{-10T} - 1\right]}{-10} \quad \{ \text{ since } \tfrac{1}{T} = \tfrac{1}{\infty} = 0 \}$$

$= 0$

So signal $e^{-5t}u(t)$ is energy signal since $0 < E < \infty$ is finite and $P = 0$.

(ii) Energy $E = \int_{-\infty}^{\infty} (2\sin 2t + 3\cos \pi t)^2 dt$

As the signal is non-periodic so $P = 0$

(b) (i) $\int_{-\infty}^{5} (5 + \cos t)\delta(t-10)dt = 0$

Since $\delta(t-10) = \begin{cases} 1 & \text{for } t = 10 \\ 0 & \text{for } t \neq 10 \end{cases}$

So multiplication of $(5+\cos t)$ and $\delta(t-10)$ is zero with the limit of integration.

(ii) $\int_{-\infty}^{\infty} (t-2)^2\, \delta(t-2)dt$

Since $\delta(t-2) = \begin{cases} 1 & \text{for } t = 2 \\ 0 & \text{for } t \neq 0 \end{cases}$

So $\int_{-\infty}^{\infty} (t-2)^2\, \delta(t-2)dt = 0$

(iii) $\int_{-\infty}^{\infty} e^{-a^2} \delta(t-10)dt$

Since $\delta(t-10) = \begin{cases} 1 & \text{for } t = 10 \\ 0 & \text{for } t \neq 10 \end{cases}$

So $\int_{-\infty}^{\infty} e^{-at^2} \delta(t-10)dt = e^{-100a}$

(c) Give differential equation $\frac{dy(t)}{dt} + ty(t) = x(t)$

For an input $x_1(t)$, let the output be $y_1(t)$ and can be described by the differential equation as

$$\frac{dy_1(t)}{dt} + ty_1(t) = x_1(t) \qquad (1)$$

Similarly for an input $x_2(t)$, let the output be $y_2(t)$

$$\frac{dy_2(t)}{dt}+ty_2(t)=x_2(t) \tag{2}$$

Multiplying eqn. (1) by a and eqn. (2) by b

$$a\frac{dy_1(t)}{dt}+aty_1(t)+b\frac{dy_2(t)}{dt}+bty_2(t)=ax_1(t)+bx_2(t)$$

$$\frac{d}{dt}\underbrace{[ay_1(t)+by_2(t)]}_{\text{weighted sum of outputs}}+t\underbrace{[ay_1(t)+by_2(t)]}_{\text{weighted sum of outputs}}=\underbrace{[ax_1(t)+bx_2(t)]}_{\text{weighted sum of inputs}} \tag{3}$$

Since weighted sum of inputs to the system produces an output which is equal to weighted sum of outputs to each of the individual inputs, therefore, the system is linear.

(d) (i) $x(t)=u(t)-u(t-2)$

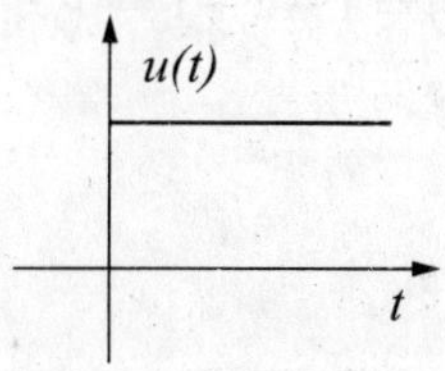

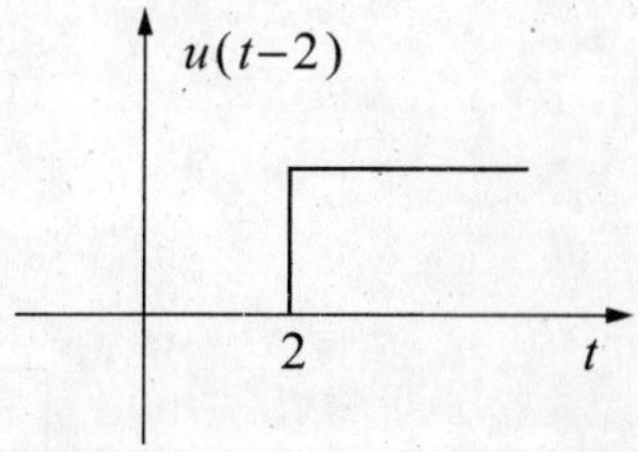

$x(t) = u(t) - u(t-2)$

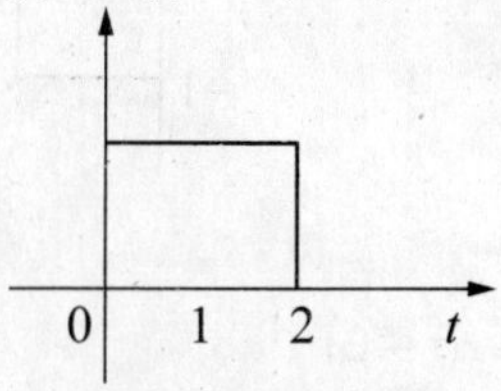

(ii) $x(t)=u(t+1)-2u(t)+u(t-1)$

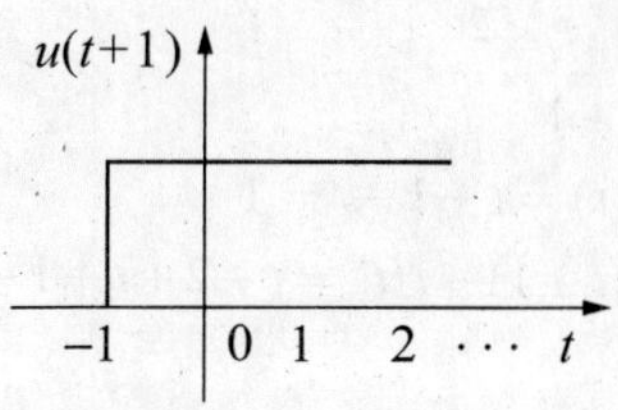

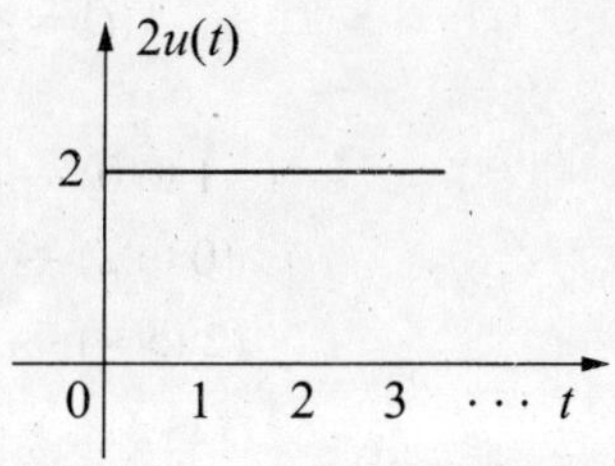

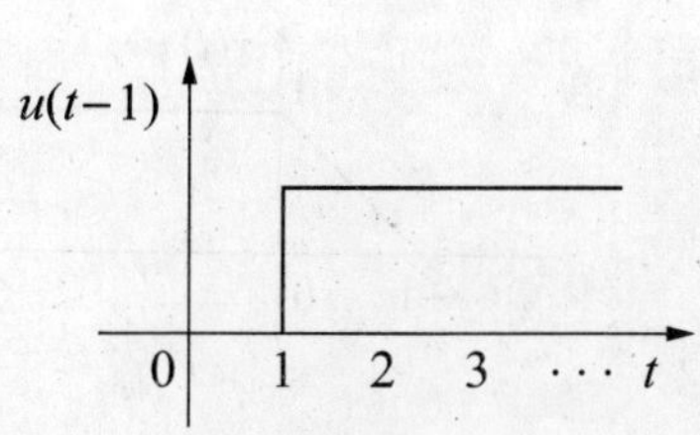

From -1 to 0; value of waveform $(1) = 1$

value of waveform $(2) = 0$

value of waveform $(3) = 0$

So $(1)+(2)+(3) = 1$

Similarly, from 0 to 1 $\quad (1) = 1$

$(2) = -2$

$(3) = 0$

So $(1)+(2)+(3) = -1$

from 0 to ∞ $\quad (1) = 1$

$(2) = -2$

$(3) = 1$

$$(1)+(2)+(3) = 1-2+1 = 0$$

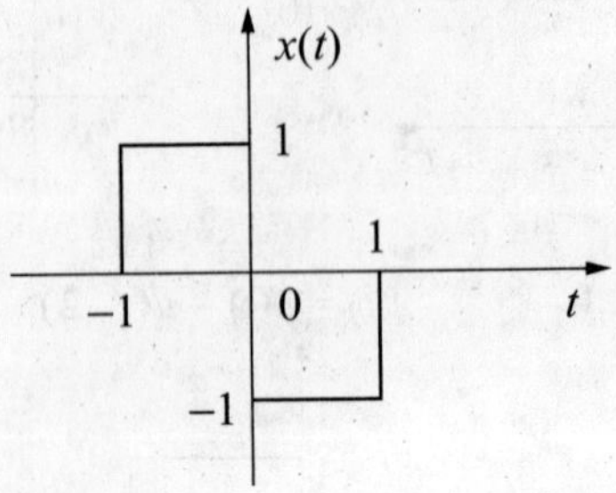

(iii) $x(t) = r(t+1) - r(t) + r(t-2)$

$$x(t) = (t+1)u(t+1) - tu(t) + (t-2)u(t-2)$$

$$x(t) = \begin{cases} t+1 & \text{for } t \geq -1 \\ t & \text{for } t \geq 0 \\ t-2 & \text{for } t \geq 2 \end{cases}$$

$(-1 \text{ to } 0) \rightarrow r(t+1) = t+1$

$(0 \text{ to } 2) \rightarrow r(t+1) - r(t) = t+1-t = 1$

$(2 \text{ to } \infty) \rightarrow r(t-2) + r(t+1) - r(t) = t-2+t+1-t = t-1$

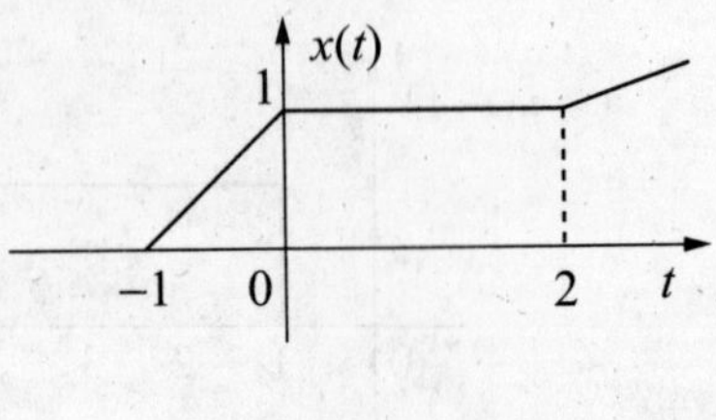

(e)

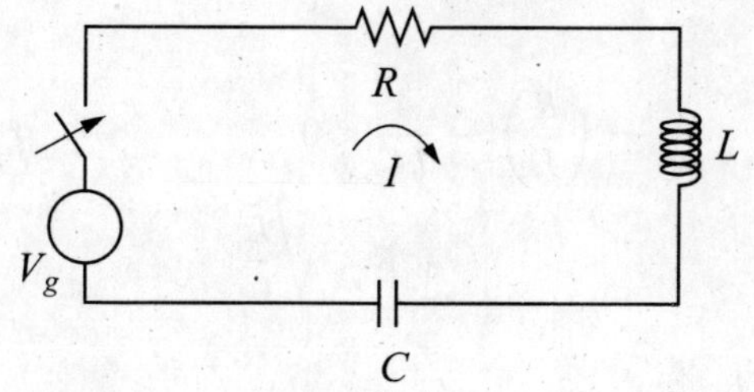

Loop equation for zero initial condition

$$iR + L\frac{di}{dt} + \frac{1}{C}\int idt = V_g$$

$$L\frac{di}{dt} + Ri + \frac{1}{C}\int idt = V_g \quad (1)$$

On differentiating eqn. (1) on both sides

$$L\frac{d^2i}{dt^2} + R\frac{di}{dt} + \frac{1}{C}i = 0$$

$$\frac{d^2i}{dt^2} + \frac{R}{L}\frac{di}{dt} + \frac{1}{LC}i = 0 \quad (2)$$

Solution of above equations $r^2 + \frac{R}{L}r + \frac{1}{LC} = 0$

$$\text{Roots are } r = \frac{\frac{-R}{L} \pm \sqrt{\left(\frac{R}{L}\right)^2 - \frac{4}{LC}}}{2}$$

$$r = \frac{-R}{2L} \pm \frac{1}{2}\sqrt{\left(\frac{R}{L}\right)^2 - \frac{4}{LC}}$$

Case 1: Roots are equal

$$\left(\frac{R}{L}\right)^2 - \frac{4}{LC} = 0$$

$$R = \sqrt[2]{\frac{L}{C}} \quad (3)$$

System is critically damped.

Case 2: Roots are real but unequal

$$\left(\frac{R}{L}\right)^2 - \frac{4}{LC} > 0$$

$$R > \sqrt[2]{\frac{L}{C}} \quad (4)$$

Overdamped.

Case 3: Roots are complex

$$\left(\frac{R}{L}\right)^2 - \frac{4}{LC} < 0$$

$$R < \sqrt[2]{\frac{L}{C}} \tag{5}$$

Underdamped.

Q2:

(a) Waveform symmetries

(i) Even symmetry: A function $x(t)$ is a even function if

$$x(t) = x(-t)$$

The sum of two or more even functions is also even function

For such functions $\int_0^T x(t)dt = \int_{\frac{-T}{2}}^{\frac{T}{2}} x(t)dt = 2\int_0^{T/2} x(t)dt$

e.g. $\cos wt, t\sin wt$

$$\cos wt = \cos(-wt)$$

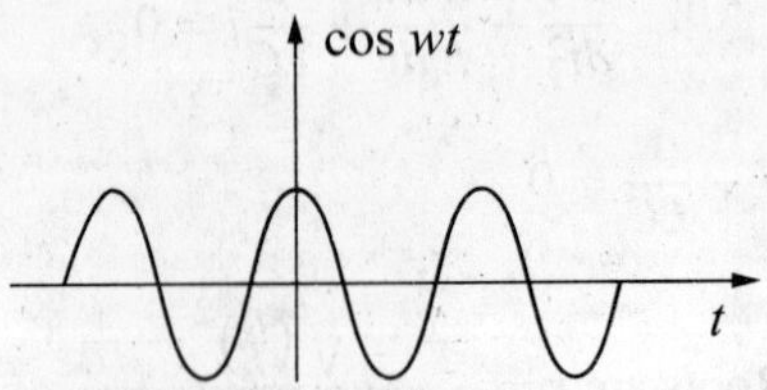

Odd symmetry: A function is said to be odd

$$x(t) = -x(-t)$$

For such functions $\int_0^T x(t)dt = \int_{-\frac{T}{2}}^{\frac{T}{2}} x(t)dt = 0$

e.g. $\sin wt$

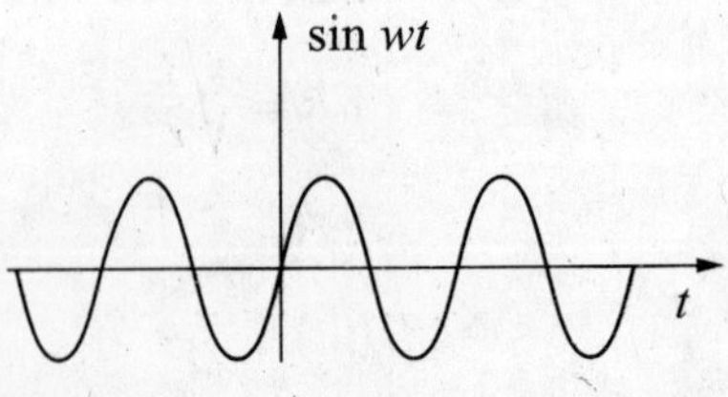

(ii) Half-wave symmetry: A half-wave symmetric function is neither completely even nor a completely odd function. For half-wave symmetric function

$$x(t) = -x\left(t \pm \frac{T}{2}\right)$$

(iii) Quarter-wave symmetry: A quarter-wave symmetric function has both even or odd as well as half-wave symmetry.

2(b)(1) $x(t) = \delta(t)$

Fourier transform of $x(t)$ is given as

$$X(jw) = \int_{-\infty}^{\infty} x(t)e^{-jwt}\,dt$$

$$X(jw) = \int_{-\infty}^{\infty} \delta(t)e^{-jwt}\,dt$$

$$\text{Since } \delta(t) = \begin{cases} 1 & \text{for } t = 0 \\ 0 & \text{for } t \neq 0 \end{cases}$$

$$X(jw) = e^{-jwt}\Big|_{t=0} = 1 \qquad (1)$$

(2) $X(jw) = 2\pi\,\delta(w)$

Inverse Fourier transform $X(jw)$ is

$$x(t) = \frac{1}{2\pi}\int_{-\infty}^{\infty} X(jw)e^{jwt}\,dw$$

$$= \frac{1}{2\pi}\int_{-\infty}^{\infty} 2\pi\,\delta(w)e^{jwt}\,dw$$

$$= \int_{-\infty}^{\infty} \delta(w)e^{jwt}\,dw = e^{jwt}\Big|_{w=0} = 1$$

We can also solve using duality property

$$x(t) \xrightarrow{FT} X(jw)$$

$$X(t) \xrightarrow{FT} 2\pi\,x(-w)$$

(c)

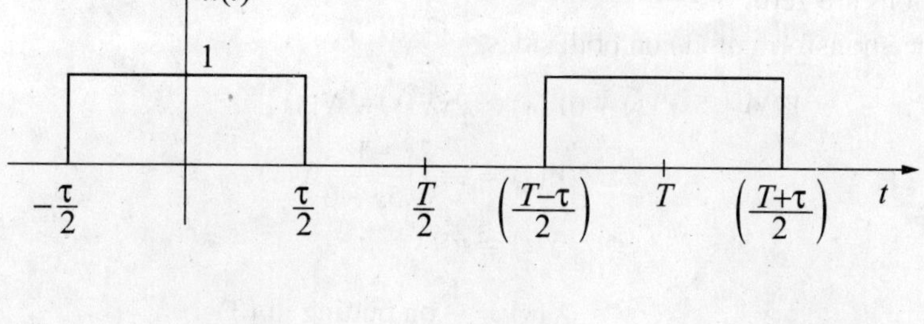

$$x(t) = \begin{cases} 1 & 0 < t < \frac{\tau}{2} \\ 0 & \frac{\tau}{2} < t < \frac{T-\tau}{2} \\ 1 & \frac{T-\tau}{2} < t < T \end{cases} \tag{1}$$

$$\text{Since} \quad x(t) = \sum_{n=-\infty}^{\infty} C_n e^{jnw_0 t}$$

Here

$$C_n = \frac{1}{T}\int_T x(t)e^{-jnw_0 t}dt$$

$$= \frac{1}{T}\left[\int_0^{\frac{T}{2}} 1e^{-jnw_0 t}dt + \int_{\frac{\tau}{2}}^{\frac{T-\tau}{2}} 0e^{-jnw_0 t}dt + \int_{\frac{T-\tau}{2}}^{T} 1e^{-jnw_0 t}dt\right]$$

$$C_n = \frac{1}{T}\left[-\frac{1}{jnw_0}\left(e^{\frac{-jnw_0\tau}{2}} - 1\right) - \frac{1}{jnw_0}\left(e^{-jnw_0 T} - e^{-jnw_0\frac{(T-\tau)}{2}}\right)\right]$$

$$w_0 = \frac{2\pi}{T} \text{ or } w_0 T = 2\pi$$

$$C_n = -\frac{1}{jn2\pi}\left[e^{-\frac{jnw_0\tau}{2}} - 1 + e^{-j2n\pi} - e^{-jn\pi} - e^{jn\frac{w_0\tau}{2}}\right]$$

$$C_n = \frac{1}{n\pi}\left[\sin\frac{nw_0\tau}{2} - (-1)^n\right] \tag{2}$$

$$C_0 = \frac{1}{T}\left[\int_0^{\frac{2}{2}} dt + \int_{\frac{T-\tau}{2}}^{T} dt\right] = \frac{1}{T}\left[\frac{\tau}{2} + T - \frac{T}{2} + \frac{\tau}{2}\right]$$

$$C_0 = \frac{1}{T}\left[\tau + \frac{T}{2}\right] = \left[\frac{\tau}{T} + \frac{1}{2}\right]$$

Q3:

(a)

$$\frac{d^2y(t)}{dt^2} + 5\frac{dy(t)}{dt} + 6y(t) = 2\frac{dx(t)}{dt} + x(t) \tag{1}$$

Given

$$\text{input} \quad x(t) = u(t) \tag{2}$$

Initial conditions are zero

Taking Laplace transform of (1) on both sides

$$s^2Y(s) + 5sY(s) + 6Y(s) = 2sX(s) + X(s)$$

$$Y(s) = \frac{2s+1}{s^2+5s+6}X(s) \tag{3}$$

$$x(t) = u(t)$$

$$X(s) = \frac{1}{s} \text{ on putting in (3)}$$

$$Y(s) = \frac{(2s+1)}{s(s+2)(s+3)}$$

By partial fraction $Y(s) = \frac{A}{s} + \frac{B}{s+2} + \frac{C}{s+3}$

$$A = s\left.\frac{2s+1}{s(s+2)(s+3)}\right|_{s=0} = \frac{1}{6}$$

$$B = (s+2)\left.\frac{2s+1}{s(s+2)(s+3)}\right|_{s=-2} = \frac{3}{2}$$

$$C = (s+3)\left.\frac{2s+1}{s(s+2)(s+3)}\right|_{s=-3} = \frac{-5}{3}$$

$$Y(s) = \frac{\frac{1}{6}}{s} + \frac{\frac{3}{2}}{s+2} - \frac{\frac{5}{3}}{s+3} \tag{4}$$

On taking inverse Laplace transform

$$y(t) = \frac{1}{6}u(t) + \frac{3}{2}e^{-2t}u(t) - \frac{5}{3}e^{-3t}u(t)$$

$$y(t) = \left(\frac{1}{6} + \frac{3}{2}e^{-2t} - \frac{5}{3}e^{-3t}\right)u(t) \tag{5}$$

(b) (1)

$$L\left[y\left(\frac{t}{a}\right)\right] = aY(s), a > 0$$

$$r(t) = y\left(\frac{t}{a}\right)$$

$$LT\,(x(t)) = \int_0^\infty x(t)e^{-st}dt$$

$$R(s) = \int_0^\infty y\left(\frac{t}{a}\right)e^{-st}dt$$

Let $\frac{t}{a} = p \quad dt = a\,dp$

$$R(s) = a\int_0^\infty y(p)e^{-asp}dp$$

$$= a\int_0^\infty y(p)e^{-(as)p}dp$$

$$\left(y\left(\frac{t}{a}\right)\right) = aY(as)$$

3 (b) (2)

$$LT\,[ty(t)] = \frac{d}{ds}Y(s)$$

$$LT\,[y(t)] = Y(s) = \int_0^\infty y(t)e^{-st}dt$$

So

$$\frac{d}{ds}Y(s) = \frac{d}{ds}\int_0^\infty y(t)e^{-st}\,dt$$

$$\frac{d}{ds}Y(s) = \int_0^\infty y(t)\frac{d}{ds}\left(e^{-st}\right)dt$$

$$= \int_0^\infty y(t)(-t)e^{-st}\,dt$$

$$= -\int_0^\infty [ty(t)]\,e^{-st}\,dt$$

$$\frac{d}{ds}Y(s) = -LT\,[ty(t)]$$

$$\text{so}\quad LT\,[ty(t)] = -\frac{d}{ds}Y(s)$$

3(c)

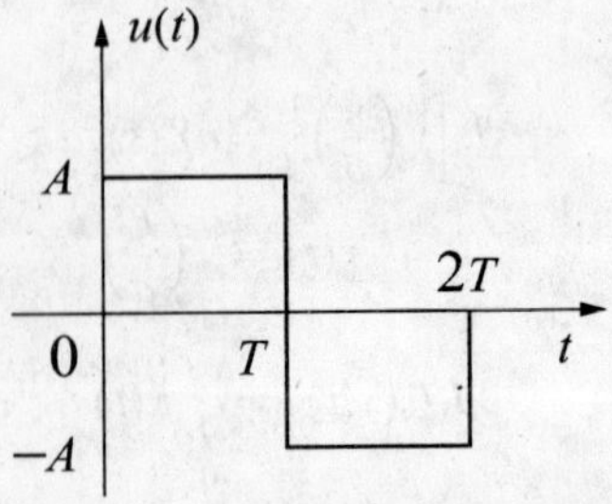

$$v(t) = A\,[u(t) - u(t-T)] - A\,[u(t-T) - u(t-2T)]$$

$$T \text{ of } v(t) = V(s) = A\left(\frac{1}{s} - \frac{e^{-sT}}{s}\right) - A\left(\frac{e^{-sT}}{s} - \frac{e^{-2sT}}{s}\right)$$

Using time shifting property

$$V(s) = \frac{A}{s}\left[1 - e^{-sT} - e^{-sT} - e^{-2sT}\right]$$

$$V(s) = \frac{A}{s}\left[1 - 2e^{-sT} - e^{-2sT}\right]$$

$$= \frac{A}{s}\left[1 - 2e^{-sT} - \left(e^{-2sT}\right)^2\right]$$

$$V(s) = \frac{A}{s}\left(1 - e^{-sT}\right)^2$$

Q4 :(a)

$$u(a)A = \begin{pmatrix} 0 & 1 \\ -6 & -5 \end{pmatrix}$$

$$(sI - A) = \begin{pmatrix} s & s-1 \\ s+6 & s+5 \end{pmatrix}$$

$$\phi(s) = (sI - A)^{-1} = \begin{pmatrix} s & s-1 \\ s+6 & s+5 \end{pmatrix}^{-1}$$

$$\begin{vmatrix} s & s-1 \\ s+6 & s+5 \end{vmatrix} = s(s+5) - (s+6)(s-1)$$

$$= s^2 + 5s - s^2 - 5s + 6 = 6$$

$$\therefore \phi(s) = (sI - A)^{-1} = \frac{1}{6}\begin{vmatrix} s+5 & -(s-1) \\ -(s+6) & s \end{vmatrix}$$

State transition matrix

$$\phi(t) = LT^{-1}[\phi(s)]$$

4-(b) Refer Question 8 of Unit 4

4-(c)

$$\begin{pmatrix} \dot{x}_1 \\ \dot{x}_2 \end{pmatrix} = \begin{pmatrix} 0 & 1 \\ -2 & -3 \end{pmatrix}\begin{pmatrix} x_1 \\ x_2 \end{pmatrix} + \begin{pmatrix} 0 \\ 2 \end{pmatrix} r$$

$$\begin{pmatrix} x_1(0) \\ x_2(0) \end{pmatrix} = \begin{pmatrix} 0 \\ 1 \end{pmatrix} r \qquad r = \text{unit step}$$

Find the state response $x(t)$.

$$\text{Ans. } A = \begin{pmatrix} 0 & 1 \\ -2 & -3 \end{pmatrix} B = \begin{pmatrix} 0 \\ 2 \end{pmatrix}$$

$$\begin{pmatrix} x_1(0) \\ x_2(0) \end{pmatrix} = \begin{pmatrix} 0 \\ 1 \end{pmatrix}$$

$$(sI - A) = \begin{pmatrix} s & s-1 \\ s+2 & s+3 \end{pmatrix}$$

$$\phi(s) = (sI - A)^{-1} = \begin{pmatrix} s & s-1 \\ s+2 & s+3 \end{pmatrix}^{-1}$$

$$|sI - A| = s(s+3) - (s+2)(s-1)$$

$$= s^2 + 3s - (s^2 + 2s - s - 2)$$

$$= 2s + 2 = 2(s+1)$$

$$\phi(s) = \frac{1}{2(s+1)}\begin{pmatrix} s+3 & -(s-1) \\ -(s+2) & s \end{pmatrix}$$

$$= \frac{1}{2}\begin{pmatrix} \frac{s+3}{s+1} & \frac{-(s-1)}{s+1} \\ \frac{-(s+2)}{s+1} & \frac{s}{s+1} \end{pmatrix}$$

$$= \frac{1}{2}\begin{pmatrix} 1 + \frac{2}{s+1} & -1\frac{+2}{s+1} \\ -1\frac{-1}{s+1} & 1 - \frac{1}{s+1} \end{pmatrix}$$

$$\text{State transition matrix} = e^{At} = \tfrac{1}{2}\begin{pmatrix} \delta(t) + 2e^{-t}r(t) & -\delta(t) + 2e^{-t}r(t) \\ -\delta(t) - e^{-t}r(t) & \delta(t) - e^{-t}r(t) \end{pmatrix}$$

The response to unit step function

$$X(t) = e^{At}X(0) + \int_0^t e^{A(t-\tau)}Br(\tau)d\tau$$

$$= e^{At}\begin{pmatrix}0\\1\end{pmatrix} + \int_0^t \frac{1}{2}\begin{pmatrix}\delta(t-\tau)+2e^{-(t-\tau)} & -\delta(t-\tau)+2e^{-(t-\tau)}\\ -\delta(t-\tau)-e^{-(t-\tau)} & \delta(t-\tau)-1e^{-(t-\tau)}\end{pmatrix}\begin{pmatrix}0\\1\end{pmatrix}r(\tau)d\tau$$

$$= e^{At}\begin{pmatrix}0\\1\end{pmatrix} + \frac{1}{2}\int_0^t \begin{pmatrix}-\delta(t-\tau)+2e^{-(t-\tau)}\\ \delta(t-\tau)-e^{-(t-\tau)}\end{pmatrix} 1d\tau$$

$$= e^{At}\begin{pmatrix}0\\1\end{pmatrix} + \frac{1}{2}\begin{pmatrix}-u(t-\tau)-2e^{-(t-\tau)}\\ u(t-\tau)+e^{-(t-\tau)}\end{pmatrix}_0^t$$

$$= \begin{pmatrix}-\delta(t)+2e^{-t}\\ \delta(t)-e^{-t}\end{pmatrix} + \begin{pmatrix}1-2e^{-t}\\ e^{-t}\end{pmatrix}$$

$$\begin{pmatrix}X_1(t)\\X_2(t)\end{pmatrix} = \begin{pmatrix}1-\delta(t)\\ \delta(t)\end{pmatrix}$$

Q5: (a)

(i) $x(k) = \delta(k)$

$$z\text{-transform}\quad X(z) = \sum_{k=-\infty}^{\infty} x(k)z^{-k}$$

$$= \sum_{k=-\infty}^{\infty} \delta(u)z^{-k}$$

$$\delta(u) = \begin{cases} 1 & \text{for } k=0\\ 0 & \text{for } k\neq 0\end{cases}$$

$$X(z) = 1.z^{-k}|_{k=0} = 1 \quad \text{Entire } z\text{-plane}$$

(ii) $x(k) = u(k)$

$$X(z) = \sum_{k=-\infty}^{\infty} x(k)z^{-k}$$

$$= \sum_{k=-\infty}^{\infty} u(k)z^{-k} = \sum_{k=0}^{\infty} z^{-k}$$

$$= 1 + z^{-1} + z^{-2} + z^{-3} + \ldots$$

$$= \frac{1}{1-z^{-1}} = \frac{z}{z-1} \qquad \text{ROC}: |z^{-1}| < 1$$

$$|z| > 1$$

(iii) $x(k) = e^{\pm\beta k}; k \geq 0$

$$X(z) = \sum_{k=-\infty}^{\infty} x(u)z^{-k}$$

$$= \sum_{k=0}^{\infty} e^{\pm\beta k}z^{-k}$$

$$= \sum_{k=0}^{\infty} \left(e^{\pm\beta} z^{-1}\right)^k = 1 + e^{\pm\beta} z^{-1} + \left(e^{\pm\beta} z^{-1}\right)^2 + \left(e^{\pm\beta} z^{-1}\right)^3 + \ldots$$

$$= \frac{1}{1 - e^{\pm\beta} z^{-1}} = \frac{z}{z - e^{\pm\beta}} \qquad \text{ROC} : |z| > e^{\pm\beta}$$

(b) (i) See the proof of initial value theorem
(ii) See the proof of final value theorem

(c) (i) $F(z) = (.4)^n u(n)$ (ii) $F(z) = -(.4)^n u(-n-1)$

BASIC SYSTEM ANALYSIS

U.P.T.U. Examination 2008-09

Time: 3 Hours **Total Marks: 100**

Note:
(i) Attempt all questions.
(ii) All questions carry equal marks.
(iii) In case of numerical problems assume data wherever not provided.

Q1: Attempt any *four* parts of the following: **(5 × 4 = 20)**

(a) Whether following continuous-time signal is periodic or not?

$$s(t) = \begin{cases} \cos(t) \text{ if } t < 0 \\ \sin(t) \text{ if } t \geq 0 \end{cases}$$

(b) Distinguish the following systems:
(i) Linear and non-linear systems
(ii) Time-invariant and time-varying systems.

(c) Input-output relationship for following continuous time systems is given as
(i) $y(t) = s(t-2) + s(2-t)$
(ii) $y(t) = \cos(3t)\ s(t)$
Whether the system is causal or not?

(d) Discuss force-current analogy used for electrical analog of the mechanical systems. Also write rules for it.

(e) Draw the force-voltage analogous of the spring mass-damper system given in Fig. 1(e).

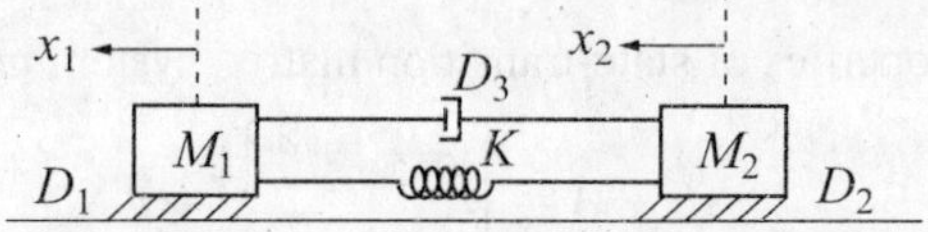

Fig. 1(e). Mechanical system

(f) What do you mean by electromechanical systems? Discuss fundamental theory of electromechanical systems with suitable examples.

2. Attempt any *two* parts of the following: **(10 × 2 = 20)**

(a) Find the continuous-time Fourier transform of the double gate function shown in Fig. 2(a) using time shifting property.

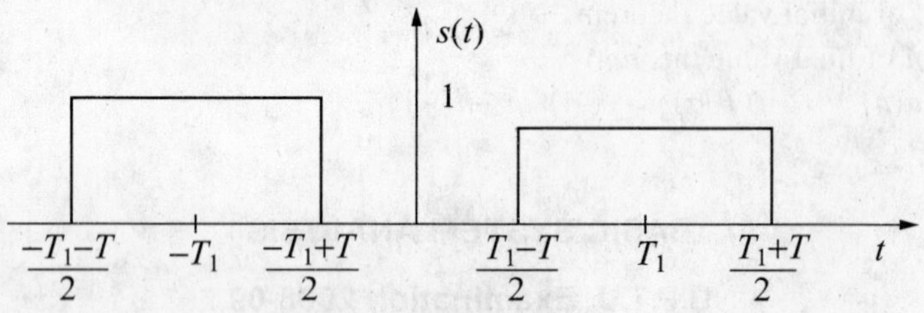

Fig. 2(a). Double gate function

(b) Find the continuous-time Fourier transform of the following functions:

(i) Impulse function

(ii) Sinusoidal function

(iii) Signum function

3. Attempt any *two* parts of the following: **(10 × 2 = 20)**

(a) Define Laplace transform. Give the relationship between Laplace transform and continuous-time Fourier transform. Find the initial and final values of the function if its Laplace is given by $F(s) = \frac{(s-2)}{s(s+1)}$

(b) Determine the inverse Laplace of the following:

(i) $\frac{s-1}{s(s+1)}$ (ii) $\frac{s^3+1}{s(s+1)(s+2)}$

(c) Laplace transform of the network is given by

$$H(s) = \frac{3s}{(s+2)(s^2+s+2)}$$

(i) Draw plot-zero plot

(ii) Find its impulse response $b(t)$

4. Attempt any *two* parts of the following: **(10 × 2 = 20)**

(a) Write various important properties of state-transition matrix, system matrix of a system is given by

$$A = \begin{bmatrix} -\frac{1}{2} & -\frac{5}{2} \\ \frac{1}{2} & -\frac{7}{5} \end{bmatrix}$$

Find the state transition matrix $\phi(t)$ of the system.

(b) State space representation of a system is given by

$$X = \begin{bmatrix} 0 & 1 \\ -2 & -3 \end{bmatrix} X + \begin{bmatrix} 0 \\ 1 \end{bmatrix} u$$

where $u(t) \Rightarrow$ unit step function and all initial conditions are zero. Find the time response of the system.

Representation of the network given in Fig. 4(c).

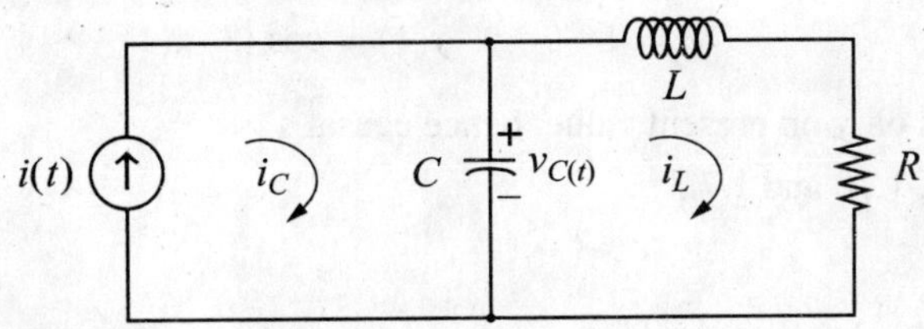

5. Attempt any *two* parts of the following: **(10 × 2 = 20)**

(a) Determine the causal discrete-time signal $f(n)$ for $f(z)$ given by $f(z) =$

$$\frac{1+2z^{-1}}{1-2z^{-1}+4z^{-2}}$$

(b) (i) Find the response of the system

$$f(n+2)-3f(n+1)+2f(n)=8(n)$$

when all the initial conditions are zero.

(ii) Find the initial and final values of the sequence $f(n)$. Given

$$F(z)=2+3z^{-1}+4z^{-2}$$

(c) Define z-transform. Give the relationship between z-transform and discrete-time Fourier transform. Also determine the inverse z-transform for $F(z) = \dfrac{z^{-1}}{1-3z^{-1}}$ with ROC : $|z| < 3$.

SOLUTIONS OF U.P.T.U. EXAMINATION 2008-09

Q1:

(a)

$$s(t) = \begin{cases} \cos t \text{ if } t < 0 \\ \sin t \text{ if } t \geq 0 \end{cases}$$

Periodic, since considering $t > 0$ and $t < 0$ separately, $s(t)$ does not repeat itself over every interval of length 2π and also has a discontinuity at the time origin that does not happen again at any other time.

(b) **See unit-1**

(c) (i) $y(t) = x(t-2) + x(2-t)$

$$\begin{aligned} \text{For } t &= -1 & y(-1) &= x(-3)+x(3) \\ t &= 0 & y(0) &= x(-2)+x(2) \\ t &= 1 & y(1) &= x(-1)+x(1) \end{aligned}$$

Since output depends on future value at $t = -1$, so non-causal.

(ii) $y(t) = \cos(3t)\, x(t)$

$$
\begin{aligned}
t &= -1 \qquad y(-1) = \cos(-3)\, x(-1) \\
t &= 0 \qquad y(0) = \cos(0)\, x(0) \\
t &= 1 \qquad y(1) = \cos(3)\, x(1)
\end{aligned}
$$

Since output depends only on present value, hence causal.

(d) See unit-1, Refer 1.9, 1.10 and 1.7.3

(e)

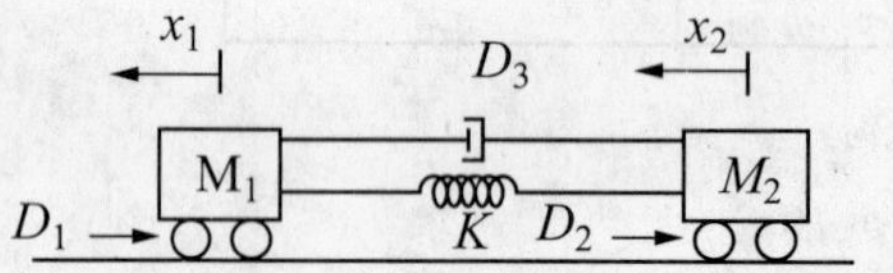

Using $f - v$ analogy for the above figure, corresponding to x_1, x_2 we have two loops in electrical circuit and $R_3(D_3)\ M(L),\ K\left(\frac{1}{C}\right)$. Then the electrical equivalent of the mechanical system is

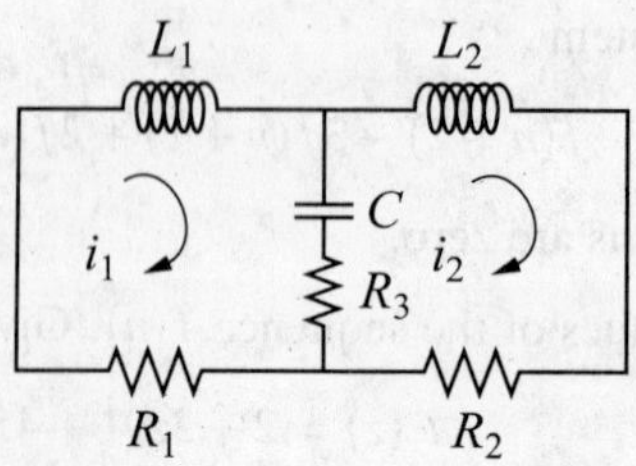

(f) Refer-1.12

2(a) Fourier transform of

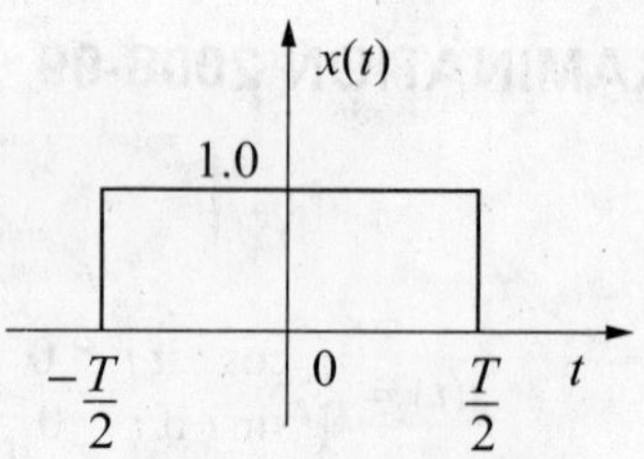

Fig. 1.

$$
x(t) = \begin{cases} 1 \text{ for } -\frac{T}{2} < t < \frac{T}{2} \\ 0, \text{ otherwise} \end{cases}
$$

$$
X(jw) = T \sin c\left(\frac{wT}{2}\right) \tag{1}
$$

Given double gate function

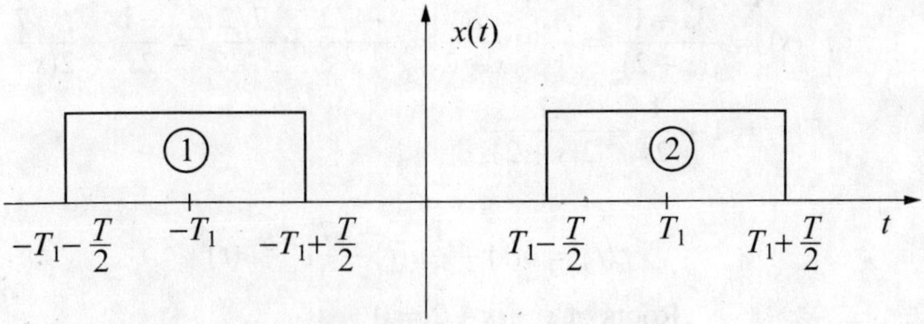

We can write waveform (1) as $x_1(t+T_1)$ and (2) as $x_1(t-T_1)$

Using time shifting property $x(t-t_0) \overset{F.T.}{\Rightarrow} e^{-jwt_0}X(jw)$

$$\begin{aligned} X(jw) &= FT[(1)] + FT[(2)] \\ &= FT[x_1(t+T_1)] + FT[x(t-T_1)] \\ &= e^{jwT_1}X_1(jw) + e^{-jwT_1}X_1(jw) \\ &= Te^{jwT_1}\sin c\left(\frac{wT}{2}\right) + Te^{-jwT_1}\sin c\left(\frac{wT}{2}\right) \\ &= 2T\sin c\left(\frac{wT}{2}\right)\cos(wT_1) \end{aligned}$$

2(b) See Unit-2

2(c) See Unit-2

3(a)

$$F(s) = \frac{s-2}{s(s+1)}$$

Initial value $f(0) = \lim_{s\to\infty}[sF(s)] = \lim_{s\to\infty}\left[\frac{s-2}{s+1}\right]$

$$f(0) = \lim_{s\to\infty}\left[\frac{1-\frac{2}{s}}{1+\frac{1}{s}}\right] = 1$$

Final value $f(\infty) = \lim_{s\to 0}[sF(s)] = \lim_{s\to 0}\left[\frac{s-2}{s+1}\right] = -2$

3(b) $F(s) = \dfrac{s-1}{s(s+1)} = \dfrac{A}{s} + \dfrac{B}{s+1} = -\dfrac{1}{s} + \dfrac{2}{s+1}$

Taking inverse Laplace transform

$$f(t) = -u(t) + 2e^{-t}u(t) = (-1+2e^{-t})\,u(t)$$

$$\text{(ii) } F(s) = \frac{s^3+1}{s(s+1)(s+2)}$$

$$= 1 - \frac{3s^2+2s-1}{s(s+1)(s+2)} = 1 - \frac{(3s-1)(s+1)}{s(s+1)(s+2)} \qquad (1)$$

$$= 1 - \frac{3s-1}{s(s+2)} = 1 - F_1(s)$$

$$F_1(s) = \frac{3s-1}{s(s+2)} = \frac{A}{s} + \frac{B}{s+2} = \frac{-1/2}{s} + \frac{7/2}{s+2} = \frac{-1}{2s} + \frac{7}{2(s+2)}$$

$$F(s) = 1 + \frac{1}{2s} - \frac{7}{2(s+2)}$$

Taking ILT

$$f(t) = \delta(t) + \frac{1}{2}u(t) - \frac{7}{2}e^{-2t}u(t)$$

3(c) $H(s) = \frac{3s}{(s+2)(s^2+s+2)}$ Roots of $s^2+s+2=0$

$\frac{-1\pm\sqrt{1-8}}{2} = \frac{-1\pm j\sqrt{7}}{2}$

Zero at $s = 0$

Poles at $s = -2, \frac{-1+j\sqrt{7}}{2}, \frac{-1-j\sqrt{7}}{2}$

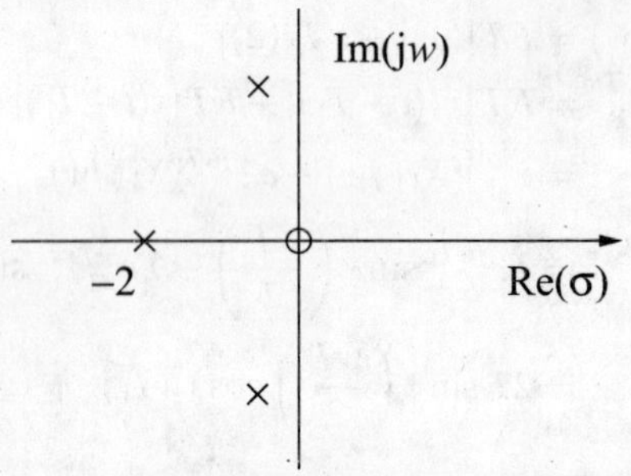

Pole zero plot

$$H(s) = \frac{A}{s+2} + \frac{Bs+c}{s^2+s+2}$$

$$= \frac{-3/2}{s+2} + \frac{\frac{3}{2}(s+1)}{s^2+s+2}$$

$$= \frac{-3/2}{s+2} + \frac{\frac{3}{2}\left(s+\frac{1}{2}+\frac{1}{2}\right)}{\left(s+\frac{1}{2}\right)^2 + 2 - \frac{1}{4}}$$

$$= \frac{-3/2}{s+2} + \frac{\frac{3}{2}\left(s+\frac{1}{2}\right)+\frac{3}{4}}{\left(s+\frac{1}{2}\right)^2 + \frac{7}{4}} = \frac{-3/2}{s+2} + \frac{\frac{3}{2}\left(s+\frac{1}{2}\right)+\frac{3}{4}}{\left(s+\frac{1}{2}\right)^2 + \left(\frac{\sqrt{7}}{2}\right)^2}$$

$$= \frac{-3/2}{s+2} + \frac{\frac{3}{2}\left(s+\frac{1}{2}\right)}{\left(s+\frac{1}{2}\right)^2 + \left(\frac{\sqrt{7}}{2}\right)^2} + \frac{3/4 \times \frac{\sqrt{7}}{2}}{\left(s+\frac{1}{2}\right)^2 + \left(\frac{\sqrt{7}}{2}\right)^2} \times \frac{1}{\frac{\sqrt{7}}{2}}$$

$$A+B=0 \quad (1)$$
$$A+2B+C=3 \quad (2)$$
$$2A+2C=0 \quad (3)$$
$$\text{So} \quad A=-B=-C$$
$$A-2A-A=3$$
$$A=-3/2\; B=C=3/2$$

Take inverse Laplace transform on both sides

$$h(t)=-3/2e^{-2t}+\frac{3}{2}e^{-t/2}\cos\frac{\sqrt{7}t}{2}+\frac{3}{42}\times\frac{2}{\sqrt{7}}e^{-t/2}\sin\left(\frac{\sqrt{7}\,t}{2}\right)$$
$$=\left[-3/2e^{-2t}+\frac{3}{2}e^{-t/2}+\frac{3}{2\sqrt{7}}e^{-t/2}\sin\left(\frac{\sqrt{7}t}{2}\right)\right]u(t)$$

4(a) Refer - 4.13.2.

$$A=\begin{pmatrix}-1/2 & -5/2\\ 1/2 & -7/5\end{pmatrix}=\frac{-1}{10}\begin{pmatrix}5 & 25\\ -5 & 14\end{pmatrix}$$
$$(sI-A)=\begin{pmatrix}s & 0\\ 0 & s\end{pmatrix}-\begin{pmatrix}\frac{-1}{2} & -5/2\\ 1/2 & -7/5\end{pmatrix}$$
$$(sI-A)=\begin{pmatrix}s+\frac{1}{2} & 5/2\\ -\frac{1}{2} & s+\frac{7}{5}\end{pmatrix}$$
$$|sI-A|=\left(s+\frac{1}{2}\right)\left(s+\frac{7}{5}\right)-\left(\frac{-1}{2}\right)\left(\frac{5}{2}\right)$$
$$=s^2+\frac{7}{10}+\frac{s}{2}+\frac{7s}{5}+\frac{5}{4}$$
$$=s^2+\frac{5s+14s}{10}+\frac{5}{4}$$
$$=s^2+\frac{21s}{10}+\frac{5}{4}$$

After that find $(sI-A)^{-1}$ and then take the Laplace inverse of $[(sI-A)^{-1}]$.

$$\phi(t)=LT^{-1}\left[(SI-A)^{-1}\right]$$

4(b) Refer Question 8 of Unit 4

4(c) Refer Question 12 of Question Bank Unit 4

5(a)

$$F(z)=\frac{1+2z^{-1}}{1-2z^{-1}+4z^{-2}}=\frac{2(z+2)}{z^2-2z+4}$$
$$=\frac{2(z+2)}{(z-1-j\sqrt{3})\,(z-1+j\sqrt{3})}$$

$$= \frac{1+2z^{-1}}{z^{-2}(z-p_1)(z-p_2)} \qquad P_1 = 1+j\sqrt{3}$$
$$\qquad\qquad\qquad\qquad\qquad p_1 = 1-j\sqrt{3}$$

$$= \frac{1+2z^{-1}}{(1-p_1z^{-1})(1-p_2z^{-1})}$$

$$= \frac{A}{1-p_1z^{-1}} + \frac{B}{1-p_2z^{-1}}$$

$$f(n) = A(p_1)^n + B(p_2)^n$$

$$A+B = 1$$
$$-p_2A - p_1B = 2$$
$$p_2A + p_1(1-A) = -2$$
$$A(p_2-p_1) + p_1 = -2$$

$$A = \frac{-2-p_1}{P_2-P_1} = \frac{2+p_1}{p_1-p_2}$$

$$A = \frac{3+j\sqrt{3}}{2j\sqrt{3}}$$

$$B = 1 - \frac{3+j\sqrt{3}}{2j\sqrt{3}}$$

$$= \frac{-3+j\sqrt{3}}{2j\sqrt{3}}$$

$$f(n) = \frac{3+j\sqrt{3}}{2j\sqrt{3}}\left(1+j\sqrt{3}\right)^n + \frac{-3+j\sqrt{3}}{2j\sqrt{3}}\left(1-j\sqrt{3}\right)^n$$

5(b) $f(n+2) - 3f(n+1) + 2f(n) = \delta n$

Taking L.T. on both sides

$$z^2F(z) - 3zF(z) + 2F(z) = \frac{\delta z}{(z-1)^2}$$

$$F(z) = \frac{\delta z}{(z-1)^2(z^2-3z+2)}$$

$$\frac{F(z)}{z} = \frac{\delta}{(z-1)^2(z-1)(z-2)} = \frac{\delta}{(z-1)^3(z-2)}$$

$$\frac{F(z)}{z} = \frac{A}{z-2} + \frac{B}{z-1} + \frac{C}{(z-1)^2} + \frac{D}{(z-1)^3}$$

$$A = (z-2)\left.\frac{\delta}{(z-2)(z-1)^3}\right|_{z=2} = \delta$$

$$B = \frac{1}{L^2}\frac{d^2}{dz^2}\left[(z-1)^3\frac{\delta}{(z-2)(z-1)^3}\right]_{z=1}$$

$$= \frac{1}{L^2}\frac{d}{dz}\left[\frac{-\delta}{(z-2)^2}\right]_{z=1}$$

$$= \frac{1}{2}\left[\frac{-16(z-2)}{(z-2)^4}\right]_{z=1} = \frac{1}{2}\left[\frac{-16}{(z-2)^3}\right]_{z=1} = \delta$$

$$C = \frac{1}{L^1}\frac{d}{dz}\left[(z-1)^3\frac{\delta}{(z-2)(z-1)^3}\right]_{z=1}$$

$$= \left.\frac{-\delta}{(z-2)^3}\right|_{z=1} = \delta$$

$$D = \left[(z-1)^3\frac{\delta}{(z-2)(z-1)^3}\right]_{z=1} = -\delta$$

$$\frac{F(z)}{z} = \frac{\delta}{z-2} + \frac{\delta}{z-1} + \frac{\delta}{(z-1)^2} - \frac{\delta}{(z-1)^3}$$

$$F(z) = \frac{\delta z}{z-2} + \frac{\delta z}{z-1} + \frac{\delta z}{(z-1)^2} - \frac{\delta z}{(z-1)^3}$$

Taking I.Z.T.

$$f(n) = \delta(2)^n u(n) + \delta u(n) + \delta n u(n) - \delta\frac{n(n-1)}{L^2}u(n)$$

$$f(n) = \delta\left[(2)^n + 1 + n - \frac{n(n-1)}{2}\right]u(n)$$

5(b)(ii)

$$F(z) = 2 + 3z^{-1} + 4z^{-2}$$

Initial value $t(0) = \lim_{z\to\infty} F(z)$

$$= \lim_{z\to\infty} 2 + 3z^{-1} + 4z^{-2} = 2$$

Final value $x(\infty) = \lim_{z\to 1}\left(1 - z^{-1}\right)X(z) = \lim_{z\to 1}\left(1 - z^{-1}\right)\left(2 + 3z^{-1} + 4z^{-2}\right)$

$$= \lim_{z\to 1}(2 + 3z^{-1} + 4z^{-2} - 2z^{-1} - 3z^{-2} - 4z^{-3})$$

$$= (2 + 3 + 4 - 2 - 3 - 4) = 0$$

5(c) See unit-5

$$F(z) = \frac{z^{-1}}{1 - 3z^{-1}}$$

$$f(n) = (3)^{n-1}u(n-1)$$

Appendices

Appendix-A

TRIGONOMETRIC RELATIONS

(1) $\cos\theta = \dfrac{e^{j\theta} + e^{-j\theta}}{2}$

(2) $\sin\theta = \dfrac{e^{j\theta} - e^{-j\theta}}{2j}$

(3) $\tan\theta = \dfrac{\sin\theta}{\cos\theta} = \dfrac{e^{j\theta} - e^{j\theta}}{j(e^{j\theta} + e^{-j\theta})}$

(4) Euler's theorem: $e^{j\theta} = \cos\theta + j\sin\theta$

(5) $\cos 2\theta = \cos^2\theta - \sin^2\theta = 1 - 2\sin^2\theta = 2\cos^2\theta - 1$

(6) $\sin 2\theta = 2\sin\theta\cos\theta$

(7) $\cos 3\theta = 4\cos^3\theta - 3\cos\theta$

(8) $\sin 3\theta = 3\sin\theta - 4\sin^3\theta$

(9) $\cos\left(\theta + \dfrac{\pi}{2}\right) = -\sin\theta$

(10) $\sin\left(\theta + \dfrac{\pi}{2}\right) = \cos\theta$

(11) $\sin(A+B) = \sin A\cos B + \cos A\sin B$

(12) $\cos(A+B) = \cos A\cos B - \sin A\sin B$

(13) $2\cos A\cos B = \cos(A-B) + \cos(A+B)$

(14) $2\sin A\sin B = \cos(A-B) - \cos(A+B)$

(15) $2\sin A\cos B = \sin(A-B) + \sin(A+B)$

Appendix-B

DERIVATIVES AND INTEGRATION

Definition: $$\frac{df(\theta)}{d\theta} = \lim_{\Delta\theta \to 0} \frac{f\left(\theta + \frac{\Delta\theta}{2}\right) - f\left(\theta - \frac{\Delta\theta}{2}\right)}{\Delta\theta}$$

Differentiation rules: $$\frac{du(x)v(x)}{dx} = u(x)\frac{dv(x)}{dx} + v(x)\frac{du(x)}{dx}$$

$$\frac{d\left(\frac{u(x)}{v(x)}\right)}{dx} = \frac{v(x)\frac{du(x)}{dx} - u(x)\frac{dv(x)}{dx}}{v^2(x)}$$

Derivative table:

$$\frac{d(x^n)}{dx} = nx^{n-1}$$

$$\frac{d}{dx}\cos x = -\sin x$$

$$\frac{d}{dx}(\sec x) = \sec x \tan x$$

$$\frac{d}{dx}(e^x) = e^x$$

$$\frac{d}{dx}(\tan^{-1} x) = \frac{1}{1+x^2}$$

$$\frac{d}{dx}(\mathrm{cosec}^{-1} x) = \frac{-1}{x\sqrt{x^2-1}}$$

$$\frac{d}{dx}(\tanh x) = \mathrm{sech}^2 x$$

$$\frac{d}{dx}(\log_e x) = \frac{1}{x}$$

$$\frac{d}{x}(\tan x) = \sec^2 x$$

$$\frac{d}{dx}(\mathrm{cosec}\, x) = -\mathrm{cosec}\, x \cot x$$

$$\frac{d}{dx}(\sin^{-1} x) = \frac{1}{\sqrt{1-x^2}}$$

$$\frac{d}{dx}(\cot^{-1} x) = \frac{-1}{1+x^2}$$

$$\frac{d}{dx}(\sinh x) = \cosh x$$

$$\frac{d}{dx}(\coth x) = -\text{cosech}^2 x$$

$$\frac{d}{dx}(\sin x) = \cos x$$

$$\frac{d}{dx}(\cot x) = -\text{cosec}^2 x$$

$$\frac{d}{dx}(a^x) = a^x \log_e a$$

$$\frac{d}{dx}(\cos^{-1} x) = \frac{-1}{\sqrt{1-x^2}}$$

$$\frac{d}{dx}(\sec^{-1} x) = \frac{1}{x\sqrt{x^2-1}}$$

$$\frac{d}{dx}(\cosh x) = \sinh x$$

$$\frac{d}{dx}(\text{sech}\, x) = -\,\text{sech}\, x \tanh x$$

$$\frac{d}{dx}(\text{cosech}\, x) = -\text{cosech}\, x \coth x$$

Definition: $\int f(x)dx = \lim_{\Delta x \to 0} \left\{ \sum_n [f(n\Delta x)]\,\Delta x \right\}$

Integration techniques: (1) let $v = u(x)$

$$\int_a^b f(x)dx = \int_{u(a)}^{v(a)} \left[\left. \frac{f(x)}{dv/dx} \right|_{x=u^{-1}(v)} \right] dv$$

(2) $\int u dv = uv - \int v du$

Indefinite and definite integrals:

$$\int x^n dx = \frac{x^{n+1}}{n+1}$$

$$\int a^x dx = a^x \log_a e$$

$$\int \tan x \, dx = \log \sec x$$

$$\int \frac{1}{x} dx = \log_e x$$

$$\int \sin x \, dx = \cos x$$

$$\int \cos x \, dx = \sin x$$

$$\int \cot x \, dx = \log \sin x$$

$$\int e^x dx = e^x$$

$$\int \sec x\,dx = \log(\sec x + \tan x) = \log\tan\left(\frac{x}{2} + \frac{\pi}{u}\right)$$

$$\int \operatorname{cosec} x\,dx = \log(\operatorname{cosec} x - \cot x) = \log\tan\frac{x}{2}$$

$$\int \sec x \tan x\,dx = \sec x$$

$$\int \operatorname{cosec} x \cot x\,dx = -\operatorname{cosec} x$$

$$\int \frac{dx}{\sqrt{a^2 - x^2}} = \sin^{-1}\frac{x}{a}$$

$$\int \frac{-dx}{\sqrt{a^2 - x^2}} = \cos^{-1}\frac{x}{a}$$

$$\int \frac{dx}{\sqrt{a^2 + x^2}} = \sinh^{-1}\frac{x}{a}$$

$$\int \frac{dx}{\sqrt{x^2 - a^2}} = \cosh^{-1}\frac{x}{a}$$

$$\int \frac{dx}{a^2 + x^2} = \frac{1}{a}\tan^{-1}\frac{x}{a}$$

$$\int \frac{-dx}{x^2 + a^2} = \frac{1}{a}\cot^{-1}\frac{x}{a}$$

$$\int \frac{dx}{a^2 - x^2} = \frac{1}{2a}\log\frac{a + x}{a - x}$$

$$\int \frac{dx}{x\sqrt{x^2 - a^2}} = \frac{1}{a}\sec^{-1}\frac{x}{a}$$

$$\int \frac{-dx}{x\sqrt{x^2 - a^2}} = \frac{1}{a}\operatorname{cosec}^{-1}\frac{x}{a}$$

$$\int \sinh x\,dx = \cosh x$$

$$\int \cosh x\,dx = \sinh x$$

$$\int \operatorname{sech}^2 x\,dx = \tanh x$$

$$\int \operatorname{cosech}^2 x\,dx = -\coth x$$

$$\int \operatorname{sech} x \tanh x\,dx = -\operatorname{sech} x$$

$$\int \operatorname{cosech} x \coth x\,dx = -\operatorname{cosech} x$$

$$\int \sqrt{a^2-x^2}\, dx = \frac{x}{2}\sqrt{a^2-x^2} + \frac{a^2}{2}\sin^{-1}\frac{x}{a}$$

$$\int \sqrt{a^2+x^2}\, dx = \frac{x}{2}\sqrt{a^2+x^2} + \frac{a^2}{2}\sinh^{-1}\frac{x}{a} = \frac{x}{2}\sqrt{a^2+x^2} + \frac{a^2}{2}\log\frac{x+\sqrt{a^2}}{a}$$

$$\int \sqrt{x^2-a^2}\, dx = \frac{x}{2}\sqrt{x^2-a^2} - \frac{a^2}{2}\cosh^{-1}$$

$$\frac{x}{a} = \frac{x}{2}\sqrt{x^2-a^2} - \frac{a^2}{2}\log\frac{x+\sqrt{x^2-a^2}}{a}$$

Theorems on definite integral:

$$\int_a^b f(x)dx = \int_a^b f(t)dt, \int_a^b f(x)dx = -\int_b^a f(x)dx$$

$$\int_a^b f(x)dx = \int_a^c f(x)dx + \int_c^b f(x)dx \quad a < c < b$$

$$\int_0^a f(x)dx = \int_0^a f(a-x)dx$$

$$\int_0^{2a} f(x)dx = \int_0^a f(x)dx + \int_0^a f(2a-x)dx$$

$$\int_0^{2a} f(x)dx = \begin{cases} 2\int_0^a f(x)dx & \text{if } f(2a-x) = f(x) \\ 0 & \text{if } f(2a-x) = -f(x) \end{cases}$$

$$\int_{-a}^{a} f(x)dx = \begin{cases} 2\int_0^a f(x)dx & \text{if } f(-x) = f(x), \text{ even function} \\ 0 & \text{if } f(-x) = -f(x), \text{ odd function} \end{cases}$$

Appendix-C

SERIES EXPANSION

Infinite series:

(1) $f(x) = \sum_{n=-\infty}^{\infty} C_n e^{jnw_0t} \quad \text{for } 0 \le t \le T \text{(Fourier)}$

$$\text{where} \quad C_n = \frac{1}{T}\int_{T}^{0} f(x)e^{-jnw_0t}dt$$

(2) $e^x = \sum_{n=0}^{\infty} \frac{x^n}{\lfloor n}$

(3) $\sin x = \sum_{n=0}^{\infty} \frac{(-1)^n x^{2n+1}}{\lfloor 2n+1}$

$$\cos x = \sum_{n=0}^{\infty} \frac{(-1)^n x^{2n}}{\lfloor 2n}$$

$$f(x) = \sum_{n=0}^{\infty} \left[\frac{f^n(a)}{\lfloor n}\right](x-a)^n \quad \text{(Taylor's series)}$$

Finite series:

$$\sum_{n=1}^{N} n = \frac{N(N+1)}{2}$$

$$\sum_{n=1}^{N} n^2 = \frac{N(N+1)(2N+1)}{6}$$

$$\sum_{n=1}^{N} n^3 = \frac{N^2(N+1)^2}{4}$$

$$\sum_{n=0}^{N} a^n = \frac{a^{N+1}-1}{a-1}$$

$$\sum_{n=N_1}^{N_2} a^n = \frac{a^{N_1}-a^{N_2+1}}{1-a}, N_2 \ge N_1$$

Appendix-D

FOURIER TRANSFORM RELATIONS

Properties:

(1) Linearity: $FT\,[ax_1(t)+bx_2(t)] = aX_1(jw)+bX_2(jw)$
a and b constant

(2) Time shifting: $FT\,[x(t-t_0)] = e^{\pm jwt_0}X(jw)$

(3) Frequency shifting: $FT\,\left[e^{\pm jw_0t}x(t)\right] = X\,[j(w\mp w_0)]$

(4) Time scaling: $FT\,[x(at)] = \dfrac{1}{|a|}X\left(\dfrac{jw}{a}\right)$ a is constant

(5) Time reversal: $FT\,[x(-t)] = X(-jw)$

(6) Differentiation in time domain: $FT\left[\dfrac{d}{dt}x(t)\right] = jwX(jw)$

(7) Differentiation in frequency domain: $FT\,[tx(t)] = j\dfrac{d}{dw}X(jw)$

(8) Integration in time domain: $FT\left[\int_{-\infty}^{t} x(\tau)d\tau\right] = \dfrac{1}{jw}X(jw)$

(9) Convolution in time domain:

$$FT\,[x_1(t)^*x_2(t)] = FT\left[\int_{-\infty}^{\infty} x_1(\tau)x_2(t-\tau)d\tau\right] = X_1(jw)X_2(jw)$$

(10) Multiplication in time domain or convolution in frequency domain:

$$FT\,[x_1(t)x_2(t)] = \frac{1}{2\pi}\int_{-\infty}^{\infty} X_1(\lambda)X_2(w-\lambda)d\lambda$$

Fourier transform pair:

(1) Unit impulse: $FT\,[\delta(t)] = 1$

(2) Unit step: $FT\,[u(t)] = \pi\delta(w)+\dfrac{1}{jw}$

(3) Signal function: $FT\,[\text{sgn}(t)] = \dfrac{2}{jw}$

(4) DC signal: $FT[1] = 2\pi\delta(w)$

(5) Decaying exponential signal: $FT\,[e^{-at}u(t)] = \dfrac{1}{a+jw}$ for $t>0$

(6) Double exponential signal: $FT\left[e^{-a(t)}, a>0\right] = \dfrac{2a}{a^2+w^2}$

(7) Sine wave signal: $FT\,[\sin w_0t] = \dfrac{\pi}{\delta}\,[\delta(w-w_0)+\delta(w-w_0)]$

(8) Cosine wave signal: $FT\,[\cos w_0 t] = \pi\,[\delta(w - w_0) + \delta(w + w_0)]$

(9) Rectangular pulse of width T: $FT\left[\text{rect}\left(\frac{t}{T}\right)\right] = T\sin\left(\frac{wT}{2}\right)$

(10) Sinc pulse with zero: $FT\,[\sin c(2wt)] = \frac{1}{2w}\text{rect}\left(\frac{t}{2w}\right)$ crossing at $\frac{n}{2w}$

(11) Impulse train or sampling function: $FT\left[\sum_{m=-\infty}^{\infty} \delta(t - mT_0)\right] = w_0 \sum_{k=-\infty}^{\infty} \delta(w - kw_0)$

Appendix-E

LAPLACE TRANSFORM RELATIONS

Properties:

(1) Linearity: $LT\left[ax_1(t)+bx_2(t)\right] = aX_1(s)+bX_2(s)$

(2) Time shifting: $LT\left[x(t-t_0)\right] = e^{-st_0}X(s)$

(3) Frequency shifting: $LT\left[e^{js_0t}x(t)\right] = X(s-s_0)$

(4) Time scaling: $LT\left[x(at)\right] = \dfrac{1}{|a|}X\left(\dfrac{s}{a}\right)$

(5) Differentiation in time domain: $LT\left(\dfrac{dx(t)}{dt}\right) = sX(s)-x(0)$

$$LT\left(\frac{d^2x(t)}{dt^2}\right) = s^2X(s)-sx(0)-\frac{dx(0)}{dt}$$

(6) Differentiation in frequency domain: $LT\left[-tx(t)\right] = \dfrac{dX(s)}{ds}$

(7) Integration in time domain:

$$LT\left[\int_0^t x(\tau)d\tau\right] = \frac{X(s)}{s}$$

$$LT\left[\int_{-\infty}^t x(\tau)d\tau\right] = \frac{X(s)}{s}+\frac{1}{s}\int_{-\infty}^0 x(\tau)d\tau$$

(8) Integration in frequency domain: $LT\left[\dfrac{x(t)}{t}\right] = \int_s^\infty X(s)ds$

(9) Convolution in time domain: $LT\left[x_1(t)^*x_2(t)\right] = X_1(s)X_2(s)$

(10) Convolution in frequency domain: $LT\left[x_1(t)x_2(t)\right] = X_1(s)^*X_2(s)$

(11) Initial value theorem: $LT\left[\lim\limits_{t\to 0} x(t)\right] = \lim\limits_{s\to\infty} sX(s)$

(12) Final value theorem: $LT\left[\lim\limits_{t\to\infty} x(t)\right] = \lim\limits_{s\to 0} sX(s)$

Table 1. Laplace transform pair

S.No	$x(t)$	$X(s)$
1.	$\delta(t)$	1
2.	1	$1/s$
3.	t	$1/s^2$
4.	t^n	$\dfrac{n}{s^n+1}$
5.	e^{-at}	$1/(s+a)$
6.	te^{-at}	$1/(s+a)^2$
7.	$t^n e^{-at}$	$n/(s+a)^{n+1}$
8.	$\sin w_0 t$	$w_0/(s^2+w_0{}^2)$
9.	$\cos w_0 t$	$s/(s^2+w_0{}^2)$
10.	$e^{-at}\sin w_0 t$	$w_0/(s+a)^2+w_0{}^2$
11.	$e^{-at}\cos w_0 t$	$\dfrac{s+a}{(s+a)^2 w_0{}^2}$

Appendix-F

Z-TRANSFORM RELATIONS

Properties:

(1) Linearity: $ZT\,[ax_1(n)+bx_2(n)] = aX_1(z)+bX_2(z)$

(2) Time shifting: $ZT\,[x(n-m)] = z^{-m}X(z)$

(3) Multiplication by exponential sequence: $ZT\,[a^n x(n)] = X(a^{-1}z)$

(4) Time reversal: $ZT\,[x(-n)] = X(z^{-1})$

(5) Differentiation in the Z-domain: $ZT\,[nx(n)] = -z\dfrac{d}{dz}X(z)$

(6) Convolution: $ZT\,[x_1(n)^* x_2(n)] = X_1(z)X_2(z)$

(7) Correlation: $ZT\,[x_1(l)^* x_2(-l)] = X_1(z)X_2(z^{-1})$

(8) Multiplication: $ZT\,[x_1(n)x_2(n)] = \dfrac{1}{2\pi j}X_1(v)X_2\left(\dfrac{z}{v}\right)v^{-1}dv$

(9) Initial value theorem: $ZT\left[\lim\limits_{h\to 0} x(n)\right] = \lim\limits_{z\to\infty} X(n)$

(10) Final value theorem: $ZT\left[\lim\limits_{n\to\infty} x(n)\right] = \lim\limits_{z\to 1}(1-z^{-1})X(z)$, if $(1-z^{-1})X(z)$ has no pole on or outside the unit circle.

(11) Parseval's relation:

$$\sum_{n=-\infty}^{\infty} x_1(n)x_2^*(n) = \frac{1}{2\pi j}\oint_c X_1(v)X_2^*\left(\frac{1}{v^*}\right)v^{-1}dv$$

Table 2. Z-transform pairs

S.No	Sequence $[x(n)]$	z-transform $[X(z)]$	ROC
1.	$\delta(n)$	1	All z
2.	$u(n)$	$\frac{1}{1-z^{-1}}$	$\|z\|>1$
3.	$-u(-n-1)$	$\frac{1}{1-z^{-1}}$	$\|z\|<1$
4.	$\delta(n-m)$	z^{-m}	All z except at $z=0$; $m>$ All z except at $z=\infty$; if $m<$
5.	$a^n u(n)$	$\frac{1}{1-az^{-1}}$	$\|z\|>\|a\|$
6.	$-a^n u(-n-1)$	$\frac{1}{1-az^{-1}}$	$\|z\|<\|a\|$
7.	$na^n u(n)$	$\frac{az^{-1}}{(1-az^{-1})^2}$	$\|z\|>\|a\|$
8.	$-na^n u(-n-1)$	$\frac{az^{-1}}{(1-az^{-1})^2}$	$\|z\|<\|a\|$
9.	$\sin w_0 n u(n)$	$\frac{(\sin w_0)z^{-1}}{1-2(\cos w_0)z^{-1}+z^{-2}}$	$\|z\|>\|$
10.	$\cos w_0 n u(n)$	$\frac{1-(\cos w_0)z^{-1}}{1-(2\cos w_0)z^{-1}+z^{-2}}$	$\|z\|>\|$
11.	$r^n \sin w_0 n u(n)$	$\frac{r(\sin w_0)z^{-1}}{1-2r(\cos w_0)z^{-1}+r^2z^{-2}}$	$\|z\|>\|r\|$
12.	$r^n \cos w_0 n u(n)$	$\frac{1-r(\cos w_0)z^{-1}}{1-2r(\cos w_0)z^{-1}+r^2z^{-2}}$	$\|z\|>\|r\|$
13.	$\frac{1}{n}$; $n>0$	$-\log_e(1-z^{-1})$	$1<\|z\|$
14.	$a^{\|n\|}$ for all n	$\frac{1-a^2}{(1-az)(1-az^{-1})}$	$\|a\|<\|z\|<\|\frac{1}{a}\|$

Index